PHYSICS OF SEMICONDUCTOR DEVICES

PHYSICS OF SEMICONDUCTOR DEVICES

by

J. P. Colinge
Department of Electrical and Computer Engineering
University of California, Davis

C. A. Colinge
Department of Electrical and Electronic Engineering
California State University

KLUWER ACADEMIC PUBLISHERS
Boston / Dordrecht / London

Library of Congress Cataloging-in-Publication Data

Colinge, Jean-Pierre.
 Physics of semiconductor devices / by J.P. Colinge, C.A. Colinge.
   p. cm.
   1. Semiconductors I. Colinge, C.A. (Cynthia A.) II. Title.

ISBN: 1-4020-7018-7 (HC)   ISBN: 0-387-28523-7 (SC)
ISBN 13: 9781402070181   ISBN 13: 9780387285238

e-ISBN: 0-306-47622-3

Printed on acid-free paper.

First softcover printing 2006.

Printed in the United States of America.

9 8 7 6 5 4 3 2 1   SPIN 11544159

springeronline.com

CONTENTS

PREFACE

This Textbook is intended for upper division undergraduate and graduate courses. As a prerequisite, it requires mathematics through differential equations, and modern physics where students are introduced to quantum mechanics. The different Chapters contain different levels of difficulty. The concepts introduced to the Reader are first presented in a simple way, often using comparisons to everyday-life experiences such as simple fluid mechanics. Then the concepts are explained in depth, without leaving mathematical developments to the Reader's responsibility. It is up to the Instructor to decide to which depth he or she wishes to teach the physics of semiconductor devices.

In the Annex, the Reader is reminded of crystallography and quantum mechanics which they have seen in lower division materials and physics courses. These notions are used in Chapter 1 to develop the Energy Band Theory for crystal structures.

An introduction to basic Matlab® programming is also included in the Annex, which prepares the students for solving problems throughout the text. Matlab® was chosen because of its ease of use, its powerful graphics capabilities and its ability to manipulate vectors and matrices. The problems can be used in class by the Instructor to graphically illustrate theoretical concepts and to show the effects of changing the value of parameters upon the result. We believe it is important for students to understand and experience a "hands-on" feeling of the consequences of changing variable values in a problem (for instance, what happens to the C-V characteristics of a MOS capacitor if the substrate doping concentration is increased? - What happens to the band structure of a semiconductor if the lattice parameter is increased? - What happens to the gain of a bipolar transistor if temperature increases?). Furthermore, some Matlab® problems make use of a basic numerical, finite-difference technique in which the "exact" numerical solution to an equation is compared to a more approximate, analytical solution such as the solution of the Poisson equation using the depletion approximation.

Chapters 1 to 3 introduce the notion of energy bands, carrier transport and generation-recombination phenomena in a semiconductor. End-of-chapter problems are used here to illustrate and visualize quantum mechanical effects, energy band structure, electron and hole behavior, and the response of carriers to an electric field.

Chapters 4 and 5 derive the electrical characteristics of PN and metal-semiconductor contacts. The notion of a space-charge region is introduced and carrier transport in these structures is analyzed. Special applications such as solar cells are discussed. Matlab® problems are used to visualize charge and potential distributions as well as current components in junctions.

Chapter 6 analyzes the JFET and the MESFET, which are extensions of the PN or metal-semiconductor junctions. The notions of source, gate, drain and channel are introduced, together with two-dimensional field effects such as pinch-off. These important concepts lead the Reader up to the MOSFET chapter.

Chapter 7 is dedicated to the MOSFET. In this important chapter the MOS capacitor is analyzed and emphasis is placed on the physical mechanisms taking place. The current expressions are derived for the MOS transistor, including second-order effects such as surface channel mobility reduction, channel length modulation and threshold voltage roll-off. Scaling rules are introduced, and hot-carrier degradation effects are discussed. Special MOSFET structures such as non-volatile memory and silicon-on-insulator devices are described as well. Matlab® problems are used to visualize the characteristics of the MOS capacitor, to compare different MOSFET models and to construct simple circuits.

Chapter 8 introduces the bipolar junction transistor (BJT). The Ebers-Moll, Gummel-Poon and charge-control models are developed and second-order effects such as the Early and Kirk effects are described. Matlab® problems are used to visualize the currents in the BJT.

Heterojunctions are introduced in Chapter 9 and several heterojunction devices, such as the high-electron mobility transistor

(HEMT), the heterojunction bipolar transistor (HBT), and the laser diode, are analyzed.

Chapter 10 is dedicated to the most recent semiconductor devices. After introducing the tunnel effect and the tunnel diode, the physics of low-dimensional devices (two-dimensional electron gas, quantum wire and quantum dot) is analyzed. The characteristics of the single-electron transistor are derived. Matlab® problems are used to visualize tunneling through a potential barrier and to plot the density of states in low-dimensional devices.

Chapter 11 introduces silicon processing techniques such as oxidation, ion implantation, lithography, etching and silicide formation. CMOS and BJT fabrication processes are also described step by step. Matlab® problems analyze the influence of ion implantation and diffusion parameters on MOS capacitors, MOSFETs, and BJTs.

The solutions to the end-of-chapter problems are available to Instructors. To download a copy of the solution manual, the Matlab® files corresponding to the end-of-chapter problems, as well as additional problems, please go to the following URL:
        http://www.wkap.nl/prod/b/1-4020-7018-7

This Book is dedicated to Gunner, David, Colin-Pierre, Peter, Eliott and Michael. The late Professor F. Van de Wiele is acknowledged for his help reviewing this book and his mentorship in Semiconductor Device Physics.

Cynthia A. Colinge   *Jean-Pierre Colinge*
California State University   *University of California*

Matlab® is a Registered Trademark of The MathWorks, Inc.

Chapter 1

ENERGY BAND THEORY

1.1. Electron in a crystal

This Section describes the behavior of an electron in a crystal. It will be demonstrated that the electron can have only discrete values of energy, and the concept of "energy bands" will be introduced. This concept is a key element for the understanding of the electrical properties of semiconductors.

1.1.1. Two examples of electron behavior

An electron behaves differently whether it is in a vacuum, in an atom, or in a crystal. In order to comprehend the dynamics of the electron in a semiconductor crystal, it is worthwhile to first understand how an electron behaves in a simpler environment. We will, therefore, study the "classical" cases of the electron in a vacuum (free electron) and the electron confined in a box-like potential well (particle-in-a-box).

1.1.1.1. Free electron

The free electron model can be applied to an electron which does not interact with its environment. In other words, the electron is not submitted to the attraction of the atoms in a crystal; it travels in a medium where the potential is constant. Such an electron is called a free electron. For a one-dimensional crystal, which is the simplest possible structure imaginable, the time-independent Schrödinger equation can be written for a constant potential V using Relationship A3.12 from Annex 3. Since the reference for potential is arbitrary the potential can be set equal to zero ($V = 0$) without losing generality.[1] The time-independent Schrödinger equation can, therefore, be written as:

$$-\frac{\hbar^2}{2m}\frac{d^2}{dx^2}\,\Psi(x)\ =\ E\,\Psi(x) \qquad (1.1.1)$$

where E is the electron energy, and m is its mass. The solution to Equation 1.1.1 is :

$$\Psi(x) = C_1\,exp(jkx) + C_2\,exp(-jkx) \qquad (1.1.2)$$

where:

$$k = \sqrt{\frac{2mE}{\hbar^2}} \quad or \quad E = \frac{\hbar^2 k^2}{2m} \qquad (1.1.3)$$

Equation 1.1.2 represents two waves traveling in opposite directions. C_1 $exp(jkx)$ represents the motion of the electron in the $+x$ direction, while C_2 $exp(-jkx)$ represents the motion of the electron in the $-x$ direction.

What is the meaning of the variable k? At first it can be observed that the unit in which k is expressed is m^{-1} or cm^{-1}; k is thus a vector belonging to the reciprocal space. In a one-dimensional crystal, however, k can be considered as a scalar number for all practical purposes. The momentum operator, p_x, of the electron, given by relationship A3.2, is:

$$p_x = \frac{\hbar}{j}\frac{\partial}{\partial x}$$

Considering an electron moving along the $+x$ direction in a one-dimensional sample and applying the momentum operator to the wave function $\Psi(x) = C_1\,exp(jkx)$ we obtain:

$$p_x\,\Psi(x) = \frac{\hbar}{j}\frac{d\Psi(x)}{dx}\ =\ C_1\,\hbar k\,exp(jkx) =\ \hbar k\,\Psi(x)$$

The eigenvalues of the operator p_x are thus given by:

$$p_x = \hbar k \qquad (1.1.4)$$

Hence, we can conclude that the number k, called the wave number, is equal to the momentum of the electron, within a multiplication factor $\hbar$. In classical mechanics the speed of the electron is equal to $v=p/m$, which yields $v = \hbar k/m$. We can thus relate the expression of the electron energy, given by Expression 1.1.3, to that derived from classical mechanics:

$$v = \hbar k/m \Rightarrow E = \frac{\hbar^2 k^2}{2m} = \frac{1}{2}mv^2 \qquad (1.1.5)$$

The energy of the free electron is a parabolic function of its momentum k, as shown in Figure 1.1. This result is identical to what is expected from classical mechanics considerations: the "free" electron can take any value of energy in a continuous manner. It is worthwhile noting that electrons with momentum k or $-k$ have the same energy. These electrons have the same momentum but travel in opposite directions.

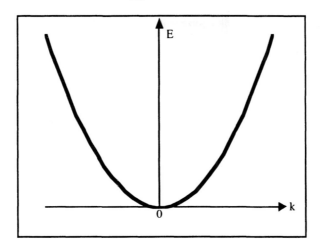

Figure 1.1: Energy vs. k for a free electron.

Another interpretation can be given to k. If we now consider a three-dimensional crystal, $\boldsymbol{k}$ is a vector of the reciprocal space. It is the called the wave vector. Indeed, the expression *exp(jkr)*, where $\boldsymbol{r}=(x,y,z)$ is the position of the electron, and represents a plane spatial wave moving in the direction of $\boldsymbol{k}$. The spatial frequency of the wave is equal to $\boldsymbol{k}$, and its spatial wavelength is equal to $\lambda = \dfrac{2\pi}{|\boldsymbol{k}|}$.

1.1.1.2. The particle-in-a-box approach

After studying the case of a free electron, it is worthwhile to consider a situation where the electron is confined within a small region of space. The confinement can be realized by placing the electron in an infinitely deep potential well from which it cannot escape. In some way the electron can be considered as contained within a box or a well surrounded by infinitely high walls (Figure 1.2). To some limited extent, the particle-in-a-box problem resembles that of electrons in an atom, where the attraction from the positively charge nucleus creates a potential well that "traps" the electrons.

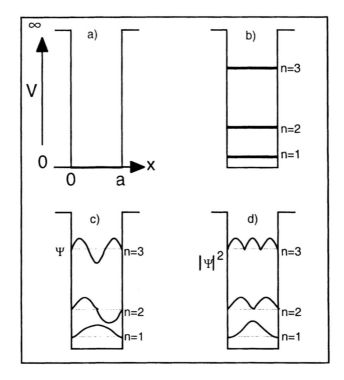

Figure 1.2: Particle in a box: a) Geometry of potential well; b) Energy levels; c) Wave functions; d) Probability density for n=1,2, and 3.

By definition the electron is confined inside the potential well and therefore, the wave function vanishes at the well edges: thus the boundary conditions to our problem are: $\Psi(x \leq 0) = \Psi(x \geq a) = 0$. Within the potential well $(0 \leq x \leq a)$, where $V = 0$, the time-independent Schrödinger equation can be written as:

$$-\frac{\hbar^2}{2m} \frac{d^2\Psi(x)}{dx^2} = E\,\Psi(x) \qquad (1.1.6)$$

which can be rewritten in the following form:

$$\frac{d^2\Psi(x)}{dx^2} + k^2\Psi(x) = 0 \;\; with \; k = \sqrt{2mE/\hbar^2} \;\;\; or \;\; E = \frac{\hbar^2 k^2}{2m} \qquad (1.1.7)$$

The solution to this homogenous, second-order differential equation is:

$$\Psi(x) = A\,sin(kx) + B\,cos(kx) \qquad (1.1.8)$$

Using the first boundary condition $\Psi(x=0) = 0$ we obtain $B = 0$. Using the second boundary condition $\Psi(a) = 0$ we obtain $A\,sin(ka) = 0$ and therefore:

$$k = \frac{n\pi}{a} \quad \text{with} \quad n = 1,2,3,... \quad\quad (1.1.9)$$

The wave function is thus given by: $\Psi_n(x) = A_n \, sin(\frac{n\pi x}{a})$ (1.1.10)

and the energy of the electron is: $E_n = \dfrac{n^2 \pi^2 \hbar^2}{2ma^2}$ (1.1.11)

This result is quite similar to that obtained for a free electron, in both cases the energy is a function of the squared momentum. The difference resides in the fact that in the case of a free electron, the wave number k and the energy E can take any value, while in the case of the particle-in-a-box problem, k and E can only take discrete values (replacing k by $n\pi/a$ in Expression 1.1.3 yields Equation 1.1.11). These values are fixed by the geometry of the potential well. Intuitively, it is interesting to note that if the width of the potential well becomes very large ($a \to \infty$) the different values of k become very close to one another, such that they are no longer discrete values but rather form a continuum, as in the case for the free electron.

Which values can k take in a finite crystal of macroscopic dimensions? Let us consider the example of a one-dimensional linear crystal having a length L (Figure 1.3). If we impose $\Psi(x=0) = 0$ and $\Psi(x=L) = 0$ as in the case of the particle-in-the-box approach, Relationships 1.1.9 and 1.1.11 tell us that the permitted values for the momentum and for the energy of the electron will depend on the length of the crystal. This is clearly unacceptable for we know from experience that the electrical properties of a macroscopic sample do not depend on its dimensions.

Much better results are obtained using the Born-von Karman boundary conditions, referred to as cyclic boundary conditions. To obtain these conditions, let us bend the crystal such that $x=0$ and $x=L$ become coincident. From the newly obtained geometry it becomes evident that for any value of x, we have the cyclical boundary conditions: $\Psi(x+L) = \Psi(x)$. Using the free-electron wave function (Expression 1.1.2), and taking into account the periodic nature of the problem, we can write:

$$\Psi(x+L) = A \, exp(jk(x+L)) = A \, exp(jkx) \, exp(jkL) = A \, exp(jkx) = \Psi(x)$$

which imposes:

$$exp(jkL) = 1 \quad \Rightarrow \quad k = \frac{2\pi n}{L} \quad\quad (1.1.12)$$

where n is an integer number.

In the case of a three-dimensional crystal with dimensions (L_x, L_y, L_z), the Born-von Karman boundary conditions can be written as follows:

$$k_x = \frac{2\pi n_x}{L_x} \ , \ k_y = \frac{2\pi n_y}{L_y} \ \text{ and } \ k_z = \frac{2\pi n_z}{L_z} \qquad (1.1.13)$$

where n_x, n_y, n_z are integer numbers.

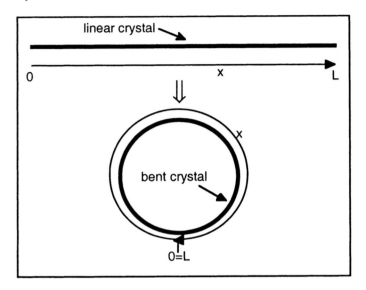

Figure 1.3: Bending of a crystal; Born-von Karman boundary conditions.

1.1.2. Energy bands of a crystal (intuitive approach)

In a single atom, electrons occupy discrete energy levels. What happens when a large number of atoms are brought together to form a crystal? Let us take the example of a relatively simple element with low atomic number, such as lithium (Z=3). In a lithium atom, two electrons of opposite spin occupy the lowest energy level (1s level), and the remaining third electron occupies the second energy level (2s level). The electronic configuration is thus $1s^2\ 2s^1$. All lithium atoms have exactly the same electronic configuration with identical energy levels. If an hypothetical molecule containing two lithium atoms is formed, we are now in the presence of a system in which four electrons "wish" to have an energy equal to that of the 1s level. But because of the Pauli exclusion principle, which states that only two electrons of opposite spins can occupy the same energy level, only two of the four 1s electrons can occupy the 1s level. This clearly poses a problem for the molecule. The problem is solved by splitting the 1s level into two levels having very close, but nevertheless different energies (Figure 1.4).

If a crystal of lithium containing N number of atoms is now formed, the system will contain N number of 1s energy levels. The same consideration is valid for the 2s level. The number of atoms in a cubic centimeter of a crystal is on the order of 5×10^{22}. As a result, each energy level is split into 5×10^{22} distinct energy levels which extend throughout the crystal. Each of these levels can be occupied by two electrons by virtue of the Pauli exclusion principle. In practice, the energy difference between the highest and the lowest energy value resulting from this process of splitting an energy level is on the order of a few electron-volts; therefore, the energy difference between two neighboring energy levels is on the order of 10^{-22} eV. This value is so small that one can consider that the energy levels are no longer discrete, but form a continuum of permitted energy values for the electron. This introduces the concept of energy bands in a crystal. Between the energy bands (between the 1s and the 2s energy bands in Figure 1.4) there may be a range of energy values which are not permitted. In that case, a forbidden energy gap is produced between permitted energy bands. The energy levels and the energy bands extend throughout the entire crystal. Because of the potential wells generated by the atom nuclei, however, some electrons (those occupying the 1s levels) are confined to the immediate neighborhood of the nucleus they are bound to. The electrons of the 2s band, on the other hand, can overcome nucleus attraction and move throughout the crystal.

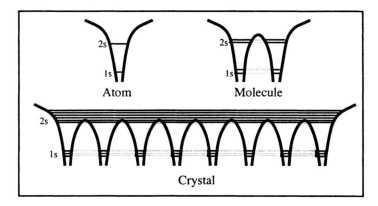

Figure 1.4: Permitted energy levels an atom, an hypothetical molecule, and a crystal of lithium.

1.1.3. Krönig-Penney model

Semiconductors, like metals and some insulators, are crystalline materials. This implies that atoms are placed in an orderly and periodic manner in the material (see Annex A4). While most usual crystalline materials are polycrystalline, semiconductor materials used in the electronics industry are single-crystal. These single crystals are almost perfect and defect-free, and their size is much greater than any of the microscopic physical dimensions which we are going to deal with in this chapter.

In a crystal each atom of the crystal creates a local potential well which attracts electrons, just like in the lithium crystal described in Figure 1.4. The potential energy of the electron depends on its distance from the atom nucleus. Electrostatics provides us with a relationship establishing the potential energy resulting from the interaction between an electron carrying a charge -q and a nucleus bearing a charge +qZ, where Z is the atomic number of the atom and is equal to the number of protons in the nucleus:

$$V(x) = \frac{-Zq^2}{4\pi\varepsilon|x|} \tag{1.1.14}$$

In this relationship x is the distance between the electron and the nucleus, $V(x)$ is the potential energy and ε is the permittivity of the material under consideration. Equation 1.1.14 ignores the presence of other electrons, such as core electrons "orbiting" around the nucleus. These electrons actually induce a screening effect between the nucleus and outer shell electrons, which reduces the attraction between the nucleus and higher-energy electrons. The energy of the electron as a function of its distance from the nucleus is sketched in Figure 1.5.

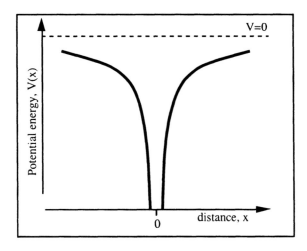

Figure 1.5: Energy of an electron as a function of its distance from the atom nucleus (*V=0* when *x=∞*). [2]

How will an electron behave in a crystal? In order to simplify the problem, we will suppose that the crystal is merely an infinite, one-dimensional chain of atoms. This assumption may seem rather coarse, but it preserves a key feature of the crystal: the periodic nature of the position of the atoms in the crystal. In mathematical terms, the expression of the periodic nature of the atom-generated potential wells can be written as:

$$V(x+a+b) = V(x) \qquad (1.1.15)$$

where $a+b$ is the distance between two atoms in the x-direction (Figure 1.6).

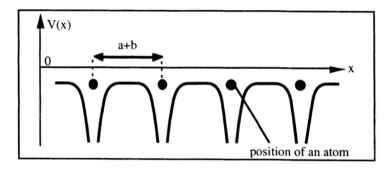

Figure 1.6: Periodic potential in a one-dimensional crystal.

The periodic nature of the potential has a profound influence on the wave function of the electron. In particular, the electron wave function must satisfy the time-independent Schrödinger equation whenever $x+a+b$ is substituted for x in the operators that act on $\Psi(x)$.[3] This condition is obtained if the wave function satisfies the Bloch theorem, which can be formulated as follows:

If $V(x)$ is periodic such that $V(x+a+b) = V(x)$,
then $\Psi(x+a+b) = \Psi(x)\, e^{jk(a+b)}$ $\qquad (1.1.16)$

A second formulation of the theorem is:
If $V(x)$ is periodic such that $V(x+a+b) = V(x)$,
then $\Psi(x) = u(x)\, e^{jkx}$ with $u(x+a+b) = u(x)$.

These two formulations are equivalent since
$\Psi(x+a+b) = u(x+a+b)\, e^{jk(x+a+b)} = u(x)\, e^{jkx}\, e^{jk(a+b)} = \Psi(x)\, e^{jk(a+b)}$

Since the potential in the crystal, $V(x)$, is a rather complicated function of x, we will use the approximation made by Krönig and Penney in 1931, in which $V(x)$ is replaced by a periodic sequence of rectangular potential wells.[4] This approximation may appear rather crude, but it preserves the periodic nature of the potential variation in the crystal while allowing a closed-form solution for $\Psi(x)$. The resulting potential is depicted in

Figure 1.7, and the following notations will be used: the inter-atomic distance is $a+b$, the potential energy near an atom is V_1, and the potential energy between atoms is V_0. Both V_1 and V_0 are negative with respect to an arbitrary reference energy, $V=0$, taken outside the crystal. We will study the behavior of an electron with an energy E lying between V_1 and V_0 ($V_0 > E > V_1$). This case is similar to a 1s electron previously shown for lithium.

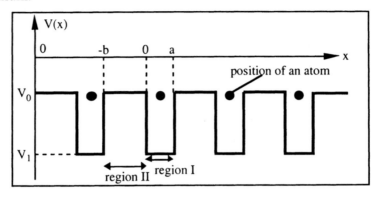

Figure 1.7: Periodic potential of the Krönig and Penney model.

In region I $(0<x<a)$, the potential energy is $V(x)=V_1$, and the time-independent Schrödinger equation can be written as:

$$\frac{\hbar^2}{2m}\frac{d^2}{dx^2}\Psi(x) + [E - V_1]\,\Psi(x) = 0 \qquad (1.1.17)$$

In region II $(-b<x<0)$, the potential energy is $V(x)=V_0$, and the time-independent Schrödinger equation becomes:

$$\frac{\hbar^2}{2m}\frac{d^2}{dx^2}\Psi(x) + [E - V_0]\,\Psi(x) = 0 \qquad (1.1.18)$$

The solution to these homogenous second-order differential equations are:

$$\Psi_I(x) = A\,exp(j\beta x) + B\,exp(-j\beta x) \quad \text{with } \beta = \sqrt{\frac{2m(E-V_1)}{\hbar^2}} \qquad (1.1.19)$$

and

$$\Psi_{II}(x) = C\,exp(\alpha x) + D\,exp(-\alpha x) \quad \text{with } \alpha = \sqrt{\frac{2m(V_0-E)}{\hbar^2}} \qquad (1.1.20)$$

Note that α and β are real numbers. The periodic nature of the crystal lattice suggests that the wave function satisfies the Bloch theorem (1.1.16) and can be written in the following form:

$$\Psi(x) = u_k(x)\,exp(jkx)$$

where $u_k(x)$ is a periodic function with period $a+b$, which imposes $u_k(x+n(a+b)) = u_k(x)$. One can thus write:

$$\Psi_I(x+n(a+b)) = \Psi_I(x) \, exp(jnk(a+b)) \qquad (1.1.21)$$

and

$$\Psi_{II}(x+n(a+b)) = \Psi_{II}(x) \, exp(jnk(a+b)) \qquad (1.1.22)$$

Boundary conditions must be used to calculate the four integration constants A, B, C and D of Equations 1.1.19 and 1.1.20. This can be done by imposing the condition that the wave function, $\Psi(x)$, and its first derivative, $d\Psi(x)/dx$, are continuous at $x=0$ and $x=a$. By doing so one obtains the following equations:

◊ $\Psi(x)$ is continuous at $x=0$. Thus $\Psi_I(0) = \Psi_{II}(0)$, which yields:

$$A+B=C+D \qquad (1.1.23)$$

◊ $d\Psi(x)/dx$ is continuous at $x=0$. Therefore, $d\Psi_I(0)/dx = d\Psi_{II}(0)/dx$:

$$j\beta(A-B)=\alpha(C-D) \qquad (1.1.24)$$

◊ $\Psi(x)$ is continuous at $x=a$ giving $\Psi_I(a) = \Psi_{II}(a)$. Using the Bloch theorem (Equation 1.1.16) at $x=a$ we have $\Psi_I(a) = \Psi_I(-b) \, exp(jk(a+b))$, which yields:

$$exp(jk(a+b)) \, [A \, exp(-j\beta b) + B \, exp(j\beta b)] = C \, exp(\alpha a) + D \, exp(-\alpha a) \qquad (1.1.25)$$

◊ $d\Psi(x)/dx$ is continuous at $x=a$ giving $d\Psi_I(a)/dx = d\Psi_{II}(a)/dx$. Using Bloch's theorem: $\Psi_I(a) = \Psi_I(-b) \, exp(jk(a+b))$ we obtain:

$$exp(jk(a+b)) \, j\beta[A \, exp(-j\beta b) - B \, exp(j\beta b)] = \alpha \, [C \, exp(\alpha a) - D \, exp(-\alpha a)] \qquad (1.1.26)$$

Equations (1.1.23) to (1.1.26) form a system of four equations with four unknowns: A, B, C and D. This system can be written in a matrix form:

$$\begin{bmatrix} 1 & 1 & -1 & -1 \\ j\beta & -j\beta & -\alpha & \alpha \\ exp(jk(a+b))exp(-j\beta b) & exp(jk(a+b))exp(j\beta b) & -exp(\alpha a) & -exp(-\alpha a) \\ exp(jk(a+b))j\beta exp(-j\beta b) & -exp(jk(a+b))j\beta exp(j\beta b) & -\alpha exp(\alpha a) & \alpha exp(-\alpha a) \end{bmatrix} \begin{bmatrix} A \\ B \\ C \\ D \end{bmatrix} = \begin{bmatrix} 0 \\ 0 \\ 0 \\ 0 \end{bmatrix}$$

$$(1.1.27)$$

In order to obtain a non-trivial solution for A, B, C and D, *i.e.* a solution different from $A=B=C=D=0$, the determinant of the 4×4 matrix must be equal to zero, which is equivalent to writing (see Problem 1.5):

$$\frac{\alpha^2 - \beta^2}{2\alpha\beta} \, sinh(\alpha a) \, sin(\beta b) + cosh(\alpha a) \, cos(\beta b) = cos(k(a+b)) \qquad (1.1.28)$$

The right-hand term of this equation depends only on E, through α and β (Expressions 1.1.19 and 1.1.20). Let us call this term $P(E)$ and rewrite Expression 1.1.28 in the following form:

$$P(E) = cos(k(a+b)) \qquad (1.1.29)$$

The right-hand side of Equation 1.1.29 is sketched as a function of energy in Figure 1.8. Because the argument in the exponential term of (1.1.16) must be imaginary, k must be real. Therefore, simultaneous solution of both left- and right-hand side of Equation 1.1.29 imposes that $-1 \leq P(E) \leq 1$. This defines permitted values of energy forming the energy bands, and forbidden values of energy constituting forbidden energy bands. This important result is the same to that intuitively unveiled in Section 1.1.2: in a crystal there are bands of permitted energy values separated by bands of forbidden energy values.

Note: In the case when the electron energy is greater than V_0, $E-V_0$ has a positive value and Equation 1.1.20 becomes:

$$\Psi_{II}(x) = C \, exp(j\alpha x) + D \, exp\,(-j\alpha x) \quad with \; \alpha = \sqrt{\frac{2m(V_0-E)}{\hbar^2}}$$

In that case the Krönig-Penney model yields an equation different from Relationship 1.1.28; however, the same general conclusion can be drawn, *i.e.*, the existence of permitted and forbidden energy bands.

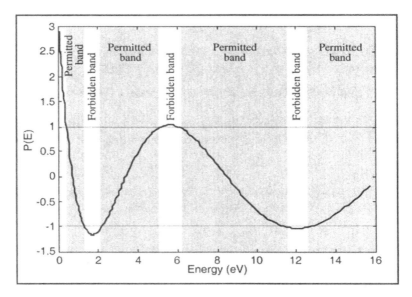

Figure 1.8: P(E) as a function of the electron energy, E, for silicon. The shaded areas correspond to the permitted energy bands, where there is a solution to Equation 1.1.29.

Using Expression 1.1.28 the *E(k)* diagram can be plotted as well. Figure 1.9 presents the energy of the electron as a function of the wave number *k*. The *E(k)* diagram for a free electron is also shown. It can be observed that the energy of the electron in a crystal coarsely represents the same dependence on k as that of a free electron. The main differences reside in the existence of forbidden energy values and curvatures of each segment of the *E(k)* curves.

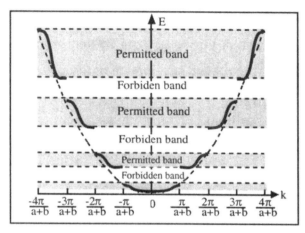

Figure 1.9: Energy versus k in a one-dimension crystal. The dotted line parabola represents the E(k) relationship for a free electron (from Figure 1.1).

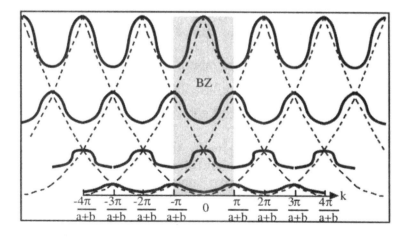

Figure 1.10: E(k) diagram of Figure 1.9, repeated with a *2π/(a+b)* period. The shaded area highlights the first Brillouin zone (BZ).[5]

Because of the periodicity of the crystal lattice (period $= a+b$), the periodicity of the reciprocal lattice (k-space) is $\frac{2\pi}{a+b}$. The $E(k)$ curve can be extended from $k = -\infty$ to $k = +\infty$ with a periodicity of $\frac{2\pi}{a+b}$, which yields the permitted energy values for the entire one-dimensional crystal (Figure 1.10).

The $E(k)$ curves shown in Figure 1.10 can be limited to k-values ranging from $\frac{-\pi}{a+b}$ to $\frac{\pi}{a+b}$ without any loss of information. This particular region of the k-space is called the *first Brillouin zone*. The second Brillouin zone extends from $\frac{-2\pi}{a+b}$ to $\frac{-\pi}{a+b}$ and from $\frac{\pi}{a+b}$ to $\frac{2\pi}{a+b}$, the third zone extends from $\frac{-3\pi}{a+b}$ to $\frac{-2\pi}{a+b}$ and from $\frac{2\pi}{a+b}$ to $\frac{3\pi}{a+b}$, etc.

Applying the Born-von Karman boundary conditions (Expression 1.1.12) to the one-dimensional crystal yields the values for k:

$$exp \; (jkN(a+b)) = 1 \Rightarrow k = \frac{2\pi n}{N(a+b)} \quad (n=0,\pm1,\pm2,\pm3,...) \qquad (1.1.30)$$

where N is the number of lattice cells in the crystal (or the number of atoms in the case of a one-dimension crystal). The length of the crystal is equal to $N(a+b)$. Since we limit our study to the first Brillouin zone, the k-values which have to be considered are given by the following relationship: $\frac{-\pi}{a+b} \leq k < \frac{\pi}{a+b}$ (the value $k = \frac{\pi}{a+b}$ is excluded because it is a duplicate of the $k = \frac{-\pi}{a+b}$ wave number). The corresponding values for n range from $-N/2$ to $(N/2-1)$. Therefore, the values of k to consider are:

$$k = \frac{2\pi n}{N(a+b)} \quad (n=0,\pm1,\pm2,\pm3,...,\pm(\frac{N}{2}-1), -N/2) \qquad (1.1.31)$$

There are thus N wave numbers in the first Brillouin zone, which corresponds to the number of elementary lattice cells. For every wave number there is a permitted energy value in *each* energy band. By virtue of the Pauli exclusion principle, each energy band can thus contain a maximum of $2N$ electrons.

The one-dimensional volume of the first Brillouin zone is equal to $2\pi/(a+b)$. Since it contains N k-values, the density of k-values in the first Brillouin zone is given by:

$$n(k) = density \; of \; k \; = \frac{number \; of \; k\text{-}values}{volume \; of \; the \; zone}$$

$$= \frac{N}{2\pi/(a+b)} = \frac{N(a+b)}{2\pi} = \frac{L}{2\pi} \qquad (1.1.32)$$

In the case of a three-dimensional crystal, energy band calculations are, of course, much more complicated, but the essential results obtained from the one-dimensional calculation still hold. In particular, there exist permitted energy bands separated by forbidden energy gaps. The 3-D volume of the first Brillouin zone is $8\pi^3 N/V$, where V is the volume of the crystal, the number of wave vectors is equal to the number of elementary crystal lattice cells, N. The density of wave vectors is given by:

$$n(k) = density\ of\ \mathbf{k} = \frac{number\ of\ \mathbf{k}\text{-}vectors}{volume\ of\ the\ zone} = \frac{NV}{8\pi^3 N} = \frac{V}{8\pi^3} \qquad (1.1.33)$$

1.1.4. Valence band and conduction band

Chemical reactions originate from the exchange of electrons from the outer electronic shell of atoms. Electrons from the most inner shells do not participate in chemical reactions because of the high electrostatic attraction to the nucleus. Likewise, the bonds between atoms in a crystal, as well as electric transport phenomena, are due to electrons from the outermost shell. In terms of energy bands, the electrons responsible for forming bonds between atoms are found in the last occupied band, where electrons have the highest energy levels for the ground-state atoms. However, there is an infinite number of energy bands. The first (lowest) bands contain core electrons such as the 1s electrons which are tightly bound to the atoms. The highest bands contain no electrons. The last ground-state band which contains electrons is called the *valence band*, because it contains the electrons that form the -often covalent- bonds between atoms.

The permitted energy band directly above the valence band is called the *conduction band*. In a semiconductor this band is empty of electrons at low temperature ($T=0K$). At higher temperatures, some electrons have enough thermal energy to quit their function of forming a bond between atoms and circulate in the crystal. These electrons "jump" from the valence band into the conduction band, where they are free to move. The energy difference between the bottom of the conduction band and the top of the valence band is called "forbidden gap" or "bandgap" and is noted E_g.

In a more general sense, the following situations can occur depending on the location of the atom in the periodic table (Figure 1.11):

A: The last (valence) energy band is only partially filled with electrons, even at $T=0K$.

B: The last (valence) energy band is completely filled with electrons at *T=0K*, but the next (empty) energy band overlaps with it (*i.e.*: an empty energy band shares a range of common energy values; $E_g < 0$).

C: The last (valence) energy band is completely filled with electrons and no empty band overlaps with it ($E_g > 0$).

In cases A and B, electrons with the highest energies can easily acquire an infinitesimal amount of energy and jump to a slightly higher permitted energy level, and move through the crystal. In other words, electrons can leave the atom and move in the crystal without receiving any energy. A material with such a property is a *metal*. In case C, a significant amount of energy (equal to E_g or higher) has to be transferred to an electron in order for it to "jump" from the valence band into a permitted energy level of the conduction band. This means that an electron must receive a significant amount of energy before leaving an atom and moving "freely" in the crystal. A material with such properties is either an *insulator* or a *semiconductor*.

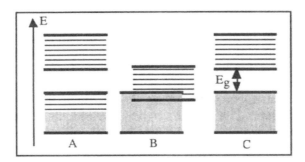

Figure 1.11: Valence band (bottom) and conduction band in a metal (A and B) and in a semiconductor or an insulator (C).[6]

The distinction between an insulator and a semiconductor is purely quantitative and is based on the value of the energy gap. In a semiconductor E_g is typically smaller than 2 eV and room-temperature thermal energy or excitation from visible-light photons can give electrons enough energy for "jumping" from the valence into the conduction band. The energy gap of the most common semiconductors are: 1.12 eV (silicon), 0.67 eV (germanium), and 1.42 eV (gallium arsenide). Insulators have significantly wider energy bandgaps: 9.0 eV (SiO_2), 5.47 eV (diamond), and 5.0 eV (Si_3N_4). In these materials room-temperature thermal energy is not large enough to place electrons in the conduction band.

Beside elemental semiconductors such as silicon and germanium, compound semiconductors can be synthesized by combining elements from column IV of the periodic table (SiC and SiGe) or by combining elements from columns III and V (GaAs, GaN, InP, AlGaAs, AlSb, GaP, AlP and AlAs). Elements from other columns can sometimes be used as well (HgCdTe, CdS,...). Diamond exhibits semiconducting properties at high temperature, and tin (right below germanium in column IV of the periodic table) becomes a semiconductor at low temperatures. About 98% of all semiconductor devices are fabricated from single-crystal silicon, such as integrated circuits, microprocessors and memory chips. The remaining 2% make use of III-V compounds, such as light-emitting diodes, laser diodes and some microwave-frequency components.

III	IV	V
B	C	N
Al	**Si**	P
Ga	Ge	As
In		Sb

Figure 1.12: Main elements used in semiconductor technology (elemental semiconductors such as Si, and compound semiconductors such as GaAs).

It is worthwhile mentioning that it is possible for non-crystalline materials to exhibit semiconducting properties. Some materials, such as amorphous silicon, where the distance between atoms varies in a random fashion, can behave as semiconductors. The mechanisms for the transport of electric charges in these materials are, however, quite different from those in crystalline semiconductors.[7]

It is convenient to represent energy bands in real space instead of k-space. By doing so one obtains a diagram such as that of Figure 1.13, where the x-axis defines a physical distance in the crystal. The maximum energy of the valence band is noted E_V, the minimum energy of the conduction band is noted E_C, and the width of the energy bandgap is E_g.

It is also appropriate to introduce the concept of a Fermi level. The Fermi level, E_F, represents the maximum energy of an electron in the material at zero degree Kelvin (*0 K*). At that temperature, all the allowed energy levels below the Fermi level are occupied, and all the energy levels above it are empty. Alternatively, the Fermi level is defined as an energy level that has a 50% probability of being filled with electrons, even though it may reside in the bandgap. In an insulator or a semiconductor, we know that the valence band is full of electrons, and the conduction band is empty at *0 K*. Therefore, the Fermi level lies somewhere in the bandgap, between E_V and E_C. In a metal, the Fermi level lies within an energy band.

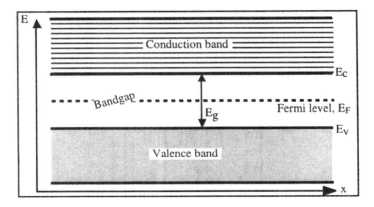

Figure 1.13: Valence and conduction band in real space.

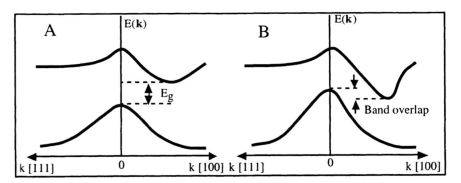

Figure 1.14: Examples of energy band extrema (minimum of the conduction band and maximum of the valence band in two crystals). In crystal A, E_g is the bandgap energy. There is no bandgap in crystal B because the conduction and the valence bands overlap.

It is impossible to represent the energy bands as a function of $k = k(k_x,k_y,k_z)$ for a three-dimensional crystal in a drawing made on a two-dimensional sheet of paper. One can, however, represent $E(k)$ along main crystal directions in k-space and place them on a single graph. For example, Figure 1.14 represents the maximum of the valence band and the minimum of the conduction band as function of k in the [100] and the [111] directions for two crystals. Crystal A is an insulator or a semiconductor ($E_g > 0$); crystal B is a metal ($E_g < 0$).

The energy band diagrams, plotted along the main crystal directions, allow us to analyze some properties of semiconductors. For instance, in Figure 1.15.B the minimum energy in the conduction band and the maximum energy in the valence band occur at the same k-values (k=0). A semiconductor exhibiting this property is called a direct-band semiconductor. Examples of direct-bandgap semiconductors include most compound elements such as gallium arsenide (GaAs). In such a semiconductor, an electron can "fall" from the conduction band into the valence band without violating the conservation of momentum law, *i.e.* an electron can fall from the conduction band to the valence band without a change in momentum. This process has a high probability of occurrence and the energy lost in that "jump" can be emitted in the form of a photon with an energy $h\nu = E_g$. In Figure 1.15.A, the minimum energy in the conduction band and the maximum energy in the valence band occur at different k-values. A semiconductor exhibiting this property is called an indirect bandgap semiconductor. Silicon and germanium are indirect-bandgap semiconductors. In such a semiconductor, an electron cannot "fall" from the conduction band into the valence band without a change in momentum. This tremendously reduces the probability of a direct "fall" of an electron from the conduction band into the valence band, as will be discussed in Chapter 3.

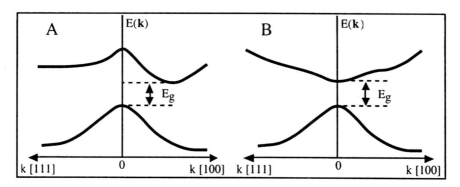

Figure 1.15: A: Indirect bandgap semiconductor, B: Direct bandgap semiconductor. [8]

1.1.5. Parabolic band approximation

For electrical phenomena, only the electrons located near the maximum of the valence band and the minimum of the conduction band are of interest. These are the energy levels where free moving electrons and missing valence electrons are found. In that case, as can be seen in Figure 1.15, the energy dependence on momentum can be approximated by a square parabolic function. Near the minimum of the conduction band one can thus write:

$$E(k)=E_{min} + A(k-k_{min})^2 \qquad (1.1.34a)$$

Near the maximum of the valence band one can write:

$$E(k)=E_{max} - B(k-k_{max})^2 \qquad (1.1.34b)$$

with A and B being constants. This approximation is called the "parabolic band approximation" and resembles the $E(k)$ relationship found for the free electron model.

1.1.6. Concept of a hole

To facilitate the understanding of electrical conduction in a solid one can make a comparison between the flow of electrical charge in the energy bands and the movement of water drops in a pipe. Let us consider (Figure 1.16.A) two pipes which are sealed at both ends. The bottom pipe is completely filled with water and the top pipe contains no water (it is filled with air). In our analogy between electricity and water, each drop of water corresponds to an electron, and the bottom and top pipes correspond to the valence and the conduction band, respectively.[9] Tilting the pipes corresponds to the application of an electric field to the semiconductor. When the filled or empty pipes are tilted, no movement or flow of water is observed, *i.e.*: there is no electric current flow in the semiconductor. Thus the semiconductor behaves as an insulator (Figure 1.16.A).

Let us now remove a drop of water from the bottom pipe and place it in the top pipe, which corresponds to "moving" an electron from the valence to the conduction band. If the pipes are now tilted, a net flow of liquid will be observed, which correspond to an electrical current flow in the semiconductor (Figure 1.16.B).

The water flow in the top pipe (conduction band) is due to the movement of the water drop (electron). In addition, there is also water flow in the bottom pipe (valence band) since drops of water can occupy the space left behind as the air bubble moves. It is, however, easier to visualize the motion of the bubble itself instead of the movement of the "valence" water.

If, in this water analogy, an electron is represented by a drop of water, a bubble or absence of water in the "valence" pipe represents what is called a *hole*. Hence, a hole is equivalent to a missing electron in the crystal valence band. A hole is not a particle and it does not exist by itself. It draws its existence from the absence of an electron in the crystal, just like a bubble in a pipe exists only because of a lack of water. Holes can move in the crystal through successive "filling" of the empty space left by a missing electron. The hole carries a positive charge $+q$, as the electron carries a negative charge $-q$ ($q = 1.6 \times 10^{-19}$ Coulomb).

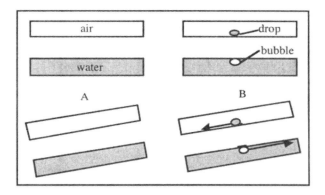

Figure 1.16: Energy bands and electrical conduction: water analogy.

1.1.7. Effective mass of the electron in a crystal

The mass m of an electron can be defined by the relationship $F=ma$ where a is the acceleration the electron undergoes under the influence of an external applied force F. The fact that the electron is in a crystal will influence its response to an applied force. As a result, the apparent, "effective" mass of the electron in a crystal will be different from that of an electron in a vacuum.

In the case of a free electron Relationship 1.1.3 can be used to find the mass of the electron [10]:

$$E = \frac{\hbar^2 k^2}{2m} \implies m = \frac{\hbar^2}{d^2 E/dk^2} \qquad (1.1.35)$$

where $m = m_o = 9.11 \times 10^{-28}$ gram is the mass of the electron in a vacuum. The mass is a constant since E is a square function of k.

Using the rightmost term of 1.1.35 as the definition of the electron mass and using Equations 1.1.28 and 1.1.29 which defines the relationship between E and k in a one-dimensional crystal, the mass of an electron within an energy band can be calculated:

$$P(E) = cos\ k(a+b)\ and\ m^* = \frac{\hbar^2}{d^2E/dk^2} \qquad (1.1.36)$$

where m^* is called the "effective mass" of the electron in a crystal. Unlike the case of a free electron the effective mass of the electron in a crystal is not constant, but it varies as a function of k (Figure 1.17).

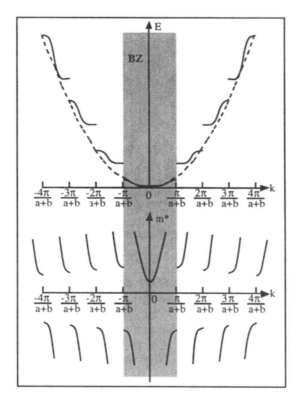

Figure 1.17: Electron energy and effective mass vs. k in a one-dimensional crystal. The first Brillouin zone (BZ) is shown in gray.

Additionally, the mass in the crystal will be different for differing energy bands. The following general observations can be made:

◊ if the electron is in the upper half of an energy band, its effective mass is negative

◊ if the electron is in the lower half of an energy band, its effective mass is positive

◊ if the electron is near the middle of an energy band, its effective mass tends to be infinite

The negative mass of electrons located in the top part of an energy band may come as a surprise, but can easily be explained using the concept of a hole. Let us consider the acceleration, a, given to an electron with charge $-q$ and negative mass, $-m^*$, by an electric field, $\mathcal{E}$. It is easy to realize that this acceleration corresponds to a hole with positive mass, $+m^*$, and positive charge, $+q$, since:

$$a = \frac{F}{-m^*} = \frac{-q\mathcal{E}}{-m^*} = \frac{q\mathcal{E}}{m^*} \quad with \quad m^* > 0 \qquad (1.1.37)$$

In the case of a three-dimensional crystal the expression of the effective mass is more complicated because the acceleration of an electron can be in a direction different from that of the applied force. In that case the effective mass is expressed by a 3×3 tensor:

$$m^* = \begin{bmatrix} m^*_{xx} & m^*_{xy} & m^*_{xz} \\ m^*_{yx} & m^*_{yy} & m^*_{yz} \\ m^*_{zx} & m^*_{zy} & m^*_{zz} \end{bmatrix}$$

with $m^*_{xx} = \dfrac{\hbar^2}{\partial^2 E/\partial k_x^2}, m^*_{xy} = \dfrac{\hbar^2}{\partial^2 E/\partial k_x \partial k_y}$, etc. $\qquad (1.1.38)$

Usually physics of semiconductor devices deals only with electrons situated near the minimum of the conduction band or holes located near the maximum of the valence band. In the case of silicon the mass of electrons near the minimum of the conduction band along the [100] k_x-direction is equal to $m^*_l = 0.97\ m_o$, and in the orthogonal directions it is $m^*_t = 0.19\ m_o$. m^*_l is called the longitudinal mass and m^*_t the transversal mass, while m_o is the mass of a free electron in a vacuum. These masses are related to the energy by the following relationship called "parabolic energy band approximation":

$$E(k) = E_c(k_m) + \frac{\hbar^2}{2m^*_l}(k_x - k_{m,x})^2 + \frac{\hbar^2}{2m^*_t}(k_y^2 + k_z^2) \qquad (1.1.39)$$

where $E_c(k_m)$ is the lowest energy state in the conduction band along the [100] or [-100] k_x-directions (Figure 1.18). In most practical cases, for the sake of simplicity, the effective mass is considered to be constant. In that case m^* is approximated by a scalar value.

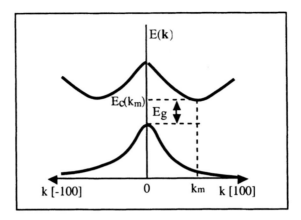

Figure 1.18: Energy bands $E(k)$ along the [100] and [-100] crystallographic directions in silicon. E_g = 1.12 eV.

In a one-dimensional case the square-law dependence of the energy on k, $E(k) = E_c(k_m) + \dfrac{\hbar^2}{2m_l^*} (k_x - k_{m,x})^2$ is illustrated by Figure 1.19.A There are two vectors k_m + dk and k_m - dk which correspond to a same energy value $E_c(k_m$ + dk). In a two-dimensional crystal (Figure 1.19.B) the locus of (k_x, k_y) values corresponding to the energy level $E_c(k_m$ + dk) is an ellipse in the (k_x, k_y) plane.

The three-dimensional case cannot be drawn on a sheet of paper, but extrapolating from the 1D and 2D cases it is easy to conceive that the k values corresponding to the energy level $E_c(k_m$ + dk) form ellipsoids in the (k_x, k_y, k_z) space (Figure 1.19.C). In a three-dimensional crystal such as silicon there are 6 equivalent crystal directions ([100], [-100], [010], [0-10], [001] and [00-1]) which present an energy minimum (conduction band minimum). The locus of k-values corresponding to a particular energy value is 6 ellipsoids (Figure 1.19.C). The center of these ellipsoids are the six k-values corresponding to the conduction band energy minima. For simplification the ellipsoids can be approximated by spheres (Figure 1.19.D), which is equivalent to equating the transverse and the longitudinal mass $(m_l^* = m_t^*)$. The energy in the vicinity of the maximum of the valence band is given by:

$$E(k) = E_v(0) - \frac{\hbar^2}{2m^*} (k_x^2 + k_y^2 + k_z^2) \qquad (1.1.40)$$

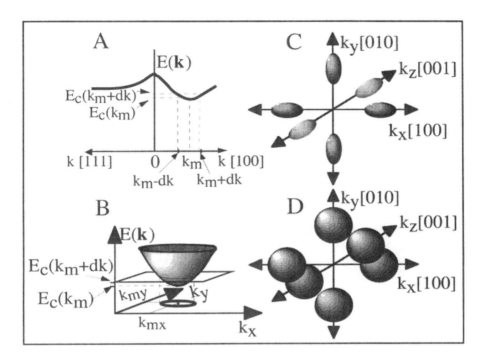

Figure 1.19: A: Values of k of equal energy. A: one-dimensional case; B: two-dimensional case; C: three-dimensional case (silicon); D: approximation of ellipsoids by spheres (silicon).

1.1.8. Density of states in energy bands

The density of permitted states in a three-dimensional crystal is given by (1.1.33). Its value is:

$$n(k)=1/8\pi^3 \qquad (1.1.41)$$

per crystal unit volume. If we define $f(k)$ as the probability that these states are occupied, then the electron density, n, in an energy band $E_n(k)$ can be calculated by integrating the product of the density of states by the occupation probability over the first Brillouin zone:

$$n = \int_{BZ} n(k)\, f(k)\, dk \qquad (1.1.42)$$

Similarly, the density of holes within an energy band is given by:

$$p = \int_{BZ} n(k)\, [1-f(k)]\, dk \qquad (1.1.43)$$

The function $n(k)$ represents the density of permitted states in an energy band. The function $f(k)$ is a statistical distribution function which is a function of the energy, $E_n(k)$. Under thermodynamic equilibrium conditions, $f(k)$ is the Fermi-Dirac distribution function defined as:[11]

Fermi-Dirac Distribution

$$f(k) = \frac{1}{1 + exp[(E_n(k)-E_F)/kT]} \qquad (1.1.44a)$$

or

$$f(E) = \frac{1}{1 + exp[(E-E_F)/kT]} \qquad (1.1.44b)$$

where E_F is an energy value called the "Fermi level", k is the Boltzmann constant, and T is the temperature in Kelvin. The Fermi-Dirac function is plotted in Figure 1.1.20 for $T > 0K$. It is worthwhile noting that $f(E) = 0.5$ if $E = E_F$, regardless of temperature. Therefore, a second definition of the Fermi level is that it is the energy level which has a 50% probability of being occupied.

In order to integrate Expressions 1.1.42 or 1.1.43 easily, the dependency of n and f on k must be transformed into a dependency on the energy, E. To do this, let us consider a unit cell of the reciprocal crystal lattice where k_x, k_y and k_z are given by Relationship 1.1.13 with $n_x=n_y=n_z=1$; the volume of this cell is equal to $k_x k_y k_z = 8\pi^3/L^3$. If the crystal has unit volume, then $L^3=1$ and the volume of a unit cell of a unit-volume crystal in k-space is equal to $8\pi^3$. In this crystal the volume of a spherical shell with a thickness dk in k-space is given by (volume of a shell of thickness dk in Figure 1.19.D):

$$\frac{4\pi}{3} [(k+dk)^3 - k^3] \cong 4\pi\, k^2\, dk \qquad (1.1.45)$$

The number of unit cells in that volume is given by the volume of the shell divided by the unit volume of the cell:

$$\frac{4\pi\, k^2\, dk}{8\pi^3} = \frac{k^2}{2\pi^2}\, dk \qquad (1.1.46)$$

The number of k vectors (and thus the number of energy levels, since there is an energy level for each k vector) is equal to the number of unit cells. Using the Pauli exclusion principle (which states that there can be only 2 electrons for each k vector), the number of electrons is given by:

$$n(k)\, dk = \frac{k^2}{\pi^2}\, dk \qquad (1.1.47)$$

Using the parabolic band approximation, $E(k)=\hbar^2 k^2/2m^*$ and using a constant effective mass, one obtains:

$$n(E) \; dE = \frac{1}{2\pi^2 \hbar^3} \; (2m^*)^{3/2} \; E^{1/2} \; dE \qquad (1.1.48)$$

This equation yields the density of states for a particle of mass m^* having an energy ranging between E and $E+dE$. In the case of electrons with a mass m_e^* located near the bottom of the conduction band, the energy is referenced to the minimum of the conduction band (E_C), which yields:

$$n(E) \; dE = \frac{1}{2\pi^2 \hbar^3} \; (2m_e^*)^{3/2} \; (E-E_C)^{1/2} \; dE \qquad (1.1.49)$$

In the case of holes with a mass m_h^* located near the top of the valence band, the energy is referenced to the maximum of the valence band (E_V), and one obtains:

$$n(E) \; dE = \frac{1}{2\pi^2 \hbar^3} \; (2m_h^*)^{3/2} \; (E_V-E)^{1/2} \; dE \qquad (1.1.50)$$

Integration of Equations 1.1.42 and 1.1.43 can now be performed. The integration can be further simplified by approximating the Fermi-Dirac (FD) distribution by the Maxwell-Boltzmann (MB) distribution. Both distributions are almost identical provided that E-E$_F$ is large enough, which is the case in typical semiconductors (*i.e.* $\frac{1}{1 + exp(u)} \cong exp(-u)$ when $u \gg 1$ (see Problem 1.10):

Maxwell-Boltzmann Distribution

$$f(E) = \frac{1}{1 + exp[(E-E_F)/kT]} \cong exp\left[-\frac{E-E_F}{kT}\right] \qquad (1.1.51)$$

$$\underset{\textit{Fermi-Dirac}}{\phantom{f(E) = \frac{1}{1 + exp[(E-E_F)/kT]}}} \quad \underset{\textit{Maxwell-Boltzmann}}{\phantom{exp\left[-\frac{E-E_F}{kT}\right]}}$$

To calculate the electron density, n, in the conduction band (CB) we replace the integral over k-values in Relationship 1.1.42 by an integral over energy:

$$n = \int_{BZ} n(k)f(k)dk = \int_{CB} n(E)f(E)dE \quad (cm-3)$$

$$= \frac{1}{2\pi^2 \hbar^3} \; (2m_e^*)^{3/2} \int_{CB} (E-E_C)^{1/2} \; exp\left[-\frac{E-E_F}{kT}\right] dE \qquad (1.1.52)$$

In a typical semiconductor the vast majority of the electrons in the conduction band have an energy close to E_C. Therefore, the lower and upper bound of the integral can thus be replaced by E_C and infinity, respectively. To integrate, a change of variables can be used where $y = (E-E_c)/kT$, which yields:

$$n = \frac{1}{2\pi^2\hbar^3}(2m_e^*)^{3/2}\int_{E_c}^{\infty}(E-E_c)^{1/2}\,exp\left[-\frac{E-E_F}{kT}\right]dE$$

$$= \frac{1}{2\pi^2\hbar^3}(2m_e^*kT)^{3/2}\,exp\left[-\frac{E_c-E_F}{kT}\right]\int_{0}^{\infty}y^{1/2}\,exp(-y)\,dy$$

$$= \frac{1}{2\pi^2\hbar^3}(2m_e^*kT)^{3/2}\,exp\left[-\frac{E_c-E_F}{kT}\right]\frac{\sqrt{\pi}}{2} \qquad (1.1.53)$$

$$\text{or:} \quad n=N_c\,exp\left[-\frac{E_c-E_F}{kT}\right] \text{ with } N_c = 2\left(\frac{2\pi m_e^*kT}{h^2}\right)^{3/2} \text{ (cm-3)} \qquad (1.1.54)$$

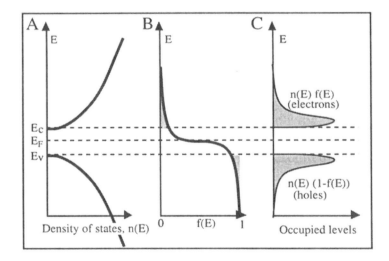

Figure 1.20: Density of states near the bottom of the conduction band and the top of the valence band (A), Fermi-Dirac function (B), and density of holes and electrons in the conduction and valence bands (C), for T≠0K. Note that at T=0K, f(E)=1 for E<E_F and f(E)=0 for E>E_F. At T=0K the valence band is completely filled with electrons (empty of holes) and there are no electrons in the conduction band.[12]

N_c is called the "*effective density of states in the conduction band*". It represents the number of states having an energy equal to E_c which, when multiplied by the occupation probability at E_c, yields the number of electrons in the conduction band. Likewise the total number of holes in the valence band can be calculated using this technique, based on Equation (1.1.43). The effective density of states for holes in the valence band is:

$$p = N_v \, exp\left[-\frac{E_F - E_v}{kT}\right] \ with \ N_v = 2\left(\frac{2\pi m_h^* kT}{h^2}\right)^{3/2} \quad (cm\text{-}3) \qquad (1.1.55)$$

The density of holes and electrons in the conduction and valence bands is shown in Figure 1.20.C for a Fermi level E_F at midpoint of E_c and E_v.

1.2. Intrinsic semiconductor

By virtue of Expressions 1.1.54 and 1.1.55 the product of the electron concentration and hole concentration in a semiconductor under thermodynamic equilibrium conditions is given by:

pn Product under Thermodynamic Equilibrium

$$pn = N_c \, exp\left[-\frac{E_c - E_F}{kT}\right] N_v \, exp\left[-\frac{E_F - E_v}{kT}\right] = N_c N_v \, exp(-E_g/kT)$$

$$= 32\left(\frac{\pi^2 k^2 m_e^* m_h^*}{h^4}\right)^{3/2} T^3 \, exp(-E_g/kT) \equiv n_i^2 \qquad (1.2.1a)$$

or: $$pn = n_i^2 \qquad (1.2.1b)$$

where n_i is called the "intrinsic carrier concentration".

A semiconductor is said to be "intrinsic" if the vast majority of its free carriers (electrons and holes) originate from the semiconductor atoms themselves. In that case if an electron receives enough thermal energy to "jump" from the valence band to the conduction band, it leaves a hole behind in the valence band. Thus, every hole in the valence band corresponds to an electron in the conduction band, and the number of conduction electrons is exactly equal to the number of valence holes:

$$p = n = n_i \qquad (1.2.2)$$

and $$E_F = \frac{E_c + E_v}{2} + \frac{3}{4} kT \, ln\left(\frac{m_h^*}{m_e^*}\right) \equiv E_i \qquad (1.2.3)$$

or, if $m_e^* = m_h^*$ (simplifying approximation): where

$$E_i = \frac{E_c + E_v}{2} \qquad (1.2.4)$$

where E_i is called the "intrinsic energy level". E_i is the energy of the Fermi level in an intrinsic semiconductor. One can generally consider that it lies right in the middle of the energy bandgap (Expression 1.2.4). n_i is the intrinsic carrier concentration (electrons or holes, $n_i=p_i$) and is a only a function of temperature and of the material through E_g. In silicon n_i is equal to 1.45×10^{10} cm^{-3} at T=300K. However, the variation of n_i with temperature is illustrated in Figure 1.21. The carrier concentration is equal to zero at $T=0K$. When temperature is raised an increasing number of electron gather sufficient thermal energy to leave the semiconductor atoms and become free to move in the conduction band. These electrons are called "free electrons". Since they can move in the crystal they can contribute to an electrical current. An equal number of "free holes" can move in the crystal and contribute to an electrical current as well.

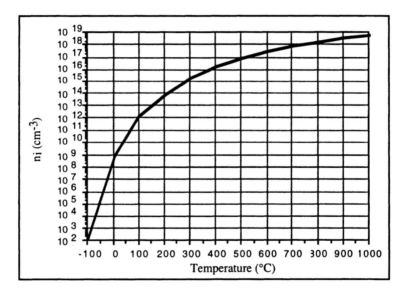

Figure 1.21: Evolution with temperature of the intrinsic carrier concentration, n_i, in silicon.

The conductivity of a material directly depends on the number of free carriers it contains (free electrons and free holes): the larger the number of carriers, the higher the conductivity. Thus, the conductivity of an intrinsic semiconductor increases with temperature (Figure 1.21).

1.3. Extrinsic semiconductor

The silicon used in the semiconductor industry has a purity level of 99.9999999%. One can, however, intentionally introduce in silicon trace amounts of elements which are close to silicon in the periodic table, such as those located in columns III (boron) or V (phosphorus, arsenic). If , for instance, an atom of arsenic is substituted for a silicon atom, it will form four bonds by sharing four electrons with the neighboring silicon atoms (Figure 1.22). The thermal energy of the crystal at room temperature is large enough to remove the loosely held fifth electron from the arsenic's outer electronic shell, such that this electron will now reside in the conduction band where it is free to move in the crystal. Arsenic atoms in silicon are called *donor* atoms because each of these atoms "donates" an electron to the crystal. The free electron can contribute to electrical conduction.

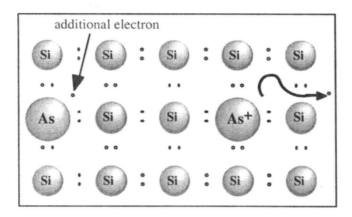

Figure 1.22: Donor impurity (arsenic in silicon). An arsenic atom introduces an extra electron in the crystal (left). An electron is released by an arsenic atom: the electron moves freely in the crystal and the arsenic atom carries a fixed positive charge (right). Note that while free electrons can move in the crystal, dopant atoms cannot.

Similarly, substituting a silicon atom with an atom from the third column of the periodic table, such as boron, will result in a missing electron (Figure 1.23). The boron atom can easily capture an electron to form a fourth bond with silicon atoms, thereby creating an immobile negatively charged boron atom. This releases a hole in the crystal, located in the valence band. This hole can move about in the crystal, thereby participating in electrical conduction. Because in silicon group III atoms create a hole which can be "filled" with an electron, these atoms are called *acceptor* atoms. Such atoms are usually introduced into the semiconductor in very small amounts (1 atom of boron per 10^6 atoms of silicon, for instance). We will see later that the introduction of even minute amounts of these impurities dramatically modify the electrical properties of a semiconductor. Atoms possessing the property of releasing or capturing electrons in a semiconductor are indiscriminately called *doping impurities, doping atoms,* or *dopants.*

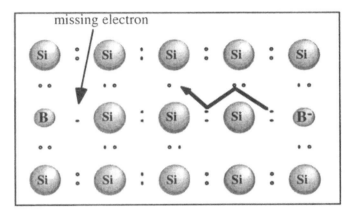

Figure 1.23: Acceptor impurity (boron in silicon). A boron atom introduces a missing electron in the crystal (left). A hole is released by a boron atom: the hole moves freely in the crystal and the boron atom carries a fixed negative charge (right). Note that while free holes can move in the crystal, dopant atoms cannot.

The introduction of a donor atom such as phosphorus (P) or arsenic (As) in silicon gives rise to a permitted energy level in the bandgap (E_d in Figure 1.24) . This level is located a few meV below the bottom of the conduction band, and at very low temperature contains the electrons which can be given by the impurity atoms to the crystal. At room temperature these electrons possess enough thermal energy (equal to kT/q = 25.6 meV) to break free from the impurity atoms and move freely in the crystal or, in other words, it can "jump" from the energy level E_d introduced by the impurity into the conduction band (Figure 1.24). When an electron moves away from a donor atom, such as arsenic (As), the atom becomes ionized (As^+) and carries a positive charge, $+q$, as shown in Figure 1.22.

Similarly, the introduction of an acceptor atom such as boron (B) in silicon gives rise to a permitted energy level in the bandgap. This level is located a few meV above the top of the valence band. At room temperature electrons in the top of the valence band possess enough thermal energy to "jump" into the energy levels created by the impurity atoms (or: valence electrons are "captured" by acceptor atoms), which gives rise to holes in the valence band. These holes are free to move in the crystal. When an electron is captured by an acceptor atom, a hole is thus released in the crystal, and the acceptor atom (boron) becomes ionized (B⁻) and carries a negative charge, -q, as shown in Figure 1.23.

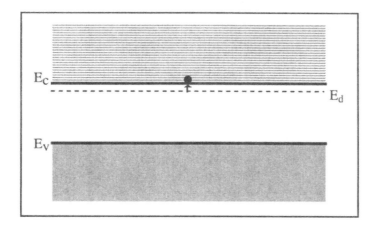

Figure 1.24: "Jump" of an electron from a donor level of energy E_d into the conduction band.

Donor and acceptor impurities are commonly introduced into semiconductors to increase electron or hole concentrations, which modifies the electrical properties of the material. The energy levels created in the bandgap by the presence of such impurities are situated close to the top of the valence band or the bottom of the conduction band. Other elements, such as gold, iron, copper and zinc introduce one or several energy levels in the bandgap of silicon. These levels are located closer to the center of the bandgap and are called "deep levels". The latter usually have a detrimental effect on semiconductors, which is why the semiconductor industry uses crystals having a very high degree of purity. The influence of deep levels on the properties of semiconductors will be discussed in Section 3.5, which is devoted to generation/recombination phenomena.

A semiconductor containing donor impurities is called an *N-type semiconductor*, since most of the carriers have a negative charge, and a semiconductor containing acceptor impurities is called a *P-type semiconductor*, since most of the carriers have a positive charge. The concentration of donor and acceptor atoms in the semiconductor are labeled N_d and N_a, respectively, and are expressed in atoms per cubic centimeters (cm^{-3}). Thus, an N-type semiconductor has more free electrons than holes, and vice-versa. However, the material itself is charge neutral due to the ionized impurities which carry a charge equal and opposite to that of the free carriers.

1.3.1. Ionization of impurity atoms

Whenever a donor (acceptor) impurity atom releases an electron (hole) it becomes ionized and carries a positive (negative) charge, $+q$ $(-q)$. If a doping atom is not ionized, it does not release a free carrier in the crystal, and therefore, does not contribute to electrical conduction. Consider a donor impurity, such as arsenic in silicon. The ionization of the arsenic atom is a reversible process which can be written as:

$$As^0 \Leftrightarrow As^+ + e^- \qquad (1.3.1)$$

where As^0 represents a non-ionized arsenic atom, and As^+ an ionized atom. Quite naturally the total impurity concentration is equal to the sum of the ionized and non-ionized impurity concentrations:

$$N_d = N_d^+ + N_d^0 \qquad (1.3.2)$$

The probability of occupancy of the donor level, E_d, can be obtained by substituting E_d for E in the Fermi-Dirac distribution function. Previously (Equation 1.1.51), the Pauli exclusion principle was taken into account for determining the probability of filling energy states. In other words, each energy level could be populated with two electrons. In this case, however, an ionized arsenic atom can receive only one electron. A correction factor, called "degeneracy factor" equal to $1/2$ must, therefore, be introduced in the Fermi-Dirac equation, which yields:

$$f(E_d) \;=\; \cfrac{1}{1 + \cfrac{1}{2}\, exp[(E_d - E_F)/kT]} \;=\; \frac{N_d^0}{N_d} \qquad (1.3.3)$$

The concentration of ionized donor atoms can be obtained using 1.3.2 and 1.3.3:

$$N_d^+ \;=\; N_d - N_d^0 = N_d\,(1 - f(E_d)) = N_d\,\cfrac{1}{1 + 2\,exp\!\left(\cfrac{E_F - E_d}{kT}\right)} \qquad (1.3.4)$$

The following example illustrate how one can determine how many donor atoms are ionized at room temperature.

Example:
Consider the following numerical example in silicon:
$E_d = E_c - 50$ meV
$E_F = (E_C + E_V)/2$ (assuming the doping concentration is very low)

$kT/q = 0.0259V$ at room temperature (T=300K)

What is the ratio of ionized donor impurities to total impurities, N_d^+/N_d?

One finds readily that $E_g/2 = 0.56eV$ and $E_F - E_d = -0.51eV$.

Therefore, using (1.3.4), $N_d^+/N_d = 0.999999996$. Thus we can conclude from this example that at room temperature, virtually all donor atoms are ionized, or in mathematical terms, $N_d^+ \cong N_d$.

In the case of acceptor impurities (boron, for example), the reversible ionization reaction is:

$$B^0 + e^- \Leftrightarrow B^- \qquad (1.3.5)$$

and we have:

$$N_a = N_a^- + N_a^0 \qquad (1.3.6)$$

Using a calculation similar to that developed for donor atoms one finds:

$$f(E_a) = \frac{1}{1 + \frac{1}{2} exp[(E_F - E_a)/kT]} = \frac{N_a^0}{N_a} \qquad (1.3.7)$$

and therefore, the probability of ionizing an acceptor is:

$$N_a^- = N_a - N_a^0 = N_a \, (1 - f(E_a)) = N_a \, \frac{1}{1 + 2 \, exp\left(\frac{E_a - E_F}{kT}\right)} \qquad (1.3.8)$$

At room temperature, virtually all acceptor atoms are ionized or, in mathematical terms, $N_a^+ \cong N_a$. Based on these derivations it is safe to assume that at room temperature every donor/acceptor atom contributes a free electron/hole to the semiconductor.

1.3.2. Electron-hole equilibrium

Consider a semiconductor crystal containing both N-type and P-type impurities. Because the crystal is charge neutral one can write:

Charge Neutrality under Thermodynamic Equilibrium

$$n + N_a^- = p + N_d^+ \qquad (1.3.9a)$$

As we have seen in the previous Section all doping impurities are ionized at room temperature, therefore, $N_a^-=N_a$ and $N_d^+=N_d$. Relationship 1.3.9a can thus be re-written in the following form:

$$n + N_a = p + N_d \Rightarrow n\text{-}p = N_d\text{-}N_a \qquad (1.3.9b)$$

Using elementary algebra one finds that $(p+n)^2 = (p\text{-}n)^2 + 4pn$. Relationship (1.3.9b) can be combined with $pn = n_i^2$ (Equation 1.2.1) to yield $(p+n)^2 = (N_d\text{-}N_a)^2 + 4n_i^2$. Since $(p+n)$ is a positive number one obtains:

$$p+n = \sqrt{(N_d\text{-}N_a)^2 + 4n_i^2} \qquad (1.3.10)$$

Combining 1.3.10 with Equation 1.3.9b one can write:

$$n = \frac{1}{2}\left[(N_d\text{-}N_a) + \sqrt{(N_d\text{-}N_a)^2 + 4n_i^2} \right] \qquad (1.3.11a)$$

and

$$p = \frac{1}{2}\left[(N_a\text{-}N_d) + \sqrt{(N_d\text{-}N_a)^2 + 4n_i^2} \right] \qquad (1.3.11b)$$

Using Relationships 1.3.11.a and 1.2.1 for an N-type semiconductor, where $N_d >> N_a$ and $N_d >> n_i$, we find that the electron and hole concentrations are given by:

Electron and Hole Concentration in N-type Semiconductor

$$n \cong N_d \quad and \quad p \cong \frac{n_i^2}{N_d} \qquad (1.3.12a)$$

Using Relationships 1.3.11.b and 1.2.1 for an P-type semiconductor, where $N_a >> N_d$ and $N_a >> n_i$, we find that the hole and electron concentrations are given by:

Electron and Hole Concentration in P-type Semiconductor

$$p \cong N_a \quad and \quad n \cong \frac{n_i^2}{N_a} \qquad (1.3.12b)$$

There are exceptions to Equations 1.3.12 a and b: at low temperatures not all impurities are ionized, and as a result, carrier freeze-out occurs: $n = N_d^+ < N_d$ and $p = N_a^- < N_a$. And at high temperature the intrinsic carrier concentration can become much larger than the concentration of carriers released by doping impurities. In that case, $N_d << n = n_i = p_i = p >> N_a$ and the semiconductor is intrinsic even though it is doped. The influence of high and low temperatures on carrier concentration is illustrated by Problem 1.12.

1.3.3. Calculation of the Fermi Level

In the case of an N-type semiconductor, combining Relationships 1.1.54 and 1.3.12a yields:

$$n = N_d = N_c \, exp\left[- \frac{E_c - E_F}{kT} \right] \tag{1.3.13a}$$

from which we find:

$$E_c - E_F = kT \, ln \left(\frac{N_c}{N_d} \right) \tag{1.3.14a}$$

Using Expression (1.2.5):

$$n_i = N_c \, exp\left[- \frac{E_c - E_i}{kT} \right] \Rightarrow E_i - E_c = kT \, ln \left(\frac{n_i}{N_c} \right)$$

one finally obtains:

$$E_i - E_F = E_c - E_F + (E_i - E_c) = kT \left(ln \left(\frac{N_c}{N_d} \right) + ln \left(\frac{n_i}{N_c} \right) \right)$$

$$\Downarrow$$

$$E_F - E_i = kT \, ln \left(\frac{N_d}{n_i} \right) \quad or \quad n = N_d = n_i \, exp\left[\frac{E_F - E_i}{kT} \right] \tag{1.3.15a}$$

Hence the Fermi level, E_F, can be calculated from Equation 1.3.15a if the doping concentration is known. In an N-type semiconductor the Fermi level is located in the upper half of the bandgap, above the intrinsic energy level, E_i. The Fermi level increases logarithmically with the donor atom concentration, N_d. It is now possible to introduce a new variable, the Fermi potential, Φ_F (unit: volt). It is defined by the following relationship:

$$-q\Phi_F = E_F - E_i \tag{1.3.16}$$

Using Equation 1.3.15a the relationship between the electron concentration and the Fermi potential can be obtained:

Fermi Potential (N-Type Semiconductor)

$$n = n_i \, exp\left[\frac{-q\Phi_F}{kT} \right] \quad or \quad \Phi_F = - \frac{kT}{q} \, ln \left(\frac{n}{n_i} \right) = - \frac{kT}{q} \, ln \left(\frac{N_d}{n_i} \right) \tag{1.3.17a}$$

For a P-type semiconductor equations 1.3.13a through 1.3.17a will use the same numbering system where the "a" is replaced by "b" in the equation. Combining Relationships 1.1.55 and 1.3.12b yields:

$$p = N_a = N_v \, exp\left[-\frac{E_F-E_v}{kT}\right] \tag{1.3.13b}$$

from which we find:

$$E_F - E_v = kT \, ln\left(\frac{N_v}{N_a}\right) \tag{1.3.14b}$$

Using Expression 1.2.5

$$n_i = N_v \, exp\left[-\frac{E_i-E_v}{kT}\right] \Rightarrow E_v - E_i = kT \, ln\left(\frac{n_i}{N_v}\right)$$

one finally obtains:

$$E_i - E_F = -(E_v - E_i) - (E_F - E_v) = - kT\left(ln\left(\frac{N_v}{N_a}\right) + ln\left(\frac{n_i}{N_v}\right)\right)$$

$$\Downarrow$$

$$E_i - E_F = kT \, ln\left(\frac{N_a}{n_i}\right) \quad or \quad p = N_a = n_i \, exp\left[\frac{E_i-E_F}{kT}\right] \tag{1.3.15b}$$

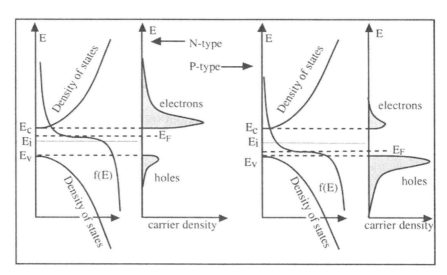

Figure 1.25: Location of the Fermi level, Fermi-Dirac distribution, f(E), and electron and hole concentration in an N-type and a P-type semiconductor. [13]

Equation 1.3.15b allows one to find the position of the Fermi level, E_F, in the bandgap. In a P-type semiconductor the Fermi level is located in the lower half of the bandgap, below the intrinsic energy level, E_i. The Fermi level decreases with increasing acceptor atom concentration, N_a.

Using Equation 1.1.16, the relationship between the Fermi potential, Φ_F, and the hole concentration can be obtained:

Fermi Potential (P-Type Semiconductor)

$$p = n_i \, exp\left[\frac{q\Phi_F}{kT}\right] \quad or \quad \Phi_F = \frac{kT}{q} \, ln\left(\frac{p}{n_i}\right) = \frac{kT}{q} \, ln\left(\frac{N_a}{n_i}\right) \qquad (1.3.17b)$$

Note that Φ_F is positive in a P-type semiconductor and negative in an N-type semiconductor. A graphical representation of electron and hole concentrations for both N- and P-type semiconductors is shown in figure 1.25. Note the position of the Fermi level, E_F, and the asymmetry of carrier densities for both types.

1.3.4. Degenerate semiconductor

We have hitherto assumed that the introduction of doping impurities in a semiconductor does not affect certain intrinsic parameters of the crystal, such as the width of the energy bandgap. As we have seen before the presence of donor doping atoms such as phosphorus or arsenic introduces a permitted energy level, E_d, in the bandgap. Typical doping concentrations are in the 10^{15} to 10^{18} atoms/cm^3 range, which is small compared to the actual number of semiconductor atoms (5×10^{22} atoms/cm^3 in silicon).

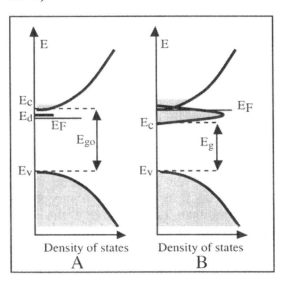

Figure 1.26: Density of states in a non-degenerate N-type semiconductor (A) and a degenerate N-type semiconductor (B). The gray areas correspond to states populated with electrons. [14]

If a very large concentration of impurities is introduced (e.g.: 10^{20} atoms/cm^3) the permitted level E_d spreads out and "degenerates" into a permitted band which overlaps with the conduction band. As a result the width of the bandgap is reduced (from E_{g0} to E_g in Figure 1.26) and the properties of the semiconductor are significantly modified. Such a semiconductor is called a "degenerate" semiconductor or a "degenerately doped" semiconductor. A degenerate semiconductor exhibits electrical properties similar to those of a metal.

1.4. Alignment of Fermi levels

Often, the doping concentration in a semiconductor is not one constant value throughout the material. Consider a piece of N-type semiconductor in which the doping concentration varies along one direction of space, x. The concentration of doping atoms is described by the function $N_d(x)$ shown in Figure 1.27.A.

Consider now that leftmost and rightmost parts of the sample are separated. According to Relationship 1.3.15a, $E_F(right) > E_F(left)$ because $N_d(right) > N_d(left)$ (Figure 1.27.B). Imagine a test energy level in the bandgap having an energy, E_T, located between $E_F(right)$ and $E_F(left)$. In the left part of the sample the test level has a low probability of being populated with an electron, because $E_T > E_F$. In the right part of the sample, on the other hand, the test level has a high probability of being populated with an electron, because $E_T < E_F$.

Let us now consider the entire sample, and in particular, focus on the middle region where the doping concentration changes abruptly. If the energy bands near $x = x_0$ stay as they are in the leftmost and rightmost parts of the sample, the test level E_T will have both a high and a low probability of being occupied by an electron, which is a contradiction in itself. The test level must have a single occupation probability. This condition can be satisfied only if E_F at the immediate left of x_0 is equal to E_F at the immediate right of x_0. And since this condition must be true for any arbitrary position along the x-axis, the Fermi level must be unique and constant throughout the sample. This is a very important property of the Fermi level, which can be enunciated the following way: *at thermodynamic equilibrium the Fermi level in a structure is unique and constant*. This property not only applies to non-homogeneously doped semiconductors, but to metal-semiconductor structures and contacts between different semiconductors. Because E_F is constant the conduction, valence, and intrinsic levels bend within a transition region around x_0 (Figure 1.27.C).

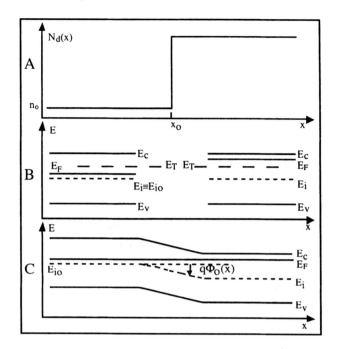

Figure 1.27: A: Inhomogenous doping profile.
B: Energy levels at the left and the right of the sample.
C: Band diagram in the complete structure.

Under thermodynamic equilibrium conditions electrons are transferred from the electron-rich right part of the sample (where the Fermi level is highest) into the electron-poor left part of the sample (where the Fermi level is lowest), through a diffusion process which will be discussed in Chapter 2. To make a comparison with fluid mechanics the alignment of the Fermi levels in the sample is similar to the alignment of the water levels in glasses of water connected together (Figure 1.28), where the transfer of electrons by a diffusion mechanism would find its equivalent in the transfer of water molecules due to a pressure differential. The diffusion process (electron transfer or water transfer) ceases when an equilibrium state is reached.

Since Relationships 1.3.13 a and b to 1.3.15 a and b are valid at any location along the x-axis, a constant Fermi level imposes a curvature of all energy bands and energy levels, E_c, E_v and E_i. However, all these levels remain parallel to one other, due to the fact that the bandgap energy is a constant of the material. The magnitude of this energy level bending reflects the presence of an internal potential, noted $\Phi_o(x)$ which, once multiplied by $-q$, is equal to the variation of the energy levels E_c, E_v and E_i between the left and the right of the sample (Figure 1.27.C). The internal potential is a real electrical potential variation due to the appearance of an electric field in the semiconductor caused by the charge imbalance resulting from the diffusion of electrons from one part of the semiconductor to the other when thermodynamic equilibrium is established. Since the electron concentration is related to $E_F\text{-}E_i$ by Relationship 1.3.15a, one can write:

$$n(x) = n_i \, exp\left[\frac{E_F\text{-}E_i(x)}{kT}\right] \tag{1.3.18}$$

or, using the notations of Figure 1.27:

$$n(x) = n_i \, exp\left[\frac{E_F\text{-}E_{io} + q\Phi_o(x)}{kT}\right] \tag{1.3.19}$$

or:

Boltzmann Relationship for Electrons

$$n(x) = n_i \, exp\left[\frac{E_F\text{-}E_{io} + q\Phi_o(x)}{kT}\right] = n_o \, exp\left[\frac{q\Phi_o(x)}{kT}\right] \tag{1.3.20a}$$

where n_o is the electron concentration in the left region of the sample, taken as reference. E_{io} is the midgap energy in the left part of the sample, also taken as reference. It is easy to show that an equivalent relationship can be derived for holes:

Boltzmann Relationship for Holes

$$p(x) = n_i \, exp\left[-\frac{E_F\text{-}E_{io} + q\Phi_o(x)}{kT}\right] = p_o \, exp\left[\frac{-q\Phi_o(x)}{kT}\right] \tag{1.3.20b}$$

Relationships 1.3.20a and 1.3.20b are called the "Boltzmann relationships". They will play an important role in the theory of the PN junction (Chapter 4).

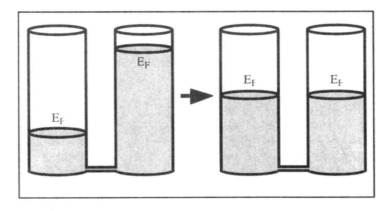

Figure 1.28: Bernoulli principle in fluids.

Important Equations

Fermi-Dirac Distribution

$$f(k) = \frac{1}{1 + exp[(E_n(k)-E_F)/kT]} \tag{1.1.44a}$$

or

$$f(E) = \frac{1}{1 + exp[(E-E_F)/kT]} \tag{1.1.44b}$$

Maxwell-Boltzmann Distribution

$$\underbrace{f(E) = \frac{1}{1 + exp[(E-E_F)/kT]}}_{Fermi\text{-}Dirac} \cong \underbrace{exp\left[-\frac{E-E_F}{kT}\right]}_{Maxwell\text{-}Boltzmann} \tag{1.1.51}$$

Free Carrier Concentration

$$n = N_c\, exp\left[-\frac{E_c-E_F}{kT}\right] \text{ with } N_c = 2\left(\frac{2\pi m_e^* kT}{h^2}\right)^{3/2} \tag{1.1.54}$$

$$p = N_v\, exp\left[-\frac{E_F-E_v}{kT}\right] \text{ with } N_v = 2\left(\frac{2\pi m_h^* kT}{h^2}\right)^{3/2} \tag{1.1.55}$$

pn Product under Thermodynamic Equilibrium

$$pn = N_c\, exp\left[-\frac{E_c-E_F}{kT}\right] N_v\, exp\left[-\frac{E_F-E_v}{kT}\right] = N_c N_v\, exp(-E_g/kT)$$

$$= 32\left(\frac{\pi^2 k^2 m_e^* m_h^*}{h^4}\right)^{3/2} T^3\, exp(-E_g/kT) \equiv n_i^2 \tag{1.2.1a}$$

or:
$$pn = n_i^2 \tag{1.2.1b}$$

Charge Neutrality under Thermodynamic Equilibrium

$$n + N_a^- = p + N_d^+ \tag{1.3.9a}$$

Electron and Hole Concentration

$$N\text{-type semiconductor:} \quad n \cong N_d \quad and \quad p \cong \frac{n_i^2}{N_d} \qquad (1.3.12a)$$

$$P\text{-type semiconductor:} \quad p \cong N_a \quad and \quad n \cong \frac{n_i^2}{N_a} \qquad (1.3.12b)$$

Fermi Potential

N-Type Semiconductor:

$$n = n_i \, exp\left[\frac{-q\Phi_F}{kT}\right] \quad or \quad \Phi_F = -\frac{kT}{q} \, ln\left(\frac{n}{n_i}\right) = -\frac{kT}{q} \, ln\left(\frac{N_d}{n_i}\right) \qquad (1.3.17a)$$

P-Type Semiconductor:

$$p = n_i \, exp\left[\frac{q\Phi_F}{kT}\right] \quad or \quad \Phi_F = \frac{kT}{q} \, ln\left(\frac{p}{n_i}\right) = \frac{kT}{q} \, ln\left(\frac{N_a}{n_i}\right) \qquad (1.3.17b)$$

Boltzmann Relationships

$$n(x) = n_i \, exp\left[\frac{E_F - E_{io} + q\Phi_o(x)}{kT}\right] = n_o \, exp\left[\frac{q\Phi_o(x)}{kT}\right] \qquad (1.3.20a)$$

$$p(x) = n_i \, exp\left[-\frac{E_F - E_{io} + q\Phi_o(x)}{kT}\right] = p_o \, exp\left[\frac{-q\Phi_o(x)}{kT}\right] \qquad (1.3.20b)$$

Problems

Problem 1.1:

Find the value of the coefficient A_n in $\Psi_n(x) = A_n \sin(\frac{n\pi x}{a})$ for a particle confined to an infinite potential well (Expression 1.1.10).

 Problem 1.2:

The wave function of a particle in a box is given by:

$$\Psi_n(x) = A_n \sin(\frac{n\pi x}{a}) \quad with \quad A_n = \sqrt{\frac{2}{a}}$$

The energy levels are given by: $E_n = \dfrac{n^2\pi^2\hbar^2}{2ma^2}$

We will use the following data:

m=9.11e-31;   % Electron mass (kg)
h=6.63e-34;   % Planck constant (J * sec)

hb=h/2/pi;   % Reduced Planck constant "$\hbar$" (J * sec)
q=1.6e-19;   % Electron charge (C)
a=1e-9;   % Width of potential well (m)

Produce a graph similar to Figure 1.2c using Matlab . The unit for energy in the plot must be electron-volts (eV). The unit for the wave function is $\sqrt{1/meter}$. Hint: it is possible to plot different units (eV and wave function unit) on the same y-axis, but if you do it as such, the wave functions and energy levels will have magnitudes with such difference that the wave functions will be much larger compared to the energy levels. Therefore, the amplitude of the wave functions must be divided by 100,000 in order to get a "nice-looking" graph.

 Problem 1.3:
Using Matlab and a finite-difference numerical method, calculate the first wave function of an electron in the four potential wells shown below. The first of those is the classical particle-in-a-box problem. Let a = 40 nm and ΔV = 10 mV.

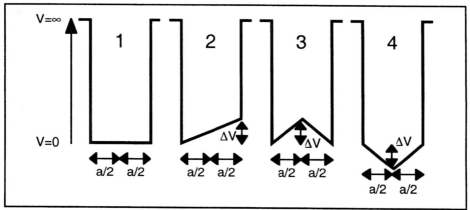

Problem Figure 1.3

Here is a description of the finite-difference technique to be used to solve this problem:

The time-independent Schrödinger equation can be written in the form:

$$\frac{-\hbar^2}{2m\,(\Delta x)^2}\,A\,\Psi\,-q\,V\,\Psi = E\,\Psi$$

where A, V, and E are n×n matrices, where n is the number of mesh points. In its discrete form, the second-derivative operator can be written:

$$\frac{d^2\Psi}{dx^2} \cong \frac{\Psi(x_{i-1}) + \Psi(x_{i+1}) - 2\Psi(x)}{(\Delta x)^2}$$

where Δx in the right-hand side of the latter expression is the constant distance between two successive mesh points. If n=6, for instance, the Schrödinger equation can be written as:

$$
\frac{-\hbar^2}{2m\,(\Delta x)^2}
\begin{bmatrix}
1 & 0 & 0 & 0 & 0 & 0 \\
1 & -2 & 1 & 0 & 0 & 0 \\
0 & 1 & -2 & 1 & 0 & 0 \\
0 & 0 & 1 & -2 & 1 & 0 \\
0 & 0 & 0 & 1 & -2 & 1 \\
0 & 0 & 0 & 0 & 1 & 1
\end{bmatrix}
\begin{bmatrix}
\Psi(x_1) \\ \Psi(x_2) \\ \Psi(x_3) \\ \Psi(x_4) \\ \Psi(x_5) \\ \Psi(x_6)
\end{bmatrix}
- q
\begin{bmatrix}
V_1 & 0 & 0 & 0 & 0 & 0 \\
0 & V_2 & 0 & 0 & 0 & 0 \\
0 & 0 & V_3 & 0 & 0 & 0 \\
0 & 0 & 0 & V_4 & 0 & 0 \\
0 & 0 & 0 & 0 & V_5 & 0 \\
0 & 0 & 0 & 0 & 0 & V_6
\end{bmatrix}
\begin{bmatrix}
\Psi(x_1) \\ \Psi(x_2) \\ \Psi(x_3) \\ \Psi(x_4) \\ \Psi(x_5) \\ \Psi(x_6)
\end{bmatrix}
= E
\begin{bmatrix}
\Psi(x_1) \\ \Psi(x_2) \\ \Psi(x_3) \\ \Psi(x_4) \\ \Psi(x_5) \\ \Psi(x_6)
\end{bmatrix}
$$

where $V_i = V(x_i)$ (vector FI in the Matlab file). The wave functions and the energy levels are found by calculating the eigenvalues of the matrix "SCH" defined as:

$$
SCH \equiv \frac{-\hbar^2}{2m\,(\Delta x)^2}\,A - q\,FI
$$

using the Matlab function [PSI,V]=eig(SCH,'nobalance'). Then the wave functions must be sorted by ascending energy values, and the wave function corresponding to the lowest energy value is finally plotted.

 Problem 1.4:
Using Matlab and a finite-difference numerical method, calculate the first wave function of an electron in the third potential well of Problem 1.3 for $\Delta V = 0, 1, 2$ and 5 mV.

Problem 1.5: Derive Equation 1.1.28 from Expression 1.1.27.

Problem 1.6:
Consider Equation 1.1.28:

$$
\frac{\alpha^2 - \beta^2}{2\alpha\beta}\,\sinh(\alpha a)\,\sin(\beta b) + \cosh(\alpha a)\,\cos(\beta b) = \cos(k(a+b))
$$

The equation can be simplified by taking the bottom of the potential wells of Figure 1.7 as reference, such that $V_1=0$. Assume that the potential wells are very narrow (a $\rightarrow$ 0), and obey the following characteristics: a $\rightarrow$ 0 and the $V_0 \times a$ product is constant and equal to an arbitrary value, δ.

Remember that $\beta = \sqrt{\dfrac{2m(E-V_1)}{\hbar^2}}$ and $\alpha = \sqrt{\dfrac{2m(V_0-E)}{\hbar^2}}$.

1: Simplify Equation 1.1.28 for the case where a $\rightarrow$ 0.

2: Using this result show that the Krönig and Penney model reduces to the free-electron model when $V_0=0$, *i.e.* when the potential wells vanish.

 Problem 1.7:
The solution to Part 1 of Problem 1.6 where a=0 is:

$$
P(E) = \frac{2mb\delta}{\hbar^2}\,\frac{\sin(\beta b)}{2\beta b} + \cos(\beta b) = \cos(kb)
$$

Since a=0 we have that a+b=b and, therefore, b is the lattice parameter. In Figures 1.8, 1.9, 1.10 and 1.18 the x-axis is k (cm^{-1}). Here we will use the dimensionless *kb* product as x-axis for 0<*kb*<16. Using that result plot P(E) and E(k) for a one-dimensional silicon crystal using the Krönig and Penney model, as well as the effective mass of the electron (normalized to m_0), as a function of *kb*, in order to obtain graphs similar to those in Figures 1.8, 1.9, 1.10 and 1.18.

Use the following data for silicon:
Silicon lattice parameter: b = 5.43 Å = 5.43×10^{-10} m
$\hbar$ = h/2π = 1.0546×10^{-34} J.s
Free electron mass: m_0 = 9.11×10^{-31} kg
V_0 = 2.18×10^{-18} V $\Rightarrow$ V_0 × a = δ = 2.725×10^{-10} eV m^{-1} = 4.352×10^{-29} J m^{-1}
1 eV = 1.6×10^{-19} J

Note: To obtain similar results for diamond, use b = 3.56 Å and V_0 = 1×10^{-17} V; for germanium, use b = 5.65 Å and V_0 = 1.4×10^{-18} V; for gray tin, use b = 6.49 Å and V_0 = 2×10^{-19} V.

Problem 1.8:
In Section 1.1.7 it was shown that the concentration of electrons in the conduction band per eV is equal to N(E) $\equiv$ n(E) f(E) where n(E) and f(E) are defined as:

$$n(E) = \frac{1}{2\pi^2\hbar^3} (2m_e^*)^{3/2} (E-E_c)^{1/2} \qquad (1.1.49)$$

$$f(E) = \frac{1}{1 + \exp[(E-E_F)/kT]} \qquad (1.1.51)$$

Using the following data for intrinsic ($E_F = E_i$) silicon at room temperature:

m_e^* = m_0 =9.11e-31; % Electron mass (kg)
h=6.63e-34; % Planck constant (J * sec)
hb=h/2/pi; % Reduced Planck constant (J * sec)
q=1.6e-19; % electron charge (C)
k=1.3805e-23 % Boltzmann constant (J/K)
Ecf=0.55*q; % E_{cf} is defined as E_C-E_F (J)
  %This is half the energy bandgap of silicon
T=300; % temperature (K)

1) Plot the electron density N(E) (unit: eV^{-1} cm^{-3}) as a function of energy for E_C < E < E_C+0.3 eV (Figure 1.20C) using Matlab.

2) Using a simple numerical integration method and Matlab, calculate the electron concentration in the conduction band: n = $\int_{E_C}^{E_C+0.3eV}$ N(E) dE (units: cm^{-3}).

Problem 1.9:

Using the program developed in Problem 1.8, where the intrinsic electron concentration, n=n$_i$, in silicon was calculated at T=300 K, calculate and plot the intrinsic carrier concentration, n$_i$, in silicon as a function of temperature, from -100°C to +1000°C, such that you produce a curve similar to that of Figure 1.21.

 Problem 1.10:

Plot the Fermi-Dirac (FD) and the Maxwell-Boltzmann (MB) distribution curves for energies ranging between 0 and 0. 5 eV at T = 300K, given that E$_F$ = 0.25 eV. Interpret the curves and comment on the appropriateness of replacing the FD distribution by the MB distribution in Relationship 1.1.52.

Problem 1.11:

As can be seen in Figure 1.25 the concentration of electrons in the conduction band reaches a maximum at some energy value ΔE above E$_C$. That energy value is independent of the position of the Fermi level. Find the value of that energy assuming T = 300 K. The answer should have the form: E = E$_C$ + ΔE (eV).

 Problem 1.12:

This Problem introduces the concept of carrier freeze-out at low temperature, as well as the effect of high temperature on total carrier concentrations. Arsenic atoms introduce a donor energy level at 0.054 eV below E$_C$ in silicon. Using the results of Problem 1.9 and Relationship 1.3.4, plot the concentration of electrons (both intrinsic and dopant electrons) in the conduction band of arsenic-doped silicon as a function of temperature (-250°C < T < 1000°C).

The arsenic doping concentration is 10^{16} cm^{-3}.

References

1   J.L. Moll, *Physics of semiconductors*, McGraw-Hill, pp. 32-52, 1964
2   J.P. McKelvey, *Solid-state and semiconductor physics*, Harper International, p. 209, 1966
3   J.M. Ziman, *Principles of the theory of solids*, 2nd Edition, Cambridge University Press, p. 15, 1972
4   R. de L. Krönig and W. G. Penney, "Quantum Mechanics of Electrons in Crystal Lattices", *Proceedings of the Royal Society* (London), Vol. A-130, p.499, 1931
5   N.W. Ashcroft, N.D. Mermin, *Solid-state physics*, Holt, Rinehart and Winston, p. 160, 1976
6   J.P. McKelvey, *Solid-state and semiconductor physics*, Harper International, p. 246, 1966
7   L.L. Kazmerski, *Polycrystalline and amorphous thin films and devices*, Academic Press, pp. 17-57, 1980
8   H.F. Wolf, *Semiconductors*, J. Wiley and Sons, p. 51, 1971
9   R.S. Muller and T.I. Kamins, *Device electronics for integrated circuits, 2nd edition*, J. Wiley & Sons, p. 9, 1986
10   J.P. McKelvey, *Solid-state and semiconductor physics*, Harper International, pp. 217-224, 1966
11   R.P. Pierret, *Advanced semiconductor fundamentals*, Modular Series on Solid-State Devices, Vol. IV, Addison-Wesley, p. 100, 1989
12   J.P. McKelvey, *Solid-state and semiconductor physics*, Harper International, p. 261, 1966
13   J.P. McKelvey, *Solid-state and semiconductor physics*, Harper International, p. 263, 1966
14   H.F. Wolf, *Semiconductors*, J. Wiley and Sons, p. 50, 1971

Chapter 2

THEORY OF ELECTRICAL CONDUCTION

In this Chapter the equations describing the movement of electric charges, as well as the relationships between charge, electric field and potential, will be derived. Electrons and holes are no longer treated separately, but are considered as macroscopic carrier populations or carrier concentrations. As a result the use of quantum mechanics is no longer required. Rather, Maxwell's equations and concepts such as the conservation of charge and the diffusion resulting from concentration gradients will be used.

2.1. Drift of electrons in an electric field

The electrons we have considered so far were found in ideal crystals with perfectly periodic potential variations. Actual crystals contain defects such as interstitials and vacancies due to displaced or missing atoms, and trace impurities. Furthermore, the atoms vibrate around their equilibrium position. The amplitude of these vibrations depends, among others, on temperature. These vibrations can be studied formally using quantum mechanics. From the study of these vibrations emerges the concept of a phonon. The phonon is a quasi-particle representing the propagation of vibration -or heat- through the crystal.[1] Both crystal imperfections and phonons can interact with electrons through the distortions they induce in the periodic potential of the crystal lattice.

The interaction between a free electron and phonons or crystal defects can be viewed as a series of collisions obeying the principles of conservation of energy and momentum.[2] As a consequence electrons are never at rest and are submitted to a perpetual random motion that can be compared to the Brownian motion of fine particles in a liquid. The trajectory of electrons is thus a series of random velocity vectors. In the absence of an applied external force all these small movements average out and the net displacement of the electron is zero, as shown in Figure

2.1.A. When an electric field is applied, on the other hand, a net drift of the electron in the opposite direction of the electric field is observed (Figure 2.1.B). It is worthwhile noting that the random thermal velocity of electrons is much larger than the velocity produced by imposing an electric field. To obtain the current flow resulting from this process one must calculate the average drift velocity of the electrons caused by the electric field.

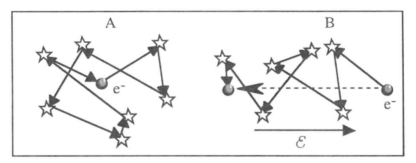

Figure 2.1: Electron motion: A: in the absence of an electric field, B: in the presence of an electric field. The stars represent collisions.

The analogy with Brownian motion in a liquid allows us to write two hypotheses concerning the motion of an electron:

◊ Each electron in the conduction band moves freely in the crystal between each collision. The average time between two collisions is called "relaxation time", and is noted τ_n. The relaxation time for electrons in a semiconductor is on the order of a tenth of a picosecond at room temperature, during which the electron can travel on the order of 10 nanometers.

◊ The direction of the electron motion after a collision is random. Collision events, are therefore, isotropic.

Among all the electrons in the conduction band there are $n(t_o)$ electrons which, at the instant $t=t_o$, undergo a collision event. Let us follow the evolution of this electron population. At $t > t_o$, some of these electrons will already have undergone new collisions. Therefore, at $t > t_o$, there is a smaller number of electrons, $n(t)$, which have not yet undergone a collision event. The population of these electrons, $n(t)$, decreases between t and $t+dt$ by an amount dn according to the following equation:

$$dn = \frac{-1}{\tau_n} n(t) \ dt \qquad\qquad (2.1.1)$$

Integrating this equation between t_0 and t the evolution of the number of electrons that have not undergone a collision since t_0 can be obtained:

$$n(t) = n(t_0) \, exp \left[\frac{-(t - t_0)}{\tau_n} \right] \qquad (2.1.2)$$

Let us now describe the influence of a time-independent electric field, $\mathcal{E}$, on an electron. The equation for the movement of a quasi-free electron with an effective mass m_e^* is:

$$F = ma \quad \Rightarrow \quad m_e^* \frac{dv}{dt} = -q\mathcal{E} \qquad (2.1.3)$$

Using Expression 2.1.3 and assuming that the effective mass is isotropic, the velocity of an electron which has not had a collision since t_0 is, at time t :

$$v(t) = v(t_0) - \frac{1}{m_e^*} q \, (t-t_0) \, \mathcal{E} \qquad (2.1.4)$$

Since the average velocity at t_0, $v(t_0)$, is equal to zero (isotropic collision events are one of our starting hypotheses) one can write:

$$v(t) = - \frac{1}{m_e^*} q \, (t-t_0) \, \mathcal{E} \qquad (2.1.5)$$

where $v(t)$ is the velocity vector at time t. This relationship is valid for the $-dn$ ($dn<0$) electrons from Relationship 2.1.1 that undergo a collision between t and dt, but which traveled collision-free from t_0 to t. Integrating Relationship 2.1.5 for t ranging from t_0 to ∞ (or for $n(t)$ ranging from $n(t_0)$ to 0) one obtains the average drift velocity, v_{dn}, of the $n(t_0)$ electron population, *i.e.* the drift velocity resulting from the application of the electric field:

$$v_{dn} = \frac{1}{n(t_0)} \int_{n(t_0)}^{0} \frac{-1}{m_e^*} q \, (t-t_0) \, \mathcal{E} \, (-dn) \qquad (2.1.6)$$

2.2. Mobility

Using Relationship 2.1.2, Equation 2.1.6 can be converted into an integral over time, which yields:

$$v_{dn} = - \frac{q}{m_e^*} \mathcal{E} \int_{t_0}^{\infty} \frac{1}{\tau_n} (t-t_0) \, e^{-(t-t_0)/\tau_n} \, dt \qquad (2.2.1)$$

and since

$$\int_{t_0}^{\infty} \frac{1}{\tau_n} (t-t_o) \, e^{-(t-t_0)/\tau_n} \, dt \;=\; \tau_n \int_0^{\infty} \frac{t-t_o}{\tau_n} \, e^{-(t-t_0)/\tau_n} \, d\!\left(\frac{t-t_o}{\tau_n}\right) = \tau_n \int_0^{\infty} y \; e^{-y} \, dy = \tau_n$$

we finally obtain:

$$v_{dn} = -\frac{q \, \tau_n}{m^*_e} \, \mathcal{E} \equiv -\mu_n \, \mathcal{E} \qquad\qquad (2.2.2)$$

where μ_n is called the *mobility* of the electrons in the conduction band. The unit for the mobility (velocity divided by an electric field) is cm^2 $volt^{-1}$ sec^{-1}. Using Relationship 2.2.2 the mobility is defined by the following relationship:

$$\mu_n = \frac{q \, \tau_n}{m^*_e} \qquad\qquad (2.2.3)$$

Mobility is proportional to the relaxation time of the electrons and inversely proportional to their effective mass. Since mobility is proportional to the relaxation time it decreases with temperature because thermal lattice vibrations -or phonons- increase with increasing temperature. Similarly, impurities and defects cause electron scattering (collisions), and therefore, mobility decreases with increasing impurity or defect concentration.

A similar derivation can made for holes in the valence band and yields:

$$v_{dp} = \frac{q \, \tau_p}{m^*_h} \, \mathcal{E} \equiv \mu_p \, \mathcal{E} \qquad\qquad (2.2.4)$$

where μ_p is the hole mobility which is defined by:

$$\mu_p = \frac{q \, \tau_p}{m^*_h} \qquad\qquad (2.2.5)$$

The actual effective mass of electrons and holes is anisotropic (see Relationship 1.1.38) and the mobility is represented by a tensor rather than by a scalar number. Because of the cubic symmetry in Si, Ge or GaAs crystals, one can, however, use a scalar expression for the effective mass, defined by:

$$\frac{1}{m^*} = \frac{1}{3}\!\left(\frac{1}{m^*_l} + \frac{2}{m^*_t}\right) \qquad\qquad (2.2.6)$$

where m^* is called "conductivity effective mass". In silicon the conductivity effective mass of electrons is equal to $m^*_e = 0.26 \; m_o$, and that of holes is $m_h^* = 0.37\ m_o$, m_o being the mass of a free electron in a vacuum.

Mobility depends on the interactions between electrons and phonons and impurities. A more thorough analysis of the scattering of electrons by phonons yields the following dependence of mobility on temperature:

$$\mu_T \propto T^{-3/2} \qquad (2.2.7)$$

and the dependence of mobility on impurity concentration, N:

$$\mu_N \propto \frac{T^{3/2}}{N} \qquad (2.2.8)$$

When the dependence on both temperature and impurities is taken into account, the mobility, μ, is given by:

$$\frac{1}{\mu} = \frac{1}{\mu_T} + \frac{1}{\mu_N} \qquad (2.2.9)$$

Equation 2.2.8 implies that the mechanism which results in the lowest mobility, will be the limiting factor for mobility. The mobility of electrons and holes in different semiconductors is shown in Figure 2.2 as a function of dopant concentration.

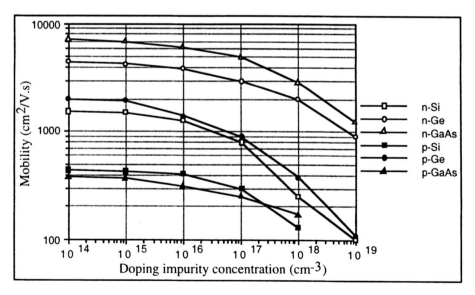

Figure 2.2: Mobility of electrons (n) and holes (p) at T=300K in different semiconductors, as a function of the impurity doping concentration.[3]

2.3. Drift current

If the electron concentration in the conduction band is equal to n, the electron drift current density is given by $J = -q \, n \, v_{dn}$ or, using Relationship 2.2.2:

$$J_n = -q \, n \, v_{dn} = q \, \mu_n \, n \, \mathcal{E} \tag{2.3.1}$$

In a similar way, the hole drift current density is given by:

$$J_p = q \, p \, v_{dp} = q \, \mu_p \, p \, \mathcal{E} \tag{2.3.2}$$

The conductivity, σ ($\Omega^{-1}\text{cm}^{-1}$), and the resistivity, ρ (Ω cm), of an homogeneously doped semiconductor are, therefore, given by:

$$\sigma = q(n \, \mu_n + p \, \mu_p) \quad \text{and} \quad \rho = \frac{1}{\sigma} \tag{2.3.3}$$

In Figure 2.3 one observes that the resistivity of a semiconductor can vary by several orders of magnitude simply by modifying the doping concentration. The resistivity can range from 10^{-4} to 10^3 Ω.cm. By comparison, the resistivity of metals is on the order of 10^{-6} Ω.cm and that of typical insulators is around 10^8 Ω.cm.

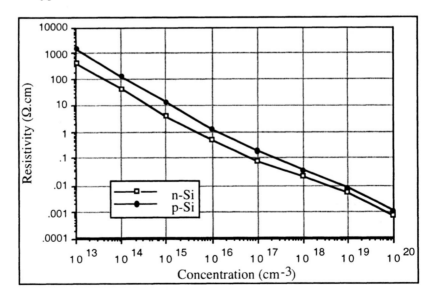

Figure 2.3: Resistivity of silicon at T=300K as a function of doping impurity concentration. [4]

2.3.1. Hall effect

According to Relationship 2.3.3 the conductivity of a semiconductor sample is given by the product of the carrier concentration and their mobility. The conductivity of the sample can readily be measured, using an ohmmeter, for instance. The carrier concentration and the mobility can be separated by performing an additional measurement based on the Hall effect.

When a magnetic field, $\mathcal{B}$, is applied perpendicular to the direction of the carrier flow in a semiconductor sample a potential difference appears in the direction perpendicular to both the current flow direction and the direction of the magnetic field (Hall effect, 1897).

Let us examine the motion of electrons in a piece of N-type semiconductor under the combined effect of a longitudinal electric field, $\mathcal{E}_L$, and of a magnetic field, $\mathcal{B}$, perpendicular to it (Figure 2.4). The current density in the y-direction, J_n, is given by Equation 2.3.1:

$$J_n = q\,\mu_n\,n\,\mathcal{E}_L = -\,q\,n\,v_{dn} \qquad (2.3.4)$$

where n is the electron concentration.

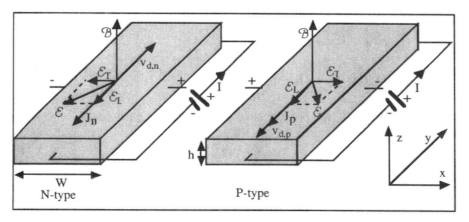

Figure 2.4: Hall effect.

Each electron in motion is submitted to a Lorentz force having a magnitude equal to $-q(v_{dn} \times \mathcal{B})$ in a direction, x, perpendicular to both the electron velocity, v_{dn}, thus also to J_n, and to $\mathcal{B}$. Since no current can flow in the x-direction a transverse electric field which exactly counteracts the Lorentz force, $\mathcal{E}_T$, is created, such that :

$$q\,\mathcal{E}_T = -q(v_{dn} \times \mathcal{B})$$

$$\Downarrow$$

$$\mathcal{E}_T = -(v_{dn} \times \mathcal{B}) = -\frac{1}{q\,n}(\mathcal{B} \times J_n) = -\mu_n\,(\mathcal{B} \times \mathcal{E}_L) \qquad (2.3.5)$$

If the width of the sample is W, a potential difference which can be measured, called "Hall voltage" will appear at the sides of the sample:

$$V_H = W\,|\mathcal{E}_T| \qquad (2.3.6)$$

If the thickness of the sample is h the current flowing in the y-direction is equal to:

$$I = h\,W\,|J_n| \qquad (2.3.7)$$

One defines the "Hall coefficient", $R_{H,n}$, which characterizes the combined effect of an electric field and a magnetic field on electrons by the following relationship: [5]

$$R_{H,n} = \frac{h\,V_H}{|I|.|\mathcal{B}|} \qquad (2.3.8)$$

Since the magnetic field is perpendicular to the direction of current flow the latter Equation can be rewritten in the following form using 2.3.4 and 2.3.5:

$$R_{H,n} = \frac{|\mathcal{E}_T|}{|J_n|\,|\mathcal{B}|} = \frac{-1}{q\,n} \qquad (2.3.9)$$

The conductivity of the N-type semiconductor is equal to $\sigma_n = q\,\mu_n\,n$. Therefore, one obtains, using Equation 2.3.9:

$$\mu_n = -R_{H,n}\,\sigma_n \qquad (2.3.10)$$

The mobility of the carriers in a sample can thus be extracted using a conductivity (or resistivity) measurement and a Hall effect measurement. Once the mobility is known, Relationship $\sigma_n = q\,\mu_n\,n$ gives access to the electron concentration. In the case of a P-type semiconductor, one finds:

$$R_{H,p} = \frac{+1}{q\,p}, \quad \sigma_p = q\,\mu_p\,p \quad \text{and} \quad \mu_p = +R_{H,p}\,\sigma_p \qquad (2.3.11)$$

In conclusion the Hall effect allows the determination of the polarity of a semiconductor (N- or P-type) through the sign of the Hall coefficient. In addition, when combined with a conductivity measurement it allows for the extraction of the majority carrier density and the majority carrier mobility.

2.4. Diffusion current

In semiconductors current can be produced due to a concentration gradient of carriers. The current in this case is called *diffusion current* and is derived below. Consider a piece of semiconductor in which, for whatever reason, there is an electron concentration gradient. By analogy with the laws of diffusion in gases or liquids one can easily conceive that electrons will diffuse from the region where their concentration is highest to the region where it is lowest. The flux of electrons, F_n, resulting from the diffusion process is directly proportional to the electron concentration gradient, dn/dx. This flux, when multiplied by $-q$, is equal to the diffusion current density of the electrons:

$$F_n = -D_n \frac{dn}{dx} \quad \Rightarrow \quad J_n = -q\,F_n = qD_n \frac{dn}{dx} \qquad (2.4.1)$$

In a similar way a hole concentration gradient gives rise to a hole diffusion current. Since each hole bears a positive charge $+q$ one can write:

$$F_p = -D_p \frac{dp}{dx} \quad \Rightarrow \quad J_p = +q\,F_p = -qD_p \frac{dp}{dx} \qquad (2.4.2)$$

D_n and D_p are constants called "diffusion coefficients" for electrons and holes, respectively. They represent the ease or the "fluidity" with which the carriers can move and diffuse in the semiconductor material.

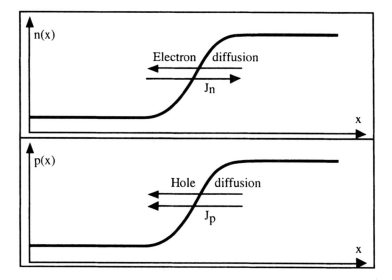

Figure 2.5: Carrier concentration gradients and the resulting diffusion currents.

2.5. Drift-diffusion equations

Based on the concepts derived in the previous sections we can now establish the drift-diffusion equations. The total hole current density in a semiconductor is composed of the sum of the drift and the diffusion components of current. Similarly, the total electron current density in a semiconductor is composed of the sum of the drift and the diffusion components of current. Using 2.3.1, 2.3.2, 2.4.1 and 2.4.2 we obtain:

$$J_p = q\,\mu_p\,p\,\mathcal{E} - qD_p\frac{dp}{dx} \qquad (2.5.1a)$$

and

$$J_n = q\,\mu_n\,n\,\mathcal{E} + qD_n\frac{dn}{dx} \qquad (2.5.1b)$$

or, in a three-dimensional case:

$$J_p = q\,\mu_p\,p\,\mathcal{E} - qD_p\,grad(p) \qquad (2.5.2a)$$

and

$$J_n = q\,\mu_n\,n\,\mathcal{E} + qD_n\,grad(n) \qquad (2.5.2b)$$

The total density of the current flowing at any point in the semiconductor is simply obtained by adding the hole and electron current densities:

$$J = J_n + J_p$$

2.5.1. Einstein relationships

The mobility and diffusion coefficient in a semiconductor are related to each other. This relationship is derived in the following section. Consider a piece of semiconductor material with a non-uniform doping concentration. Let the doping atoms be arsenic in silicon and for the sake of simplicity we will consider a one-dimensional case. The doping impurities are N-type and their concentration is $N_d(x)$, as shown in Figure 2.6. Assuming all doping impurities are ionized, we have that $n(x) = N_d(x)$. The presence of an electron concentration gradient gives rise to an electron diffusion current. The electrons diffusing to the left "leave behind" positively charged arsenic atoms. These atoms occupy substitutional sites in the crystal lattice, and unlike electrons, cannot move. Because of the increased number of electrons in the left-hand part of the sample and the presence of positive charges in the right-hand part an internal electric field develops locally. This electric field tends to "recall" the electrons towards their place of origin. This electric field and the associated potential drop are noted $\mathcal{E}_o(x) = -d\Phi_o(x)/dx$, where the subscript zero implies an internal or "built-in" field under thermal equilibrium.

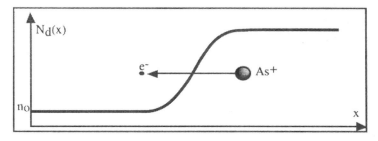

Figure 2.6: Non-uniform doping profile in an N-type silicon sample.

With no external bias applied to the sample there is no current flow and the force of the internal electric field exactly balances the diffusion force. Using the drift-diffusion equation 2.5.1b we can write:

$$J_n = q\,\mu_n\,n\,\mathcal{E}_o + qD_n\frac{dn}{dx} = 0 \qquad (2.5.3)$$

Recalling that $n(x) = n_o\,exp\left[\dfrac{q\Phi_o(x)}{kT}\right]$ (Expression 1.3.20a), and since by

definition $\mathcal{E}_o = -d\Phi_o/dx$, one obtains:

$$q\,\mu_n\,n\,\frac{d\Phi_o}{dx} = qD_n\frac{dn}{dx}$$
$$\Downarrow$$

$$q\,\mu_n\,n_o\,exp\left[\frac{q\Phi_o(x)}{kT}\right]\frac{d\Phi_o}{dx} = qD_n\frac{dn}{d\Phi_o}\frac{d\Phi_o}{dx}$$

$$= qD_n\frac{q}{kT}n_o\,exp\left[\frac{q\Phi_o(x)}{kT}\right]\frac{d\Phi_o}{dx}$$
$$\Downarrow$$

Einstein Relationships

For electrons:

$$D_n = \frac{kT}{q}\mu_n \qquad (2.5.4a)$$

For holes:

$$D_p = \frac{kT}{q}\mu_p \qquad (2.5.4b)$$

Relationships 2.5.4 a and b are called "Einstein relationships". They show that diffusion coefficients and mobilities represent the same thing, within a multiplication constant, kT/q. The value kT/q has the dimension of a voltage, and is called "thermal voltage". It is equal to 25.9 mV at room temperature and is frequently noted "U_T" or "V_T". Thus if the mobility is known the diffusion coefficient can be calculated.

2.6. Transport equations

The transport equations are a set of five equations that govern the behavior of semiconductor materials and devices. In the previous section we have related the flow of current to drift and diffusion mechanisms. The first two transport equations are the drift-diffusion equations given by Relationships 2.5.2a and 2.5.2b and are repeated below:

Drift-Diffusion Equations

$$J_p = q\,\mu_p\,p\,\mathcal{E} - qD_p\,grad(p) \tag{2.6.1a}$$
and
$$J_n = q\,\mu_n\,n\,\mathcal{E} + qD_n\,grad(n) \tag{2.6.1b}$$

or, in one-dimensional problems:

$$J_p = q\,\mu_p\,p\,\mathcal{E} - qD_p\frac{dp}{dx}$$
and
$$J_n = q\,\mu_n\,n\,\mathcal{E} + qD_n\frac{dn}{dx}$$

From Maxwell's equations $\nabla.\mathcal{D} = \rho$ and $\mathcal{D} = \varepsilon\,\mathcal{E}$, where $\mathcal{D}$ is the displacement field, and using the relationship between electric field and potential $\mathcal{E}(x) = -\nabla\Phi$ one readily obtains the Poisson equation:

$$- \nabla.\mathcal{E} = \nabla.\nabla\Phi(x,y,z) \equiv \nabla^2\Phi(x,y,z)$$

$$= -\frac{\rho(x,y,z)}{\varepsilon_s} = -\frac{q}{\varepsilon_s}(p - n + N_d^+ - N_a^-) \tag{2.6.2}$$

where ε_s is the permittivity of the semiconductor and ρ is the local charge density (C/cm^3) in the semiconductor. If all the doping atoms are ionized, which is the case at room temperature, one obtains:

Poisson's Equation

$$\nabla^2\Phi(x,y,z) = \frac{\partial^2\Phi}{\partial x^2} + \frac{\partial^2\Phi}{\partial y^2} + \frac{\partial^2\Phi}{\partial z^2}$$

$$= -\frac{q}{\varepsilon_s}\left[(p(x,y,z) - n(x,y,z) + N_d(x,y,z) - N_a(x,y,z))\right] \tag{2.6.3a}$$

or, in short:

$$\nabla^2\Phi = -\frac{\rho}{\varepsilon_s} \tag{2.6.3b}$$

and for one-dimensional problems:

$$\frac{d^2\Phi}{dx^2} = -\frac{\rho}{\varepsilon_s} \tag{2.6.4}$$

The permittivity of a material is given by the product of its relative permittivity or dielectric constant, κ, multiplied by the permittivity of vacuum where the permittivity of vacuum is equal to 8.854×10^{-14} F/cm. For example, silicon, which has a dielectric constant of 11.7, has a permittivity of 1.036×10^{-12} F/cm.

In the previous derived Equations 2.6.1a and b, and 2.6.4, steady-state was assumed, *i.e.*, there was no time dependence of any of the variables. Another set of equations which describe the evolution of carrier concentration with time can be derived. However, the local carrier concentration may vary for the following reasons:

◊ External forces can be applied to a region of the semiconductor material such that carriers are either added to or removed from that region (*i.e.* carrier injection in a PN junction).

◊ The width of the bandgap in a semiconductor is small enough to allow for electrons to "jump" from the valence band into the conduction band and reciprocally. In addition, electrons can also "jump" from the conduction or valence band into permitted energy levels located inside the bandgap. These levels arise from the presence of trace impurity elements or crystalline defects. If, for instance, an electron jumps from the valence band into the conduction band, it becomes free to move in the crystal. At the same time, a free hole is created in the valence band, which is free to move as well. Such an event is called "carrier pair generation" or, more simply, "generation". An electron can also "fall" from the conduction band into the valence band. In this process called "recombination" both a free electron and a free hole are lost. More complex generation/recombination processes can occur as well, in which permitted energy states within the bandgap are involved. The net, intrinsic, generation/recombination rates for electrons and holes are noted U_n and U_p, respectively. Generation/recombination mechanisms will be analyzed in more detail in Chapter 3. The generation/recombination rates, U_n and U_p, are taken as positive in the case of recombination, and negative in case of generation.

◊ An external source energy can increase the hole and electron concentration. If enough energy is transferred to an electron in the valence band, it can "jump" into the conduction band, a process by which a free electron-hole pair is created. The external generation rates for electrons and holes are noted G_n and G_p, respectively (unit: cm^{-3} sec^{-1}). A typical example where external generation is useful is the conversion of sun light into electrical energy in a solar cell.

A clear distinction should be made between the *intrinsic* generation/recombination rates U_n and U_p, and the *extrinsic* generation rates G_n and G_p:

◊ The *intrinsic* generation/recombination rates express the rate at which free electrons and holes are created or annihilated within a unit volume of the semiconductor material in the absence of any outside influence. U_n and U_p are positive if recombination dominates over generation, *i.e.* if more free electrons and holes disappear by spontaneous recombination than free electrons and holes are created within the material by thermal energy. U_n and U_p are negative if there is more intrinsic carrier generation than recombination. If the rates of spontaneous generation and recombination are equal, both U_n and U_p are equal to zero. In other words, U_n = (free electron intrinsic recombination rate *minus* free electron intrinsic generation rate) and U_p = (free hole intrinsic recombination rate *minus* free hole intrinsic generation rate).

◊ The *extrinsic* generation rates express the rate at which free electrons and holes are created by an outside source of energy, such as light illumination. Extrinsic generation involves only generation (*i.e.* no recombination) events.

To derive the equations describing the variation of the number of carriers due to generation/recombination events we will consider a differential volume of semiconductor material (Figure 2.7). The cross-sectional area of the volume under consideration is A with length dx. An electron current density $J_n(x)$ (unit: Amps/cm^2) enters the volume and a current density $J_n(x+dx)$ flows out of it.

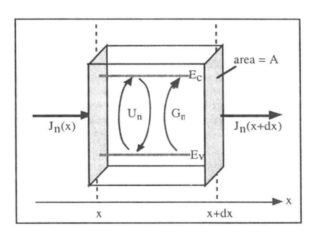

Figure 2.7: Elementary volume used for the derivation of the continuity equations. [6]

For one-dimensional current flow in the x-direction the variation of the number of free electrons in the volume Adx as a function of time is given by the number of electrons entering the volume, minus the number of electrons flowing out of the volume, plus the number of electrons generated minus the number of electrons recombined:

$$A\,\frac{\partial n}{\partial t}\,dx = A\left(\frac{J_n(x)}{-q} - \frac{J_n(x+dx)}{-q}\right) + A\,(G_n - U_n)\,dx \qquad (2.6.5)$$

$J_n(x+dx)$ can be developed in series, which yields: $J_n(x+dx) = J_n(x) + \dfrac{dJ_n(x)}{dx} dx + \ldots$

Using the latter result Equation 2.6.5 can then be rewritten to obtain the continuity equation for electrons:

$$\frac{\partial n}{\partial t} = \frac{1}{q} \frac{\partial J_n(x)}{\partial x} + (G_n - U_n) \tag{2.6.6a}$$

A similar calculation, made for holes would yield:

$$\frac{\partial p}{\partial t} = -\frac{1}{q} \frac{\partial J_p(x)}{\partial x} + (G_p - U_p) \tag{2.6.6b}$$

Extending Expressions 2.6.6a and 2.6.6b to three dimensions one obtains the continuity equations:

Continuity Equations

$$\frac{\partial n}{\partial t} = \frac{1}{q} \, \text{div} \, \boldsymbol{J}_n + (G_n - U_n) \tag{2.6.7a}$$

and

$$\frac{\partial p}{\partial t} = -\frac{1}{q} \, \text{div} \, \boldsymbol{J}_p + (G_p - U_p) \tag{2.6.7b}$$

or, in one-dimensional problems:

$$\frac{\partial n}{\partial t} = \frac{1}{q} \frac{\partial J_n}{\partial x} + (G_n - U_n) \tag{2.6.6a}$$

and

$$\frac{\partial p}{\partial t} = -\frac{1}{q} \frac{\partial J_p}{\partial x} + (G_p - U_p) \tag{2.6.6b}$$

The set of equations composed of the *drift-diffusion equations*, the *Poisson equation*, and the *continuity equations* is called the "transport equations". The transport equations allows one to derive most properties of semiconductor devices.

2.7. Quasi-Fermi levels

At thermodynamic equilibrium, and in the absence of applied external forces, the equilibrium carrier concentrations are a function of the internal potential $\Phi_o(x,y,z)$ in the semiconductor. The carrier concentrations are related to the internal potential by the Boltzmann relationships 1.3.20a and 1.3.20b. These can be rewritten in the following form:

$$n(x,y,z) = n_i \, exp\left[\frac{E_F - E_{io}}{kT}\right] exp\left[\frac{q\Phi_o(x,y,z)}{kT}\right] \tag{2.7.1}$$

$$p(x,y,z) = n_i \, exp\left[-\frac{E_F - E_{io}}{kT}\right] exp\left[\frac{-q\Phi_o(x,y,z)}{kT}\right] \qquad (2.7.2)$$

and the *pn* product is given by:

$$p(x,y,z) \, n(x,y,z) = n_i^2 \qquad (2.7.3)$$

Under thermodynamic equilibrium conditions the Fermi level, E_F, is unique for both electrons and holes.

Under non-equilibrium conditions, however, this is no longer the case. For instance when excess carriers are continuously injected into the semiconductor material or if light is continuously shone on it, the relationship between the internal potential $\Phi(x,y,z)$ and the electron and hole concentrations, $n(x,y,z)$ and $p(x,y,z)$ becomes more complicated. The Boltzmann relationships, however, are still valid if one introduces the notion of "quasi-Fermi levels". Quasi-Fermi levels are also called "imref", which means "*im*aginary *ref*erence", and quite conveniently, corresponds to the word "Fermi" spelled backwards. Instead of a single Fermi level common to both types of carriers let us define an electron quasi-Fermi level, $E_{Fn}(x,y,z)$, and a hole quasi-Fermi level, $E_{Fp}(x,y,z)$. The Boltzmann relationships can be rewritten in the following form:

$$n(x,y,z) = n_i \, exp\left[\frac{E_{Fn}(x,y,z) - E_{io}}{kT}\right] exp\left[\frac{q\Phi(x,y,z)}{kT}\right] \qquad (2.7.4)$$

$$p(x,y,z) = n_i \, exp\left[-\frac{E_{Fp}(x,y,z) - E_{io}}{kT}\right] exp\left[\frac{-q\Phi(x,y,z)}{kT}\right] \qquad (2.7.5)$$

and the *pn* product is equal to:

$$p(x,y,z) \, n(x,y,z) = n_i^2 \, exp\left[\frac{E_{Fn}(x,y,z) - E_{Fp}(x,y,z)}{kT}\right] \qquad (2.7.6)$$

From Equation 2.6.1b we know that the electron current density is given by:

$$J_n = q \, \mu_n \, n \, \mathcal{E} + qD_n \, grad(n) \qquad (2.7.7)$$

Taking the derivative of Expression 2.7.4 we can write:

$$grad(n) = n_i \, exp\left[\frac{E_{Fn} - E_{io}}{kT}\right] exp\left[\frac{q\Phi}{kT}\right] \frac{1}{kT}\left(grad(E_{Fn}) + q \, grad(\Phi)\right)$$

$$= \frac{n}{kT} \, [grad(E_{Fn}) + q \, grad(\Phi)] \qquad (2.7.8)$$

Introducing the result of Equation 2.7.8 into Relationship 2.7.7 one obtains:

$$J_n = q\,\mu_n\,n\,\mathcal{E} + qD_n\,\frac{n}{kT}\,[grad(E_{Fn}) + q\,grad(\Phi)] \qquad (2.7.9)$$

Using the Einstein Relationship $D_n = \dfrac{kT}{q}\,\mu_n$ we finally obtain:

$$J_n = n\,\mu_n\,grad(E_{Fn}) \qquad (2.7.10a)$$

A similar calculation, made for holes, would yield:

$$J_p = p\,\mu_p\,grad(E_{Fp}) \qquad (2.7.10b)$$

The two last relationships show that, *in the most general case, the current is not linked to the gradient of the internal potential, Φ_o, but to the gradient of the quasi-Fermi levels.* Under thermodynamic equilibrium conditions and in the absence of external forces, however, $E_{Fn} = E_{Fp} = E_F$ = a constant, and therefore, $J_n = J_p = 0$, and $pn = n_i^2$.

Important Equations

Einstein Relationships

For electrons:

$$D_n = \frac{kT}{q}\,\mu_n \qquad (2.5.4a)$$

For holes:

$$D_p = \frac{kT}{q}\,\mu_p \qquad (2.5.4b)$$

Drift-Diffusion Equations

$$J_p = q\,\mu_p\,p\,\mathcal{E} - qD_p\,grad(p) \qquad (2.6.1a)$$

and

$$J_n = q\,\mu_n\,n\,\mathcal{E} + qD_n\,grad(n) \qquad (2.6.1b)$$

or, in one-dimensional problems:

$$J_p = q\,\mu_p\,p\,\mathcal{E} - qD_p\,\frac{dp}{dx} \qquad (2.5.1a)$$

and

$$J_n = q\,\mu_n\,n\,\mathcal{E} + qD_n\,\frac{dn}{dx} \qquad (2.5.1a)$$

Poisson's Equation

$$\nabla^2 \Phi(x,y,z) = \frac{\partial^2 \Phi}{\partial x^2} + \frac{\partial^2 \Phi}{\partial y^2} + \frac{\partial^2 \Phi}{\partial z^2}$$

$$= -\frac{q}{\varepsilon_s}\left[(p(x,y,z) - n(x,y,z) + N_d(x,y,z) - N_a(x,y,z))\right] \quad (2.6.3a)$$

or, in short:

$$\nabla^2 \Phi = -\frac{\rho}{\varepsilon_s} \quad (2.6.3b)$$

and in one-dimensional problems:

$$\frac{d^2 \Phi}{dx^2} = -\frac{\rho}{\varepsilon_s} \quad (2.6.4)$$

Continuity Equations

$$\frac{\partial n}{\partial t} = \frac{1}{q} \, div \, J_n + (G_n - U_n) \quad (2.6.7a)$$

and

$$\frac{\partial p}{\partial t} = -\frac{1}{q} \, div \, J_p + (G_p - U_p) \quad (2.6.7b)$$

or, in one-dimensional problems:

$$\frac{\partial n}{\partial t} = \frac{1}{q} \frac{\partial J_n}{\partial x} + (G_n - U_n) \quad (2.6.6a)$$

and

$$\frac{\partial p}{\partial t} = -\frac{1}{q} \frac{\partial J_p}{\partial x} + (G_p - U_p) \quad (2.6.6b)$$

Problems

Problem 2.1:
A sample of gallium arsenide (GaAs) is doped with 10^{10} silicon atoms per cm^3. Ninety-five percent of the silicon atoms replace arsenic atoms and the remaining five percent replace gallium atoms. T=300K. The intrinsic carrier concentration, n_i, is equal to 9×10^6 cm^{-3}.

Calculate the electron and hole concentration as well as the position of the Fermi level referenced to the intrinsic level E_i.

Problem 2.2:
A silicon sample has a length of 1 μm. The N-type doping concentration varies linearly from $N_d(x=0) = 10^{16}$ cm^{-3} to $N_d(x=1 \ \mu m) = 10^{17}$ cm^{-3}. The electron mobility is 1000 $cm^2/V.s$

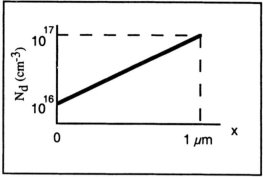

Problem Figure 2.2.

1: Assume that no external bias is applied to the sample. Calculate analytically the internal electric field, $\mathcal{E}_0(x)$ (unit: V/cm) and calculate the numerical value of the electric field at x=0.5 μm.

2: Assume that an external bias is applied in order to cancel out the electric field at x=0.5 μm. What is the current density in the sample? (unit: A cm^{-2}).

Problem 2.3:
Electromagnetics provides us with the following relationships:

$$\nabla.\mathcal{D} = \rho$$
$$\mathcal{D} = \varepsilon\,\mathcal{E}$$
$$J = \sigma\,\mathcal{E}$$
$$\nabla.J = -\,d\rho/dt$$

Consider a piece of <u>intrinsic</u> silicon ($n_i = 1.45\ 10^{10}$ cm^{-3}) of infinite size.
At time t=t_0, an arbitrary distribution of charge is injected into the sample:
$$\rho_0 = \rho(x,y,z,t=t_0)$$
Show that excess charge will vanish exponentially as a function of time, and that the time constant is: $\tau = \varepsilon/\sigma$. where σ is the conductivity of the silicon.
Calculate the time constant τ if μ_n=1417 cm^2/V.s , μ_p=471 cm^2/V.s , $\varepsilon_{si} = 11.7$ ε_0 , $\varepsilon_0 = 8.854 \times 10^{-14}$ Farad/cm.

 Problem 2.4:
A piece of P-type silicon is connected to ground on its right side. On its left side there is a metal electrode which is separated from the silicon by a thin layer of air (air is an insulator!). The potential of the left electrode is V > 0 V (Problem Figure 2.4a). Because of the positive potential on the left electrode, holes near x=0 will be pushed away to the right, leaving ionized acceptor impurities. Hint: This is similar to a parallel-plate capacitor.

As a result, a charge density equal to $\rho = q(p(x) - n(x) - N_a^-)$ appears in the left portion of the silicon sample. Since it is very difficult to solve Poisson's equation analytically for such a charge density, the so-called "depletion approximation" where the charge density is assumed to be equal to $\rho = - qN_a$ over a given distance, w, can be used. Beyond w, the silicon remains neutral. In other words, we have: $\rho = -q\,N_a$ for $0 < x < w$, and $\rho = 0$ for $x > w$ (Problem Figure 2.4b).

In the neutral part of the sample the potential and the electric field are equal to zero ($V = 0$ and $\mathcal{E} = 0$ for $x \geq w$).

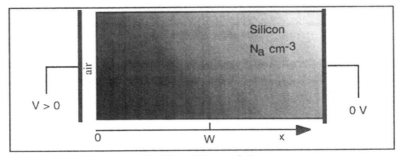

Problem Figure 2.4a.

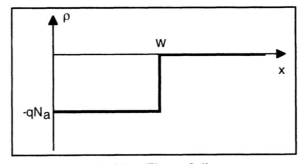

Problem Figure 2.4b.

1) Using the depletion approximation find the analytical expression of the potential $\Phi(x)$ and the electric field $\mathcal{E}(x)$ in the sample for $0<x<2w$.

2) Express w as a function of $\Phi(x=0)$.

3) Find the analytical expression of the electron and hole concentration, $n(x)$ and $p(x)$ $0<x<2w$.

4) Using the following data:

$$N_a = 5\times10^{15} \text{ cm}^{-3}$$
$$n_i = 1.45\times10^{10} \text{ cm}^{-3}$$
$$kT/q = 25.9 \text{ mV}$$
$$q = 1.6\times10^{-19} \text{ C}$$
$$\varepsilon = \text{permittivity of silicon} = 11.7\times8.854\times10^{-14} \text{ F cm}^{-1}$$

a: Plot $\rho(x)$, $\mathcal{E}(x)$, $\Phi(x)$, $n(x)$ and $p(x)$ for $0 < x < 2w$, for the two separate values of $\Phi(x=0)$ where $\Phi(x=0) = \Phi_F$ (one set of curves) and $\Phi(x=0) = 2\Phi_F$ (a second set of curves).

From Relationship 1.3.17b we know that $\Phi_F = \dfrac{kT}{q} \ln\left(\dfrac{N_a}{n_i}\right)$.

b: Plot $n(x)$ and $p(x)$. For the y-axis choose either a linear or a logarithmic scale, whichever is most appropriate. Explain your results.

5) The one-dimensional Poisson equation is given by Relationship (2.6.3a). Assuming N_d is equal to zero we have:

$$\frac{d^2\Phi(x)}{dx^2} = -\frac{q}{\varepsilon_s}(p(x) - n(x) - N_a^-) \qquad (1)$$

In its discrete form, the second-derivative operator can be written:

$$\frac{d^2\Phi}{dx^2} \cong \frac{\Phi(x_{i-1}) + \Phi(x_{i+1}) - 2\Phi(x_i)}{(\Delta x)^2} \qquad (2)$$

such that equation (1) can be written:

$$A\,\Phi = R(\Phi) \qquad (3)$$

where A is a $t \times t$ matrix, t being the number of mesh points, and where:

$$R(\Phi) = -(\Delta x)^2 \frac{q}{\varepsilon_s}(p(x) - n(x) - N_a^-)$$

$$= -(\Delta x)^2 \frac{q}{\varepsilon_s}\left(N_a \exp\left[\frac{-q\Phi(x)}{kT}\right] - \frac{n_i^2}{N_a}\exp\left[\frac{q\Phi(x)}{kT}\right] - N_a\right)$$

Δx is the constant distance between two successive mesh points. If $t = 6$, for instance, the Poisson equation can be written as:

$$
\begin{bmatrix}
1 & 0 & 0 & 0 & 0 & 0 \\
1 & -2 & 1 & 0 & 0 & 0 \\
0 & 1 & -2 & 1 & 0 & 0 \\
0 & 0 & 1 & -2 & 1 & 0 \\
0 & 0 & 0 & 1 & -2 & 1 \\
0 & 0 & 0 & 0 & 0 & 1
\end{bmatrix}
\begin{bmatrix}
\Phi(x_1) \\
\Phi(x_2) \\
\Phi(x_3) \\
\Phi(x_4) \\
\Phi(x_5) \\
\Phi(x_6)
\end{bmatrix}
= R\left(
\begin{bmatrix}
\Phi(x_1) \\
\Phi(x_2) \\
\Phi(x_3) \\
\Phi(x_4) \\
\Phi(x_5) \\
\Phi(x_6)
\end{bmatrix}
\right)
$$

The boundary conditions are $\Phi(x_1) = R(x_1) = \Phi(x=0)$ and $\Phi(x_6) = R(x_6) = \Phi(x=2w)$.

<u>For the problem use fifty mesh points (t=50), rather than six.</u>

Since the left and right terms of the latter equation are both functions of the potential, iterations must be used until acceptable accuracy is reached (see Annex 5). Chose an appropriate criterion for convergence.

Plot $\rho(x)$, $\mathcal{E}(x)$, $\Phi(x)$, $n(x)$ and $p(x)$ for $0 < x < 2w$. Plot $n(x)$ and $p(x)$ as well. Plot the curves obtained in part 4 of this problem with those obtained here (*i.e.* $\rho(x)$ from part 4 and part 5 on one graph, $\mathcal{E}(x)$ from part 4 and part 5 on one graph, etc.) and discuss the accuracy / appropriateness of using the depletion approximation.

<u>Problem 2.5:</u>
We have a sine wave-like charge distribution in a semiconductor between points a and b. The charge is equal to zero everywhere else. Calculate the electric field and potential from x=0 to x>b. $\mathcal{E}$ and Φ are both equal to zero for x=0. Between a and b the charge is given by the following expression:

$$\rho = - A \sin(\frac{\pi(x-a)}{b-a})$$

Problem Figure 2.5.

References

1   Ch. Kittel, *Introduction to solid-state physics*, 6th Edition, J. Wiley and Sons, p. 81, 1986
2   J.M. Ziman, *Principles of the theory of solids*, 2nd Edition, Cambridge University Press, p. 60, 1972
3   S.M. Sze, *Physics of semiconductor devices*, J. Wiley and Sons, p. 29, 1981
4   S.M. Sze, *Physics of semiconductor devices*, J. Wiley and Sons, p. 32, 1981
5   J.P. McKelvey, *Solid-state and semiconductor physics*, Harper International, p. 235, 1966
6   R.S. Muller and T.I. Kamins, *Device electronics for integrated circuits*, J.Wiley and Sons, pp. 219-221, 1986

Chapter 3

GENERATION/RECOMBINATION PHENOMENA

3.1. Introduction

As mentioned earlier there are electrons in the conduction band and holes in the valence band of a semiconductor, as long as the temperature is above zero Kelvin. An electron in the conduction band is free to move in the crystal. It can also "jump" into a "vacant seat" in the covalent bond network (Figure 3.1). This "vacant seat" is, of course, nothing but a hole. By doing this the electron releases energy. Such a phenomenon in which a free electron and a free hole both disappear is called a recombination event.

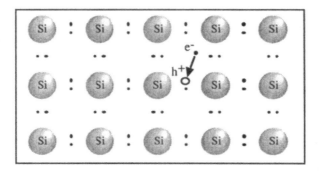

Figure 3.1: Recombination of an electron (e^-) with a hole (h^+).

Conversely, an electron can free itself from a covalent bond if enough energy is made available. By doing this it "jumps" from the valence band into the conduction band and becomes free to move in the crystal. A free hole is also created in that process, which is called "generation of an electron-hole pair" (Figure 3.2).

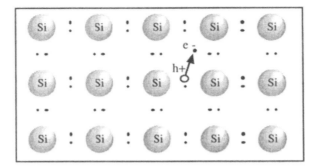

Figure 3.2: Generation of an electron-hole pair.

Under thermodynamic equilibrium, generation and recombination events exactly balance one another, such that the electron and hole equilibrium concentrations remain constant with respect to time. Using an external source of energy such as illumination with light, one can, however, increase the carrier concentration and reach a state of non-equilibrium.

3.2. Direct and indirect transitions

In a semiconductor such as gallium arsenide (GaAs) the conduction band minimum (where free electrons are located) occurs at the same k-value (k is the wave vector) as the valence band maximum. The wave vector represents the momentum of the carriers. As shown in Figure 3.3 the value of that momentum is zero. Therefore, when an electron from the conduction band recombines with a hole in the valence band the law of conservation of momentum is obeyed. A semiconductor where the minimum of the conduction band and the maximum of the valence band occur at the same k-value is called a direct-bandgap semiconductor, and the "jump" of an electron from the conduction band into the valence band is called "band-to-band recombination".

Since momentum is conserved in this example of a recombination event, recombination requires nothing more than an electron with $k=0$ and a hole with $k=0$. Since most electrons occupy the conduction band at or near $k=0$, recombination is a very likely mechanism. When a recombination event takes place the law of conservation of energy also implies that a quantum of energy is released in the form of a photon. The energy of that photon is such that $h\nu=E_g$, where h is Planck's constant, ν is the frequency of the photon, and E_g is the bandgap energy. In most direct-bandgap semiconductors the photons emitted by recombination events have an energy corresponding to visible or near-infrared light. A recombination event where photons are emitted is called "radiative recombination" and is exploited in devices such as light-emitting diodes.

The relationship between the photon wavelength, λ, and the bandgap energy, E_g, is:

$$E = h\nu \; and \; \nu = c/\lambda \;\; \Rightarrow \;\; \lambda \; (\mu m) = \frac{1.24 \; (eV \; \mu m)}{E_g(eV)} \tag{3.2.1}$$

where ν, h and c are the photon frequency, Planck's constant and the speed of light, respectively.

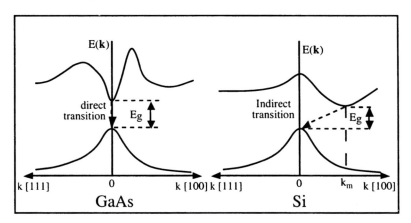

Figure 3.3: Band-to-band recombination in a direct-bandgap semiconductor (left) and in an indirect-bandgap semiconductor (right).

In silicon and germanium the minimum of the conduction band and the maximum of the valence band do not occur at a same k-value. A semiconductor where this is the case is called an "indirect-bandgap semiconductor". When recombination takes place in such a material an electron with a momentum $k=[k_m,0,0]$ recombines with a hole having a momentum $k=0$ (Figure 3.3). This can occur only if an appropriate momentum is transferred to the electron (or the hole) such that conservation of momentum is observed. This can happen through collision with a phonon or with several phonons. Since a precise value of momentum $(-k_m$ in Figure 3.3) must be transferred to the electron, band-to-band recombination is an extremely unlikely process in indirect-bandgap semiconductors. As a result there is no radiative recombination in silicon and germanium, and these materials cannot emit light. Rather recombination takes place via trap levels at various k-values within the band gap.

Gallium arsenide emits photons with a wavelength of 0.8 μm, which corresponds to near-infrared, almost visible light. To fabricate semiconductor devices producing visible light more complex semiconductor materials are used, usually based on a combination of the elements of columns III and V of the periodic table, such as Ga, Al, P, As, and N. Such semiconductors are called "III-V semiconductors".

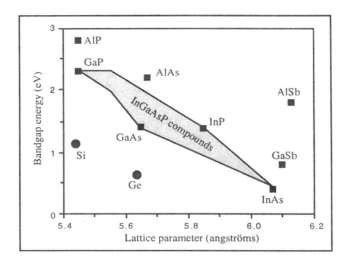

Figure 3.4: Energy bandgap in Si, Ge, and some III-V semiconductors. [1]

The main parameter that governs the electrical and optical properties of semiconductors is the bandgap energy, shown in Figure 3.4 as a function of the crystal lattice parameter. The use of ternary compound semiconductors, such as $Ga_xAl_{1-x}As$, or that of quaternary compounds, such as $Ga_xIn_{1-x}As_yP_{1-y}$ allows one to tailor the bandgap energy in order to produce a desired light wavelength. The fabrication of a semiconductor material with an "engineered" bandgap energy is obtained, for example, by adjusting the x and y coefficients during the growth of a $Ga_xIn_{1-x}As_yP_{1-y}$ crystal.

Semiconductors are transparent to photons that carry an energy, $h\nu$, smaller than the bandgap energy. Germanium, for instance, is used instead of glass to make infrared (IR) lenses for wavelengths larger than 2 μm since its bandgap energy is larger then the energy of 2 μm IR photons. Photons with an energy equal or greater than the semiconductor bandgap energy, on the other hand, can be absorbed to generate electron-hole pairs. Figure 3.5 shows the absorption coefficients in some semiconductors, as a function of wavelength. The absorption coefficient is a measure of the distance a light wave travels into the material before it is absorbed.

In addition to band-to-band recombination mechanisms, a free electron can recombine with a free hole through "recombination centers" located within the energy bandgap. These are permitted energy levels introduced by contaminants, impurity atoms or crystal defects. A recombination center acts as a catalyst that enables an electron to recombine at k values differing from the k_m of the conduction band. This is especially true in indirect-bandgap semiconductors such as silicon or germanium, where band-to-band recombination events are very unlikely to occur.

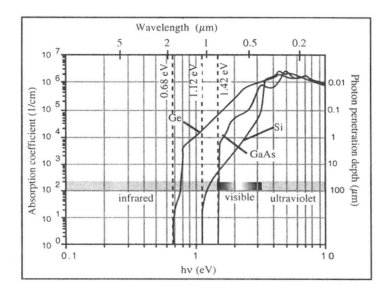

Figure 3.5: Absorption coefficient as a function of photon energy (and wavelength) in Ge, Si and GaAs, which have energy bandgaps of 0.68, 1.12 and 1.42 eV, respectively.[2]

3.3. Generation/recombination centers

Semiconductor crystals are of the highest purity and quality, but they are not perfect. They contain some crystal defects such as interstitials (excess semiconductor atoms in the crystal lattice), vacancies (missing semiconductor atoms in the crystal lattice) and dislocations (imperfections in the crystal structure), as well as traces of impurity elements such as metallic atoms or oxygen. These defects and impurities give rise to permitted levels within the energy bandgap. Let us consider one of these levels, having an energy E_t within the bandgap. This permitted level can receive an electron from the conduction band (case A in Figure 3.6), lose an electron to the valence band (case C), receive an electron from the valence band (case D), or lose an electron to the conduction band (case B). A level that is neutral if filled by an electron and positive if empty is called a "donor level", and a level that is neutral if empty and negative if filled by an electron is called an "acceptor level". Permitted levels inside the bandgap are called generation-recombination centers, or, in short, "recombination centers". In Figure 3.6 transitions A and C correspond to recombination events, and transitions B and D correspond to generation events. Since these transitions involve energies smaller than that of the bandgap they are much more likely to occur than band-to-band transitions, especially in indirect-bandgap semiconductors like silicon or germanium.

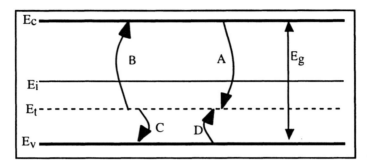

Figure 3.6: Electron transitions via a recombination center at energy E_t.

It is important to note that the terms G_n and G_p in the continuity equations 2.6.7a and 2.6.7b represent electron-hole pair generation events caused by an *external* source of energy, such as, for instance, sunlight penetrating the semiconductor. Natural, intrinsic generation in a semiconductor arising at any temperature above zero Kelvin, is encompassed in the *intrinsic* recombination-generation rate terms of the continuity equations, U_n and U_p. Using the notations of Figure 3.6 it can easily be established that $U_n = A - B$ and $U_p = C - D$. If U_n (or U_p) is positive a net recombination of carriers is taking place. If it is negative, a net generation of carriers is observed.

The energy released by a recombination event can give rise to different phenomena:

◊ In a band-to-band radiative recombination event, the energy is released in the form of a photon.

◊ In an Auger recombination event the energy released is transferred to another electron (or hole), which becomes excited to a higher energy level.

◊ In an indirect recombination event via an energy level within the bandgap, energy is transferred to the crystal lattice in the form of heat (or phonons).

Recombination of carriers takes place not only within the bulk of a semiconductor crystal, but at its surface as well. The surface is indeed a place where the periodicity of the crystal lattice is interrupted, and where contact with another substance (air, SiO_2, metal,...) is made.

◊ Within the bulk of the crystal a recombination-generation rate, or, in short, a recombination rate, is defined. The recombination rate for electrons is noted

U_n and that for holes, U_p. U_n and U_p are accounted for in the continuity equations 2.6.7a and 2.6.7b and represent the number of holes and electrons created or annihilated by intrinsic generation/recombination processes per cm^3 and per second.

◊ In a similar manner, at the surface of a semiconductor crystal a surface recombination velocity is defined. The surface recombination rate for electrons is noted S_n and that for holes, S_p. S_n and S_p are the boundary conditions for the continuity equations and represent the number of holes and electrons created or annihilated by intrinsic generation/recombination processes at the surface of a semiconductor crystal per cm^2 and per second.

When an electron is accelerated to high speeds (e.g. by an intense electric field) it can obtain an amount of kinetic energy equal to or larger than the bandgap energy, E_g. That energy can be released through a collision event in such a manner that an additional electron-hole pair is created. Therefore, instead of having a single, high-energy, free electron, we now have two free electrons and a hole. This generation mechanism is called "generation by impact ionization". If the "original" electron current is I, and an additional electron current $M{\times}I$ is created by the impact ionization mechanism, then the total electron current is equal to $(M+1){\times}I$. The M and $(M+1)$ coefficients are both called "multiplication factors".

3.4. Excess carrier lifetime

We have seen that, at thermodynamic equilibrium, the generation rate and the recombination rate are equal, such that $U_n = U_p = 0$ and $S_n = S_p = 0$. If, for some reason, the carrier concentrations are different from their equilibrium value, generation/recombination mechanisms will tend to force them back to equilibrium. Actually, U_n, U_p, S_n and S_p are directly proportional to how much the carrier concentrations depart from equilibrium:

Generation/Recombination Rate and Surface Recombination Rate

Generation/Recombination rate:

$$U_n = \frac{n-n_o}{\tau_n} \quad and \quad U_p = \frac{p-p_o}{\tau_p} \qquad (3.4.1)$$

Unit for τ: second

Surface Recombination Rate:

$$S_n = s_n\,(n-n_o) \quad and \quad S_p = s_p\,(p-p_o) \qquad (3.4.2)$$

S = surface recombination rate; unit: $cm^{-2}\ s^{-1}$
s = surface recombination velocity; unit: $cm\ s^{-1}$

In Expressions 3.4.1 and 3.4.2, n (or p) represents the electron (or hole) concentration and n_o (or p_o) represents the electron (or hole) equilibrium concentration. Thus, for example, if the electron concentration is higher than its equilibrium value, recombination events will reduce the number of electrons. Conversely, if the electron concentration is below its equilibrium value, generation events will take place. By definition τ_n and τ_p are the lifetime of the excess (or missing) electrons or holes, respectively, the equilibrium concentrations being taken as a reference. The meaning of "lifetime" is the "average" time span that excess free electrons (or holes) will "survive" before recombining, or the average time that missing electrons will be "missing" before being "re-generated" through a generation event. A similar reasoning applies to excess and missing holes. In the case of silicon the carrier lifetime ranges between 10^{-9} seconds in heavily contaminated material with many recombination centers and 10^{-3} seconds in high-purity material. In gallium arsenide, where fast band-to-band recombination takes place, the carrier lifetime is on the order of 10^{-8} seconds.

Surface recombination velocity (s_n and s_p) ranges from 10^2 to 10^5 cm/sec in silicon, depending on the cleanliness and passivation of the crystal surface. When the semiconductor surface is in contact with a metal the surface recombination velocity can be considered as infinite at the contact, which in practice means that $n=n_o$ and $p=p_o$ at the surface.

Example:
Let us consider the following example which illustrates the physical meaning of the excess carrier lifetime.

Consider a semiconductor with homogenous (constant) doping concentration which is illuminated with light such that there is an homogenous (constant) external generation rate, G, of electron-hole pairs throughout the sample. The generation is a direct, band-to-band generation, such that $G = G_n = G_p$. As a result of the external generation process the excess electron and hole concentrations are equal, *i.e.*, the generation of any electron corresponds to the generation of a hole: $n-n_o=p-p_o$. Assume a direct, band-to-band recombination mechanism where $U=U_n=U_p$.

Using Expression 3.4.1 and since $n-n_o=p-p_o$, one can write: $\tau_n=\tau_p=\tau$.

If no external bias is applied and there is no concentration gradient of carriers, there is no current flow ($J_n=J_p=0$) and the continuity Equations 2.6.7a and b become:

$$\frac{\partial n}{\partial t} = G_n - U_n = G-U = G - \frac{n-n_o}{\tau} \qquad (3.4.3a)$$

and

$$\frac{\partial p}{\partial t} = G_p - U_p = G-U = G - \frac{p-p_o}{\tau} \qquad (3.4.3b)$$

What is analytical expression for the electron and hole concentration as a function of time in the semiconductor? Under steady-state (constant illumination) conditions we have that $\partial n/\partial t = \partial p/\partial t = 0$, and thus $G = U$, which yields:

$$n = n_o + \tau G \quad \text{and} \quad p = p_o + \tau G \tag{3.4.4}$$

Assume the external generation source is suddenly removed (turn the light off) at $t=t_o$. The excess carriers will recombine to reach, after an infinite time span, their equilibrium concentration ($n=n_o$ and $p-p_o$). Using Expressions 3.4.3a and 3.4.3b an analytical expression for the carrier concentrations as a function of $t \geq t_o$ can be found:

$$\frac{dn}{dt} = -\frac{n-n_o}{\tau} \quad \text{and} \quad \frac{dp}{dt} = -\frac{p-p_o}{\tau} \tag{3.4.5}$$

which yields for electrons:

$$\int_{n(t_o)}^{n(t)} \frac{dn}{n-n_o} = -\int_{t_o}^{t} \frac{dt}{\tau} \Rightarrow \ln\left[\frac{n(t)-n_o}{n(t_o)-n_o}\right] = \frac{-(t-t_o)}{\tau}$$

$$\Rightarrow n(t) = n_o + [n(t_o) - n_o]\ exp(-(t-t_o)/\tau)$$

$$\Rightarrow n(t) = n_o + \tau G\ exp(-(t-t_o)/\tau) \tag{3.4.6a}$$

In a similar way one obtains the time-dependent hole concentration:

$$p(t) = p_o + \tau G\ exp(-(t-t_o)/\tau) \tag{3.4.6b}$$

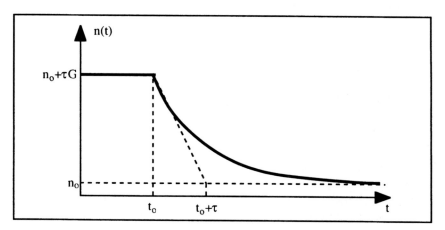

Figure 3.7: Evolution of the electron concentration with time.[3]

From this example we can see that the carrier lifetime, τ, is a constant with which the concentration of carriers, whether above or below its equilibrium value, tends to return to equilibrium.

3.5. SRH recombination

In the previous example the recombination of excess carriers was assumed to be caused by a band-to-band recombination process. In many instances, and in particular, in the case of silicon, generation/recombination events take place through recombination centers located in the energy bandgap. Such recombination events are called SRH (Shockley-Read-Hall) recombination events.

An analytical expression for the recombination rate for electrons and holes, U_n and U_p, can be determined when there are recombination centers at an energy E_t within the bandgap. Consider the case of electron generation/recombination with the assumption that the recombination centers are of the acceptor type. The centers, are therefore, neutral or negatively charged. Let N_t be the density of the recombination centers and n_t (with $n_t \leq N_t$) the concentration of electrons occupying the centers.

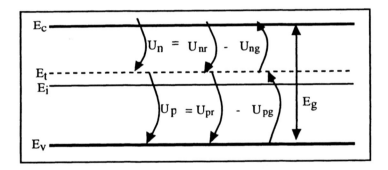

Figure 3.8: Recombination via an acceptor recombination center.[4]

To simplify the problem the electron generation/recombination rate, U_n, is split into two terms, U_{nr} and U_{ng}, which represent recombination and thermal generation, respectively. The recombination rate due to the centers, U_{nr}, is proportional to the concentration of electrons in the conduction band, n, and to the concentration of empty (or neutral) recombination centers, N_t-n_t. One can thus write:

$$U_{nr} = v_{th} \, \sigma_n \, n \, (N_t - n_t) = v_{th} \, \sigma_n \, n \, N_t \left(1 - \frac{n_t}{N_t} \right)$$

$$= v_{th} \, \sigma_n \, n \, N_t \, (1 - f(E_t)) \qquad (3.5.1)$$

where v_{th} is the thermal velocity of electrons, defined by the relationship $v_{th} = \sqrt{3kT/m^*}$ (cm/sec), and σ_n is called the "electron capture cross section" (cm^2). The capture cross section is a measure of how close an electron must be to a center in order to be captured by it, while the thermal velocity is the average speed of electrons due to "Brownian-like" or random motion at a given temperature ($\frac{1}{2} m^* v_{th}^2 = \frac{3}{2} kT$, where kT is the thermal energy).[5] Note that $f(E_t)$ is the probability that a center with energy E_t is occupied by an electron. The function $f(E_t)$ is the Fermi-Dirac distribution evaluated at the energy of the center, at thermodynamic equilibrium.[6]

The thermal generation rate, U_{ng}, is the process by which electrons can "jump" from the recombination centers into the conduction band. It is proportional to the concentration of centers occupied by an electron, $n_t = N_t f(E_t)$:

$$U_{ng} = e_n N_t f(E_t) \qquad (3.5.2)$$

where e_n is a proportionality coefficient which represents the probability of electron emission by the generation/recombination centers.

In a similar manner the recombination rate for holes between the recombination center and the valence band is given by:

$$U_{pr} = v_{th}\, \sigma_p\, p\; n_t = v_{th}\, \sigma_p\, p\, N_t f(E_t) \qquad (3.5.3)$$

The thermal generation rate, U_{pg}, is the process by which holes can "jump" from neutral recombination centers into the valence band. It is proportional to the concentration of centers not occupied by an electron, $n_t = N_t (1 - f(E_t))$:

$$U_{pg} = e_p N_t (1 - f(E_t)) \qquad (3.5.4)$$

where e_n is a proportionality coefficient which represents the probability of hole emission by the generation/recombination centers.

We are now going to calculate the proportionality coefficients e_n and e_p. When the semiconductor is in thermodynamic equilibrium the generation and the recombination rates are equal to zero:

$$U_n = U_p = 0 \qquad (3.5.5)$$

The number of negatively charged centers, *i.e.* filled centers, is given by the relationship $n_t = N_t f(E_t)$, or:

$$n_t = \frac{N_t}{1 + exp\left[\dfrac{E_t - E_F}{kT}\right]} \qquad (3.5.6)$$

Again at thermodynamic equilibrium we must have that $U_n = U_{nr} - U_{ng} = 0$. Using the Boltzmann Relationship 2.7.1 in the absence of an internal potential ($\Phi_o=0$), $U_{nr}=U_{ng}$ can be written in the following form:

$$v_{th}\ \sigma_n\ n\ N_t\ (1 - f(E_t)) = e_n\ N_t\ f(E_t)$$

Using $n = n_i\ exp\left[\dfrac{E_F-E_i}{kT}\right]$ the previous relationship becomes:

$$v_{th}\ \sigma_n\ n_i\ exp\left[\frac{E_F-E_i}{kT}\right]\left(N_t - \frac{N_t}{1 + exp\left[\frac{E_t-E_F}{kT}\right]}\right) = e_n\ \frac{N_t}{1 + exp\left[\frac{E_t-E_F}{kT}\right]}$$

$$\Downarrow$$

$$e_n = v_{th}\ \sigma_n\ n_i\ exp\left[\frac{E_t-E_i}{kT}\right] \tag{3.5.7}$$

Similarly, the hole coefficient, e_p, can be written as:

$$e_p = v_{th}\ \sigma_p\ n_i\ exp\left[\frac{E_i-E_t}{kT}\right] \tag{3.5.8}$$

Let us now use the continuity equation for electrons trapped in the generation/recombination centers, under steady-state conditions to derive an expression for the generation/recombination rate:

$$\frac{\partial n_t}{\partial t} = U_n - U_p + G_n - G_p = 0 \tag{3.5.9}$$

Since external generation creates the same amount of electrons and holes, we have $G=G_n=G_p$, which, by virtue of 3.5.9, yields $U_n = U_p$. Using Equations 3.5.1 to 3.5.4, one obtains:

$$U_n = U_p \Rightarrow U_{nr} - U_{ng} = U_{pr} - U_{pg}$$

$$\Downarrow$$

$$v_{th}\ \sigma_n\ n\ N_t\ (1 - f(E_t)) - e_n\ N_t\ f(E_t) = v_{th}\ \sigma_p\ p\ N_t\ f(E_t) - e_p\ N_t\ (1 - f(E_t)) \tag{3.5.10}$$

Solving Equation 3.5.10 for $f(E_t)$ yields:

$$f(E_t) = \frac{\sigma_n\ n + \sigma_p\ n_i\ exp\left[\frac{E_i-E_t}{kT}\right]}{\sigma_n\left(n + n_i\ exp\left[\frac{E_t-E_i}{kT}\right]\right) + \sigma_p\left(p + n_i\ exp\left[\frac{E_i-E_t}{kT}\right]\right)} \tag{3.5.11}$$

Based on the previous relationships we can now calculate the generation/recombination rate:

$$U = U_n = U_p = U_{nr} - U_{ng} = v_{th}\, \sigma_n\, n\, N_t\, (1 - f(E_t)) - e_n\, N_t\, f(E_t)$$

$$\Downarrow$$

$$U = \frac{\sigma_n \sigma_p v_{th} N_t\, (pn - n_i^2)}{\sigma_n \left(n + n_i\, exp\left[\dfrac{E_t - E_i}{kT}\right] \right) + \sigma_p \left(p + n_i\, exp\left[\dfrac{E_i - E_t}{kT}\right] \right)}$$

$$\Downarrow$$

$$U = \frac{pn - n_i^2}{\tau_p \left(n + n_i\, exp\left[\dfrac{E_t - E_i}{kT}\right] \right) + \tau_n \left(p + n_i\, exp\left[\dfrac{E_i - E_t}{kT}\right] \right)} \tag{3.5.12}$$

with τ_n and τ_p defined as:

$$\tau_n = \frac{1}{N_t v_{th} \sigma_n} \quad and \quad \tau_p = \frac{1}{N_t v_{th} \sigma_p} \tag{3.5.13}$$

where τ_n and τ_p are called "lifetime" of electrons and holes in the steady-state regime, respectively. Looking at Relationship 3.5.12 we find that the recombination rate, U, is directly proportional to $pn - n_i^2$. The recombination rate represents a "force" which tends to bring the pn product back to its equilibrium value, n_i^2. One observes that:

$$
\begin{array}{|l}
U = 0 \;\; if \;\; pn = n_i^2 \;\; \text{(equilibrium)} \\[4pt]
U > 0 \;\; if \;\; pn > n_i^2 \;\; \text{(recombination)} \\[4pt]
U < 0 \;\; if \;\; pn < n_i^2 \;\; \text{(generation)}
\end{array}
$$

It is worthwhile noting that the recombination rate is highest when the recombination centers have an energy close to E_i, e.g. when they are located close to midgap. The physical meaning of this observation is the following: consider the recombination of an electron in the conduction band with a hole in the valence band through a recombination center having an energy E_t. The recombination process requires the capture of the electron by a center (U_{nr}) followed by the emission of the electron from the center into the valence band (or the jump of a hole from the valence band into the center). If E_t is significantly larger than E_i the probability of an electron in the conduction band being captured by the center is high, simply because that process involves a small energy variation: E_c-E_t. The probability of the center capturing a hole from the valence band, U_{pr}, on the other hand, is low because the energy difference E_t - E_v is large. Thus, in this example, the term U_{pr} appears to be the limiting factor to the overall recombination rate. In the case of a center having an energy less than E_i, the capture of an electron by the center will be the limiting factor. It is, of course, when the energy of the center is close or equal to E_i that the processes limiting the recombination rate are minimized. Therefore, recombination centers near midgap yield the highest recombination rates.

Assuming that the hole and the electron capture cross sections are equal, $\sigma_n = \sigma_p = \sigma_o$, Relationship 3.5.12 can be written in the following form:

Recombination Rate

$$U = \frac{pn - n_i^2}{\tau_o \left(p + n + 2n_i \cosh\left[\frac{E_t-E_i}{kT}\right] \right)} \qquad (3.5.14)$$

with:

$$\tau_o = \frac{1}{N_t v_{th} \sigma_o} \qquad (3.5.15)$$

3.5.1. Minority carrier lifetime

Certain semiconductor devices operate by the injection of minority carriers. The lifetime of the minority carriers is important for the efficiency of these devices. In most cases the minority carrier concentrations are orders of magnitude lower than majority carrier concentrations. Let us consider Equation 3.5.12 in a case where the excess carrier concentrations, $\delta n = n - n_o$ and $\delta p = p - p_o$ are small compared to the equilibrium concentrations: $\delta n \ll n_o$ and $\delta p \ll p_o$. This condition is called "low-level injection". One can write:

$$pn = (p_o + \delta p)(n_o + \delta n) \Rightarrow pn - n_i^2 \cong p_o\delta n + n_o\delta p \qquad (3.5.16)$$

Relationship 3.5.12 can be rewritten:

$$U = \frac{p_o\delta n + n_o\delta p}{\tau_p \left(n + n_i \exp\left[\frac{E_t-E_i}{kT}\right] \right) + \tau_n \left(p + n_i \exp\left[\frac{E_i-E_t}{kT}\right] \right)} \qquad (3.5.17)$$

and for centers where the recombination rate is highest (*i.e.* for $E_t \cong E_i$):

$$U = \frac{p_o \delta n + n_o \delta p}{\tau_p \left(n + n_i\right) + \tau_n \left(p + n_i\right)} \qquad (3.5.18)$$

or, since $n = n_o + \delta n$ and $p = p_o + \delta p$ and since $\delta n \ll n_o$ and $\delta p \ll p_o$:

$$U = \frac{p_o \delta n + n_o \delta p}{\tau_p \left(n_o + n_i\right) + \tau_n \left(p_o + n_i\right)} \qquad (3.5.19)$$

Recombination Rate

In the case of a **P-type semiconductor**, we have $p_o > n_i > n_o$, and therefore:

$$U = \frac{\delta n}{\tau_n} = \frac{n - n_o}{\tau_n} \qquad (3.5.20)$$

In the case of an **N-type semiconductor**, we have $n_o > n_i > p_o$, and therefore:

$$U = \frac{\delta p}{\tau_p} = \frac{p - p_o}{\tau_p} \qquad (3.5.21)$$

An important conclusion can be drawn from Expressions 3.5.20 and 3.5.21: the lifetime of excess carriers is equal to that of the *minority carriers* (the electron lifetime in a P-type semiconductor and the hole lifetime in an N-type semiconductor). This may not appear very intuitive, but there is a sound physical reason for it. Consider a P-type semiconductor, where the hole concentration is much higher than that of electrons. In order for a recombination event to take place, both a free electron and a free hole are needed. Free holes are plentiful, while electrons are scarce and rare. Therefore, recombination events will be limited by the number of available electrons, which are minority carriers in this case, and the lifetime of excess carriers will be decided by the value of the electron lifetime. A similar process takes place in an N-type semiconductor, where the excess carrier lifetime is governed by the recombination rate of holes.

3.6. Surface recombination

Recombination of excess carriers occurs not only within the bulk of a semiconductor crystal, but at the surface of the crystal as well. The periodicity of the atoms is interrupted at the surface of the crystal, and the surface acts as an interface between the semiconductor and another material. As a result the recombination rate at the surface is different (and usually higher) than in the bulk of the semiconductor. We can define the surface recombination rate for electrons and holes, S_n and S_p, as the number of carriers disappearing per unit area and per second at the semiconductor surface due to recombination mechanisms. Therefore, S_n and S_p can be used as the boundary conditions for the continuity Equations 2.6.8a and 2.6.8b.

A formal derivation of the surface recombination rate yields an expression similar to Equation 3.5.12:

$$S = \frac{\sigma_n \sigma_p v_{th} N_{st} (p_s n_s - n_i^2)}{\sigma_n \left(n_s + n_i \, exp\left[\dfrac{E_{st}-E_i}{kT}\right]\right) + \sigma_p \left(p_s + n_i \, exp\left[\dfrac{E_i-E_{st}}{kT}\right]\right)} \qquad (3.6.1)$$

where p_s and n_s (unit: cm^{-3}) are the electron and hole concentrations at the surface, respectively, N_{st} is the concentration of surface recombination centers (cm^{-2}), and E_{st} is their energy. As in the case of bulk recombination the most efficient recombination centers are those located at midgap energy, and if we assume that $\sigma_n \cong \sigma_p = \sigma_0$, 3.6.1 yields:

$$S = \frac{\sigma_0 v_{th} N_{st} (p_s n_s - n_i^2)}{p_s + n_s + 2 n_i \, cosh\left[\dfrac{E_{st}-E_i}{kT}\right]} \quad \text{(s}^{-1}\text{ cm}^{-2}\text{)} \qquad (3.6.2)$$

The *pn* product at the surface can be written as:

$$p_s n_s = (p_0 + \delta p_s)(n_0 + \delta n_s) \Rightarrow p_s n_s - n_i^2 \cong p_0 \delta n_s + n_0 \delta p_s$$

Using a derivation similar to Equations 3.5.16 to 3.5.21 the surface recombination can be expressed as a function of minority carrier concentration at the surface (Expression 3.4.2). The recombination rate at the surface of the crystal is larger than inside the crystal. The surface recombination rate can be introduced in the continuity in the following way:

$$A \frac{J_n(x_0)}{-q} = A \, S_n = A \, s_n \, (n_s - n_0) \qquad (3.6.3)$$

for electrons, and:

$$A \frac{J_p(x_0)}{q} = A \, S_p = A \, s_p \, (p_s - p_0) \qquad (3.6.4)$$

for holes.

In some cases, such as at a metal-semiconductor contact, the surface recombination rate can be infinite. This implies $n_s = n_0$ and $p_s = p_0$ in equations 3.6.3 and 3.6.4; *i.e.* infinite surface recombination implies an equilibrium concentration at the surface.

Important Equations

Generation/Recombination Rate and Surface Recombination Rate

Generation/ Recombination rate:

$$U_n = \frac{n-n_o}{\tau_n} \quad and \quad U_p = \frac{p-p_o}{\tau_p} \tag{3.4.1}$$

Unit for τ: second

Surface Recombination Rate:

$$S_n = s_n (n-n_o) \quad and \quad S_p = s_p (p-p_o) \tag{3.4.2}$$

S = surface recombination rate; unit: $cm^{-2}\,s^{-1}$
s = surface recombination velocity; unit: $cm\,s^{-1}$

Generation/Recombination Rate (as a function of trap energy)

$$U = \frac{pn - n_i^2}{\tau_o \left(p + n + 2n_i \cosh\left[\dfrac{E_t - E_i}{kT}\right] \right)} \tag{3.5.14}$$

Recombination Rate (minority carriers)

P-type semiconductor:

$$U = \frac{\delta n}{\tau_n} = \frac{n-n_o}{\tau_n} \tag{3.5.20}$$

N-type semiconductor:

$$U = \frac{\delta p}{\tau_p} = \frac{p-p_o}{\tau_p} \tag{3.5.21}$$

Problems

Problem 3.1:

Consider an N-type silicon sample with $N_d = 10^{16}$ cm^{-3}. The dimensions of the sample can be found in Problem Figure 3.1. The carrier lifetime (electrons and holes) is 1 µs. The mobilities are $\mu_n = 625$ cm^2/Vs and $\mu_p = 200$ cm^2/Vs.

1) What is the resistance of the sample (in ohms).
2) The silicon sample is contaminated by metallic impurities which give rise to 10^{10} recombination levels per cubic centimeter. As a result, the carrier lifetime is reduced to 100 ns. These recombination centers are located at the center of the bandgap ($E_t = E_i$). What is the resistance of the sample (in ohms)?

3) The sample is illuminated with light, which gives rise to a uniform external generation $G_n=G_p=G=10^{22}$ cm^{-3}s^{-1} uniformly throughout in the sample. What is the resistance of the sample (in ohms)?

4) The concentration of metallic impurities is now doubled, while the sample remains illuminated. What is the resistance of the sample (in ohms)?

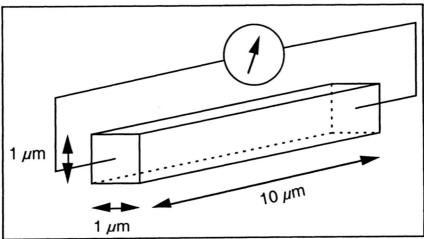

1 μm

10 μm

1 μm

Problem Figure 3.1.

Problem 3.2:

Let us consider a semi-infinite semiconductor sample on which light is shone at room temperature (Problem Figure 3.2):

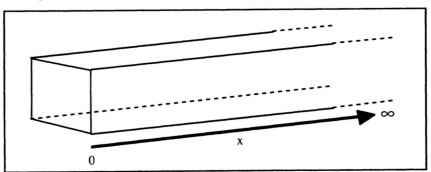

0   x   ∞

Problem Figure 3.2.

We have uniform external generation throughout the sample: $G=G_n=G_p$.

Electron and hole recombination rates are equal: $U_n=U_p=U$.

We will assume electrical neutrality everywhere such that $n-n_o=p-p_o$ where n_o and p_o are the electron and hole equilibrium concentrations, respectively. Also assume $\mu_n=\mu_p$ and that $\mathcal{E}(x=\infty)=0$. The electron and hole concentrations at equilibrium (no light) are n_o and p_o. At $x=0$, the surface recombination velocity is: $s=s_n=s_p$. Surface recombination imposes the following boundary conditions:

$$D_n \left.\frac{dn}{dx}\right|_{x=0} = s\,[n(0) - n_o]$$

and

$$D_p \left.\frac{dp}{dx}\right|_{x=0} = s\,[p(0) - p_o]$$

To simplify the expressions we will use the following notations whenever applicable: $L^2 = D\,\tau$ and $P = \frac{s\tau}{L}$. L is the "diffusion length" of minority carriers. P is a dimensionless number used to make the equations easier to manipulate.

Since $p_o = N_a^-$ and $n_o = N_d^+$, and since an equal amount of electrons and holes are photogenerated. Poisson's equation yields:

$$- \operatorname{div} \mathcal{E} = -\frac{\rho}{\varepsilon_s} = -\frac{q}{\varepsilon_s}\left((p-p_o) - (n-n_o) + p_o - n_o + N_d^+ - N_a^- \right) = 0$$

$$\Downarrow$$

$$\mathcal{E}(x) \text{ is a constant}$$

and thus, since $\mathcal{E}(\,x\to\infty) = 0$:

$$\mathcal{E}(x) = 0$$

Also, since $U_n = U_p = U$ and $n - n_o = p - p_o$ we have that $\tau_n = \tau_p = \tau$ based on $U_n = \frac{n - n_o}{\tau_n}$ and $U_p = \frac{p - p_o}{\tau_p}$.

Question:
Calculate the carrier concentrations (electrons and holes) as a function of x.
Sketch $n - n_o$ as a function of x for:
1: $s = 0$
2: $s = \infty$ (or, in other words, $n(0) = n_o$ and $p(0) = p_o$)
3: $0 < s < \infty$ (s is finite).

Problem 3.3:
An infinitely long ($-\infty < x < \infty$) piece of semiconductor is half covered by an opaque layer (Problem Figure 3.3). One shines light on the sample, such that an homogeneously uniform generation of carriers is produced for $x < 0$. The semiconductor is N-type. The electric field in the photon excited region $x < 0$ is equal to zero because there are equal numbers of holes and electrons generated, and hence $\rho(x) = 0$. We will note $\tau D_p = L_p^2$ and assume $\mathcal{E} = 0$ across the entire sample.

Find an expression for the hole current at $x = 0$. Sketch the current amplitude as a function of x.

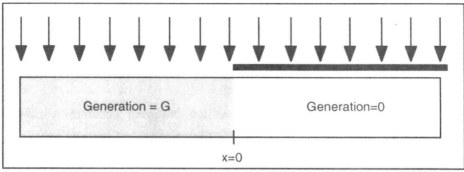

Problem Figure 3.3.

 Problem 3.4:

Consider a semi-infinite sample of silicon (Problem Figure 3.4). The cross-section area of the sample is 1 cm^2. The sample is P-type with an impurity concentration N_a (cm^{-3}).

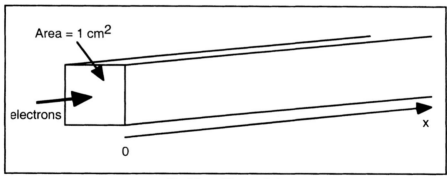

Problem Figure 3.4.

A current of electrons is continuously injected into the sample at x=0. We assume steady state, such that $\partial n/\partial t = 0$ and there is no external generation ($G_n = G_p = 0$). The electric field, $\mathcal{E}$, is equal to zero everywhere in the sample. Because of recombination the electron concentration will decrease as x is increased. Far from x=0 (*i.e.* at x=∞) the electron concentration is equal to the equilibrium electron concentration: $n(x=\infty) = n_0 = \dfrac{n_i^2}{N_a}$. The recombination rate is given by expression 3.4.1.

1) Find an analytical expression for the electron concentration n(x) as a function of x. If somewhere in the calculation you encounter the product $D_n\tau_n$, replace it by L_n^2 (in other words, $L_n^2 = D_n\tau_n$). Now assume the electron current, I, at x=0, is equal

to 5 nA. It is evenly distributed across the cross-section area of the sample, such that $J_n(x=0)$ (Ampere/cm^2) = I (Ampere), since the area is 1 cm^2.

2) Using the following data and Matlab, plot n(x) for $0 \leq x \leq 15$ micrometers.

q=1.6e-19;	%Electron charge (C)
kTq=0.0259;	%kT/q (V)
Na=1e16;	% P-type doping concentration (cm-3)
ni=1.45e10;	% Intrinsic carrier concentration (cm-3)
mu=800;	% Electron mobility (cm2/V/s)
tau=1e-9;	% Electron lifetime (s)
Ln=sqrt(Dn*tau);	% Diffusion length (cm)
n0=ni^2/Na;	% Electron equilibrium concentration
I=5e-9;	% Electron current at x=0
Jn0=I;	%Electron current density at x=0

References

1   S.M. Sze, *Physics of semiconductor devices*, J. Wiley & Sons, p. 706, 1981
2   S.M. Sze, *Physics of semiconductor devices*, J. Wiley & Sons, p. 750, 1981
3   A.S. Grove, *Physics and technology of semiconductor devices*, J. Wiley & Sons, pp. 119-127, 1967
4   R.S. Muller and T.I. Kamins, *Device electronics for integrated circuits*, J. Wiley and Sons, p. 222, 1986
5   R.F. Pierret, *Advanced semiconductor fundamentals*, Modular Serie on Solid-State Devices, Vol. VI, Addison Wesley Publishing Company, p. 158, 1989
6   A.S. Grove, *Physics and technology of semiconductor devices*, J. Wiley & Sons, pp. 129-134, 1967

Chapter 4

THE PN JUNCTION DIODE

4.1. Introduction

A PN junction is formed when a P-type and an N-type semiconductor are in contact. If the N-and P-type regions are made out of the same semiconductor material (e.g. N-type silicon and P-type silicon), the junction is a homojunction. If the semiconductor materials are different (e.g. N-type silicon and P-type germanium), the junction is a heterojunction. Heterojunctions are dealt with in Chapter 9.

A diode is a semiconductor device consisting of a single PN junction (Figure 4.1). Unlike a resistor, it has a highly non-linear current-voltage characteristics and is often used as a rectifying element. Some diodes can emit light (light-emitting diodes), and others can emit laser light (laser diodes). The proper combination of two PN junctions produces a bipolar transistor, a device capable of amplifying electric signals.

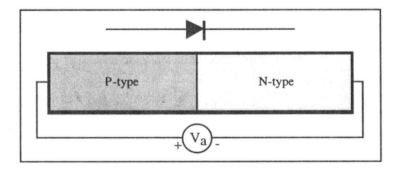

Figure 4.1: PN junction and symbol representing a diode.

The PN junction presents the following property: It allows current flow in one bias direction, but not in the other bias direction. Hence it rectifies the current. The sign convention used in this chapter is shown in Figure

4.1. The applied voltage, V_a, is positive if the potential applied to the P-side is higher than that on the N-side. As illustrated in Figure 4.2 current flows through the diode if V_a is positive, and does not if V_a is negative. If $V_a > 0$ the junction is said to be forward biased, and if $V_a < 0$ it is reverse biased.

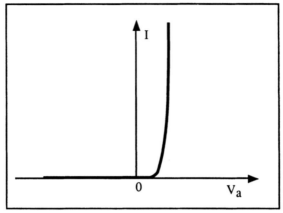

Figure 4.2: Current-voltage characteristics of a PN junction.

Experimental measurements show that the current in a PN junction, I, obeys the following equation:

$$I = I_s \left(exp \left[\frac{qV_a}{kT} \right] - 1 \right)$$
(4.1.1)

where I_s is a constant and V_a is the voltage applied to the diode.

An analogy of the diode is a valve which controls liquid flow (Figure 4.3). When a pressure differential is applied in the forward direction, the valve opens and allows the liquid flow. If the pressure differential is applied in the reverse direction, the valve closes, and no liquid flows, except for a few drops if the valve is imperfect and somewhat "leaky".

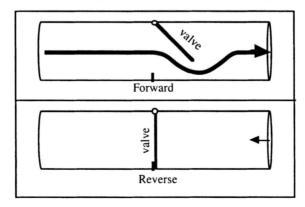

Figure 4.3: Fluid mechanics analogy of a pn junction to a valve.

4.2. Unbiased PN junction

We now consider a PN junction at thermodynamic equilibrium, *i.e.* in the absence of an applied bias ($V_a=0$). Let us first focus on the P-type and the N-type region taken separately, as if there were two separate pieces of semiconductor material. For simplicity, doping concentrations in both pieces are constant, and equal to N_d (cm^{-3}) in the N-type region, and N_a (cm^{-3}) in the P-type region. The energy band diagram of the two pieces of semiconductor are shown in Figure 4.4.

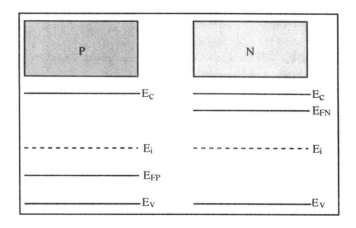

Figure 4.4: Energy band diagram in the N- and P-type regions taken separately.

Using Expressions 1.3.15a and 1.3.15b one can write:

$$E_{FN} - E_i = kT \ln\left(\frac{N_d}{n_i}\right) \quad \text{in the N-type region, and}$$

$$E_i - E_{FP} = kT \ln\left(\frac{N_a}{n_i}\right) \quad \text{in the P-type region.}$$

Let us now build the PN junction by connecting the P-type region to the N-type region. The surface where the contact is made is called the "metallurgical junction". A junction where the doping concentration "abruptly" switches from P-type to N-type (at the metallurgical junction) is called a step junction. We already know from Section 1.4 that the Fermi level is unique and constant in a structure under equilibrium: electrons instantly diffuse from the electron-rich N-type region into the electron-poor P-type region, and holes from the P-type material diffuse into the N-type region. As a result of the charge displacement an internal built-in potential called junction potential, Φ_o, is formed at the junction, as shown in Figure 4.5.

Within a multiplication factor $-q$ the junction potential is equal to the curvature of the energy bands:

$$E_{FN} - E_{FP} = q\Phi_0 = kT \ln\left(\frac{N_d}{n_i}\right) + kT \ln\left(\frac{N_a}{n_i}\right) = kT \ln\left(\frac{N_a N_d}{n_i^2}\right) \qquad (4.2.1)$$

and thus:

$$\Phi_0 = \frac{kT}{q} \ln\left(\frac{N_a N_d}{n_i^2}\right) \qquad (4.2.2)$$

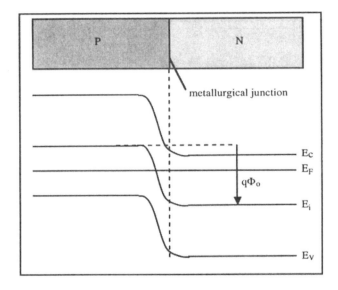

Figure 4.5: PN junction and corresponding energy band diagram.[1]

When electrons diffuse from the N-type region into the P-type material, they "leave behind" the ionized donor atoms they originated from. These atoms occupy substitutional sites in the crystal lattice and cannot move within the crystal. The region where these positively charged ions are located constitutes a space-charge region called a "depletion region" because it is depleted of electrons (Figure 4.6).

The positive charge in the depletion region attracts electrons such that at equilibrium, the force of diffusion pushing electrons into the P-type region is exactly balanced by the force of the built-in electric field that "recalls" the electrons back into the N-type region. Similarly, the diffusion of holes from the P-type into the N-type region gives rise to a depletion region in the P-type material. This region is depleted of holes and bears a negative charge because of the presence of ionized, negatively charged acceptor atoms. There are several names for the depletion region located around the metallurgical junction; it can be called the "depletion region", the "space-charge region" or the "transition region".

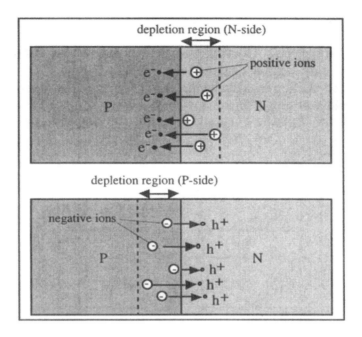

Figure 4.6: Creation of depletion regions by the diffusion of electrons and holes.

The electric field and the potential variation in the space-charge region can be calculated using the Poisson equation (Expression 2.6.2). For a one-dimensional junction the problem simplifies to:

$$\frac{d^2 \Phi(x)}{dx^2} = -\frac{q}{\varepsilon_s} (p - n + N_d^+ - N_a^-) \qquad (4.2.3a)$$

Using the Boltzmann Relationships 1.3.20a and 1.3.20b we obtain:

$$\frac{d^2 \Phi(x)}{dx^2} = -\frac{q}{\varepsilon_s} \left\{ p_o \, exp\left[\frac{-q\Phi(x)}{kT}\right] - n_o \, exp\left[\frac{q\Phi(x)}{kT}\right] + N_d^+ - N_a^- \right\} \qquad (4.2.3b)$$

with $N_d^+ = N_d$ and $N_a^+ = N_a$.

Equation 4.2.3b cannot be solved analytically and a close-form solution for the potential cannot be explicitly found. It can, however, be simplified by using the "depletion approximation". The depletion approximation assumes that the space charge is composed only of ionized doping impurities, and that the contribution of free carriers to the local charge is negligible. Furthermore, the carrier depletion in the space-charge regions is assumed to be complete. In other words, there are no free electrons in the depletion region on the N-type side, and no free holes in the depletion region on the P-type side. As a result, the charge densities in the depletion regions are equal to qN_d in the N-type material, and $-qN_a$ in the P-type material. The depletion regions extent to a distance l_{no} on the N-type side, and a distance $-l_{po}$ on the P-type side, where the metallurgical junction is taken as the origin (Figure 4.7). Additionally, the electric field and potential are shown in Figure 4.7, which can also be derived from Poisson's equation with the appropriate boundary conditions.

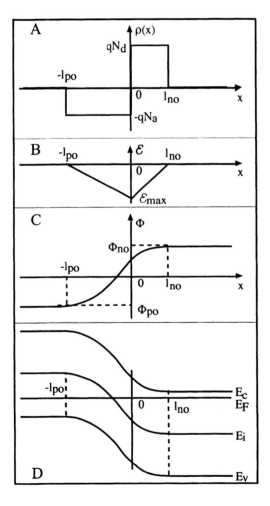

Figure 4.7: Charges (A), electric field (B), potential (C) and energy bands (D) in a PN junction.[2]

With the depletion approximation, a closed-form analytical expression can be found for the electric field $\mathcal{E}(x)$, the potential $\Phi(x)$, as well as for l_{po} and l_{no} by utilizing Poisson's equation and Gauss' law. The value of the charge density $\rho(x)$ can be expressed for four separate regions and are given by:

$$\rho(x) = \begin{array}{llll} 0 & \text{for} & -\infty < x < -l_{po} & \text{(quasi-neutral region)} \\ -qN_a & \text{for} & -l_{po} < x < 0 & \text{(space-charge region)} \\ qN_d & \text{for} & 0 < x < l_{no} & \text{(space-charge region)} \\ 0 & \text{for} & l_{no} < x < \infty & \text{(quasi-neutral region)} \end{array}$$

We will assume that charge neutrality exists in the quasi-neutral regions. Therefore, the electric field is zero in these regions. Using all the above assumptions the Poisson equation can be integrated a first time to yield the electric field:

for $-\infty < x < -l_{po}$: $\qquad\qquad \mathcal{E}(x)=0$

for $-l_{po} < x < 0$: $\qquad \dfrac{d^2\Phi(x)}{dx^2} = -\dfrac{d\mathcal{E}(x)}{dx} = \dfrac{q}{\varepsilon_s}N_a$ with $\mathcal{E}(-l_{po})=0$

$$\Downarrow$$

$$\mathcal{E}(x) = -\frac{qN_a}{\varepsilon_s}(x + l_{po}) \qquad\qquad (4.2.4)$$

for $0 < x < l_{no}$: $\qquad \dfrac{d^2\Phi(x)}{dx^2} = -\dfrac{d\mathcal{E}(x)}{dx} = -\dfrac{q}{\varepsilon_s}N_d$ with $\mathcal{E}(l_{no})=0$

$$\Downarrow$$

$$\mathcal{E}(x) = -\frac{qN_d}{\varepsilon_s}(l_{no} - x) \qquad\qquad (4.2.5)$$

and, for $l_{no} < x < \infty$ one obtains: $\qquad \mathcal{E}(x)=0$

The electric field is continuous at $x=0$ by imposing Gauss' law, which yields:

$$-\frac{qN_a}{\varepsilon_s}l_{po} = -\frac{qN_d}{\varepsilon_s}l_{no} \Rightarrow N_a l_{po} = N_d l_{no} \qquad\qquad (4.2.6)$$

Relationship 4.2.6 reiterates charge neutrality in the device, since it states that the total negative charge in the depletion region on the N-side of the junction, $-qN_d l_{no}$, is equal, in absolute value, to the total positive charge on the P-side, $qN_a l_{po}$. The potential distribution is obtained by integrating the Poisson equation a second time. In the P-type and N-type quasi-neutral regions the potentials are Φ_{po} and Φ_{no}, respectively. Using these as boundary conditions yields:

for $-\infty < x < -l_{po}$: $\qquad\qquad \Phi_0(x)=\Phi_{po}$

for $-l_{po} < x < 0$:
$$-\mathcal{E}(x) = \frac{d\Phi(x)}{dx} = \frac{qN_a}{\varepsilon_s}(x + l_{po})$$
$$\Downarrow$$
$$\Phi_o(x) = \frac{qN_a}{2\varepsilon_s}(x + l_{po})^2 + \Phi_{po} \qquad (4.2.7)$$

for $0 < x < l_{no}$:
$$-\mathcal{E}(x) = \frac{d\Phi(x)}{dx} = \frac{qN_d}{\varepsilon_s}(l_{no} - x)$$
$$\Downarrow$$
$$\Phi_o(x) = \Phi_{no} - \frac{qN_d}{2\varepsilon_s}(l_{no} - x)^2 \qquad (4.2.8)$$

for $l_{no} < x < \infty$: $\Phi_o(x) = \Phi_{no}$

The potential is a continuous function at $x=0$. Combined with 4.2.2 this condition gives an alternate expression for the junction potential, Φ_o:

$$\frac{qN_a}{2\varepsilon_s}l_{po}^2 + \Phi_{po} = \Phi_{no} - \frac{qN_d}{2\varepsilon_s}l_{no}^2$$
$$\Downarrow$$

Junction Potential
$$\Phi_o = \Phi_{no} - \Phi_{po} = \frac{qN_a}{2\varepsilon_s}l_{po}^2 + \frac{qN_d}{2\varepsilon_s}l_{no}^2 = \frac{kT}{q}\ln\left(\frac{N_a N_d}{n_i^2}\right) \qquad (4.2.9)$$

The electric field has a single maximum value at $x=0$. Its expression can be obtained using 4.2.4 or 4.2.5:

Maximum electric field
$$\mathcal{E}_{max} = -\frac{qN_a}{\varepsilon_s}l_{po} = -\frac{qN_d}{\varepsilon_s}l_{no} \qquad (4.2.10)$$

Using Expressions 4.2.6 and 4.2.9 the width of the depletion regions, l_{po} and l_{no}, can be expressed as a function of the junction potential:

Width of Depletion Regions
$$l_{po} = \sqrt{\frac{2\varepsilon_s}{q}\frac{\Phi_o N_d}{N_a(N_a+N_d)}} \qquad (4.2.11a)$$
and
$$l_{no} = \sqrt{\frac{2\varepsilon_s}{q}\frac{\Phi_o N_a}{N_d(N_a+N_d)}} \qquad (4.2.11b)$$

The two depletion regions form the "transition region", which contains both ionized acceptor and donor impurities. The width of the transition region is given by:

$$l_{no} + l_{po} = \sqrt{\frac{2\varepsilon_s}{q} \frac{\Phi_o{}^\circ(N_a+N_d)}{N_aN_d}}{}_\circ \qquad (4.2.12)$$

Actual PN junctions are strongly asymmetrical, which means that one side is doped much more heavily than the other. Consider the example of a PN$^+$ junction, with $N_a=10^{15}$ cm^{-3} and $N_d=10^{20}$ cm^{-3}. Since $N_d \gg N_a$, one obtains:

$$l_{po} = \sqrt{\frac{2\varepsilon_s}{q}\frac{\Phi_o}{N_a}} \gg l_{no} = \sqrt{\frac{2\varepsilon_s}{q}\frac{\Phi_o{}^\circ N_a}{N_d^2}}{}_\circ \qquad (4.2.13)$$

and, therefore,

$$l_{no} + l_{po} \cong l_{po} \qquad (4.2.14)$$

Comment: In a strongly asymmetrical junction, the width of the transition region is virtually equal to the width of the depletion region with the lowest doping concentration.

Example:
Calculate Φ_o, l_{no} and l_{po} in a silicon PN junction with $N_a=10^{15}$ cm^{-3} and $N_d=10^{19}$ cm^{-3}.

$$\varepsilon_{si} = \kappa_{si}\,\varepsilon_o = 11.7 \times 8.854\times10^{-14} \text{ F/cm}$$
$$q = 1.6\times10^{-19} \text{ C}$$
$$\frac{kT}{q} = 26 \text{ mV at room temperature (T = 300 K)}$$
$$n_i = 1.45\times10^{10} \text{ cm}^{-3} \text{ at room temperature}$$

$$\Phi_o = \frac{kT}{q} \ln\left(\frac{N_aN_d}{n_i^2}\right) = 0.82 \text{ V}$$

$$l_{po} = \sqrt{\frac{2\varepsilon_{si}}{q}\frac{\Phi_o{}^\circ N_d}{N_a{}^\circ(N_a+N_d)}}{}_\circ = 1.03 \text{ m}$$

$$l_{no} = \sqrt{\frac{2\varepsilon_{si}}{q}\frac{\Phi_o{}^\circ N_a}{N_d{}^\circ(N_a+N_d)}}{}_\circ = 0.000103 \text{ m} = 0.103 \text{ nm}$$

It can easily be seen that $l_{no} \ll l_{po}$

4.3. Biased PN junction

If no bias is applied to a PN junction the built-in junction potential is equal to Φ_o, as we have seen in the previous Section. The drift current generated by this potential variation is exactly equal and of opposite sign to the diffusion current caused by the carrier concentration gradients, such that the net current flow (drift + diffusion) is equal to zero. The potential variation $\Phi(x)$ actually acts as a barrier which prevents further diffusion of electrons into the P-type region and holes into the N-type region, once equilibrium has been established. That is why Φ_o is sometimes referred to as a "potential barrier" which the carriers must overcome in order to diffuse.

Consider the case when an external bias, V_a, is applied to the junction. V_a is considered positive if the potential of the P-type region is higher (more positive) than that of the N-type region. We will assume that the current flowing through the device is small enough such that the potential drops across the quasi-neutral regions are negligible. As a consequence, the external applied potential, V_a, is supported entirely by the transition region, and the internal potential, Φ, is equal to:

$$\Phi = \Phi_n - \Phi_p = \Phi_o - V_a \tag{4.3.1}$$

Noting that $-l_p$ and l_n are the edges of the transition region (Figure 4.8), the distribution of charges in the structure are:

$$\rho(x) = \quad\begin{array}{llll} 0 & \text{for} & -\infty < x < -l_p & \text{(quasi-neutral region)} \\ -qN_a & \text{for} & -l_p < x < 0 & \text{(space-charge region)} \\ qN_d & \text{for} & 0 < x < l_n & \text{(space-charge region)} \\ 0 & \text{for} & l_n < x < \infty & \text{(quasi-neutral region)} \end{array}$$

The Poisson equation can be solved just as it was in Equations 4.2.4 to 4.2.12, by replacing l_{no}, l_{po} and Φ_o by l_n, l_p and $(\Phi_o - V_a)$, respectively. The result is:

$$l_p = \sqrt{\frac{2\varepsilon_s}{q} \frac{(\Phi_o - V_a)\, N_d}{N_a\,(N_a + N_d)}} \tag{4.3.2}$$

and

$$l_n = \sqrt{\frac{2\varepsilon_s}{q} \frac{(\Phi_o - V_a)\, N_a}{N_d\,(N_a + N_d)}} \tag{4.3.3}$$

The total width of the transition region is equal to:

$$l_n + l_p = \sqrt{\frac{2\varepsilon_s}{q} \frac{(\Phi_o - V_a)\,(N_a + N_d)}{N_a N_d}} \tag{4.3.4}$$

It is worth noting that the width of the transition region increases when a reverse bias is applied ($V_a < 0$) and that it decreases when a forward bias ($V_a > 0$) is applied (Figure 4.8).

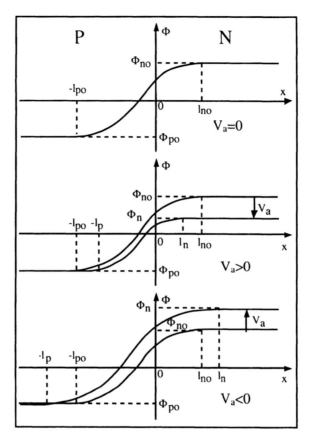

Figure 4.8: Potential in a PN junction for $V_a=0$, $V_a>0$ and $V_a<0$.

4.4. Current-voltage characteristics

As we have seen in the previous Section the potential drop across the transition region is equal to $\Phi_o - V_a$, where V_a is the applied voltage. Therefore, if V_a is positive, the potential barrier in the junction is lower than its equilibrium value, Φ_o. As a result the diffusion and electric field forces are no longer equal and of opposite sign. Diffusion acting on the carriers is only partially compensated by the force resulting from the junction potential variation, and therefore, holes can flow from the P-type region into the N-type semiconductor and electrons can flow from the N-type region into the P-type semiconductor. The resulting currents are shown in Figure 4.9. The holes injected into the N-type region are excess minority carriers (current "1" in Figure 4.9). These carriers diffuse into the N-type quasi-neutral region an average distance called the "diffusion length" before recombining with the majority carriers (electrons). Since each recombination event consumes an electron, a resulting electron current appears in the N-type region where electrons are continuously supplied by the external contact (current "2" in Figure 4.9). Similarly, the electrons injected into the P-type region (current "3" in Figure 4.9) are excess minority carriers which recombine with holes in the P-type region. Since each recombination event consumes a hole, a resulting hole current appears in the P-type region (current "4" in Figure 4.9). It is worth noting that current "1" is equal to current "2" and that current "3" is equal to current "4", in Figure 4.9.

If the junction is reverse-biased ($V_a<0$) the amplitude of the potential barrier is increased beyond its equilibrium value, Φ_o. Diffusion of holes in the N-type region and diffusion of electrons in the P-type region are reduced and net current, resulting from the drift of holes from the N-type region into the P-type region and the drift of electrons from the P-type region into the N-type region, is observed. The magnitude of this current, however, is extremely small since it involves only *minority* carriers in the vicinity of the edges of the transition region.

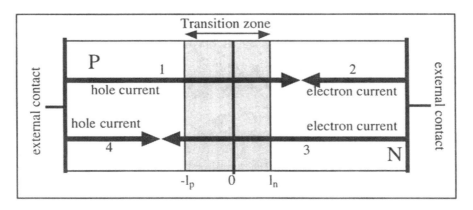

Figure 4.9: Forward-biased PN junction; 1: holes injected from the P-type region into the N-type region; 2: electrons recombining with the holes injected in the N-type region; 3: electrons injected from the N-type region into the P-type region; 4: holes recombining with the electrons injected in the N-type region.

A derivation of the current-voltage characteristics of the PN junction based on the currents of majority carriers would prove quite difficult. These are the hole current in the P-type material and the electron current in the N-type region, noted currents "4" and "2" in Figure 4.9, respectively. We know, however, that current "2" is equal to current "1" and current "4" is equal to current "3". Currents "1" and "3" are a result of minority carrier injection (holes in the N-type material and electrons in the P-type material) and the sum of these two components is equal to the total current in the device. The derivation of the modeling equations for the PN junction will, therefore, make use of currents "1" and "3".

Ultimately we want to have an equation for the PN junction which describes the current as a function of applied voltage.

4.4.1. Derivation of the ideal diode model

The notations used in this section are shown in Table 4.1:

Table 4.1: Notations used in this section

Symbol	Physical meaning	Unit
D_n	electron diffusion coefficient	$cm^2\ s^{-1}$
D_p	hole diffusion coefficient	$cm^2\ s^{-1}$
L_n	electron diffusion length	cm
L_p	hole diffusion length	cm
n_n	electron concentration in the N-type region	cm^{-3}
n_{no}	equilibrium electron concentration in the N-type region	cm^{-3}
n_p	electron concentration in the P-type region	cm^{-3}
n_{po}	equilibrium electron concentration in the P-type region	cm^{-3}
p_n	hole concentration in the N-type region	cm^{-3}
p_{no}	equilibrium hole concentration in the N-type region	cm^{-3}
p_p	hole concentration in the P-type region	cm^{-3}
p_{po}	equilibrium hole concentration in the P-type region	cm^{-3}
n'_p	excess electron concentration in the P-type region	cm^{-3}
p'_n	excess hole concentration in the N-type region	cm^{-3}
τ_n	electron lifetime	s
τ_p	hole lifetime	s

To simplify the PN junction model we will use the following starting assumptions:

1. Low-level injection assumption (or "weak injection"): the concentration of minority carriers injected in a quasi-neutral region is low compared to the majority carrier concentration.

2. The Boltzmann relationships 2.7.1 and 2.7.2 are valid in the quasi-neutral regions as well as in the transition region.

3. Current flow in the quasi-neutral regions is due to a diffusion mechanism (no potential drop, and therefore, no electric field is assumed in those regions).

4. The quasi-neutral regions of the diode are infinitely long.

5. Generation/recombination phenomena are neglected in the transition region.

Starting assumption #1- Low-level injection assumption (or "weak injection"). The *excess* concentration of minority carriers, p'_n and n'_p injected in a quasi-neutral region is low compared to the majority carrier concentration:

$$p'_n (x) \equiv p_n(x) - p_{no} \ll n_{no} \quad \text{in the N-type quasi-neutral region} \quad (4.4.1)$$
$$n'_p (x) \equiv n_p(x) - n_{po} \ll p_{po} \quad \text{in the P-type quasi-neutral region} \quad (4.4.2)$$

As a result of the low-level injection condition the concentration of majority carriers is not modified by the injection of minority carriers:

$$n_n(x) = n_{no} = N_d \quad \text{in the N-type quasi-neutral region} \qquad (4.4.3)$$
$$p_p(x) = p_{po} = N_a \quad \text{in the P-type quasi-neutral region} \qquad (4.4.4)$$

Starting assumption #2- The Boltzmann relationships 2.7.1 and 2.7.2 are valid in the quasi-neutral regions as well as in the transition region.

Considering the depletion region on the N-type side ($0 \leq x \leq l_n$) one can write:

$$n(x) = n_i \, exp\left[\frac{E_{FN}-E_{io}}{kT}\right] exp\left[\frac{q\Phi(x)}{kT}\right] \qquad (4.4.5)$$

From Relationship 4.4.3 we know that $n_n(x) = n_{no} = N_d$ *for* $l_n \leq x \leq \infty$. Since the potential in the N-type quasi-neutral region is equal to Φ_n we can write:

$$n_{no} = n_i \, exp\left[\frac{E_{FN}-E_{io}}{kT}\right] exp\left[\frac{q\Phi_n}{kT}\right] \qquad (4.4.6)$$

Expression 4.4.6 can be substituted into 4.4.5 to give:

$$n(x) = n_{no} \, exp\left[\frac{q(\Phi(x) - \Phi_n)}{kT}\right]$$

Under the assumption that the Boltzmann relationships are valid in the transition region, the latter equation can be evaluated at $x = -l_p$ where $\Phi(x) = -\Phi_p$:

$$n_p(-l_p) = n_{no} \, exp\left[\frac{q(\Phi_p-\Phi_n)}{kT}\right]$$

Since $\Phi_p - \Phi_n = V_a - \Phi_o$ we can write:

$$n_p(-l_p) = n_{no} \, exp\left[\frac{-q\Phi_o}{kT}\right] exp\left[\frac{qV_a}{kT}\right]$$

From this we now have an expression for the minority carrier concentration at the edge of the transition region which is a function of the applied voltage. The equilibrium junction potential is defined by:

$$\Phi_0 = \Phi_p - \Phi_n = \frac{kT}{q} ln\left(\frac{N_a N_d}{n_i^2}\right) = \frac{kT}{q} ln\left(\frac{p_{po} n_{no}}{n_i^2}\right)$$

Combining the two latter equations yields:

$$n_p(-l_p) = n_{no} \frac{n_i^2}{p_{po} n_{no}} exp\left[\frac{qV_a}{kT}\right] = \frac{n_i^2}{p_{po}} exp\left[\frac{qV_a}{kT}\right]$$

Since, by definition, $n_{po} = \dfrac{n_i^2}{p_{po}} = \dfrac{n_i^2}{N_a}$ we finally obtain:

$$n_p(-l_p) = n_{po} \, exp\left[\frac{qV_a}{kT}\right] \qquad (4.4.7)$$

A similar calculation, carried out for holes at the N-side edge of the transition region, would yield:

$$p_n(l_n) = p_{no} \, exp\left[\frac{qV_a}{kT}\right] \qquad (4.4.8)$$

As a result of Expression 4.4.7 the concentration of *excess* electrons at the P-side edge of the transition region is equal to:

$$n'_p(-l_p) = n_p(-l_p) - n_{po} = n_{po}\left[exp\left(\frac{qV_a}{kT}\right) - 1\right] \qquad (4.4.9)$$

Similarly the concentration of *excess* holes at the N-side edge of the transition region is given by:

$$p'_n(l_n) = p_n(l_n) - p_{no} = p_{no}\left[exp\left(\frac{qV_a}{kT}\right) - 1\right] \qquad (4.4.10)$$

Starting assumption #3- Current flow in the quasi-neutral regions is due to a diffusion mechanism (no potential drop, and therefore, no electric field is assumed in those regions).

$$J_p = -qD_p \frac{dp}{dx} \quad \text{in the N-type quasi-neutral region} \qquad (4.4.11)$$

and

$$J_n = qD_n \frac{dn}{dx} \quad \text{in the P-type quasi-neutral region} \qquad (4.4.12)$$

Let us now write the Continuity Equation 2.6.6b for holes in the N-type quasi-neutral region, with the assumption that there is no generation from an external source:

$$\frac{\partial p_n}{\partial t} = -\frac{1}{q}\frac{\partial J_p}{\partial x} - \frac{p_n - p_{no}}{\tau_p} \tag{4.4.13}$$

and, replacing J_p by its value in Equation 4.4.11:

$$\frac{\partial p_n}{\partial t} = D_p \frac{\partial^2 p_n}{\partial x^2} - \frac{p_n - p_{no}}{\tau_p} \tag{4.4.14}$$

Assuming steady-state conditions ($\partial p/\partial t = 0$) the following differential equation is obtained:

$$D_p \frac{d^2 p_n}{dx^2} = \frac{p_n - p_{no}}{\tau_p} \tag{4.4.15}$$

which admits the solution:

$$p_n(x) = p_{no} + A\ exp(-x/L_p) + B\ exp\ (x/L_p) \tag{4.4.16}$$

where A and B are integration constants, and L_p is called the *diffusion length* of holes, defined by:

$$L_p = \sqrt{D_p \tau_p} \tag{4.4.17}$$

A similar calculation, made for electrons in the P-type quasi-neutral region, would yield:

$$n_p(x) = n_{po} + C\ exp(-x/L_n) + D\ exp\ (x/L_n) \tag{4.4.18}$$

where C and D are integration constants, and L_n is called the *diffusion length* of electrons, defined by:

$$L_n = \sqrt{D_n \tau_n} \tag{4.4.19}$$

Starting assumption #4- Consider a "long-base diode", *i.e.* a diode where the length of the quasi-neutral regions is much larger than the diffusion length of the minority carriers, L_n and L_p. From a mathematical point of view this condition is equivalent to assuming that the length of the quasi-neutral regions is infinite.

Using Expression 4.4.8 and $p_n(\infty) = p_{no}$ (thermodynamic equilibrium far from the junction) as boundary conditions for Equation 4.4.16 one obtains:

$$p_n(\infty) = p_{no} \Rightarrow B = 0 \tag{4.4.20}$$

$$p_n(l_n) = p_{no}\ exp\left[\frac{qV_a}{kT}\right] = p_{no} + A\ exp\left(\frac{-l_n}{L_p}\right)$$

which yields:

$$A = p_{no}\left[exp\left(\frac{qV_a}{kT}\right) - 1\right] exp\left(\frac{l_n}{L_p}\right) \tag{4.4.21}$$

Once the integration constants A and B are known the concentration of holes in the quasi-neutral N-type region can be derived from Equation 4.4.16:

$$p_n(x) = p_{no} + p_{no} \left[exp\left(\frac{qV_a}{kT}\right) - 1 \right] exp\left(\frac{-(x-l_n)}{L_p}\right) \qquad (4.4.22)$$

The hole diffusion current in the quasi-neutral N-type region is, therefore, equal to:

$$J_p = -qD_p \frac{dp_n}{dx} = \frac{qD_p p_{no}}{L_p} exp\left(\frac{-(x-l_n)}{L_p}\right) \left[exp\left(\frac{qV_a}{kT}\right) - 1 \right] \qquad (4.4.23)$$

Similarly, the electron diffusion current in the quasi-neutral P-type region is given by:

$$J_n = qD_n \frac{dn_p}{dx} = \frac{qD_n n_{po}}{L_n} exp\left(\frac{x+l_p}{L_n}\right) \left[exp\left(\frac{qV_a}{kT}\right) - 1 \right] \qquad (4.4.24)$$

Since the diode considered here is a one-dimensional device with two access terminals the current flowing through it is constant and independent of the position x. One can, however, observe that the hole current density given by Expression 4.4.23 decreases when the value of x is increased (with $l_n < x < \infty$). This occurs because the holes, which are minority carriers in the N-type region, recombine with electrons, which are majority carriers. Since an electron must be supplied for every recombination event in which a hole disappears the current steadily transforms from a hole current into an electron current as x is increased. Similarly the electron current in the P-type region disappears to the benefit of a hole current as x (with $-\infty < x < -l_p$) is decreased. The net current density in the device is given by:

$$J = constant = J_n(x) + J_p(x), \text{ for any value of } x \qquad (4.4.25)$$

The minority carrier concentrations in the quasi-neutral regions and the hole and electron current densities are shown as a function of position, x, for $V_a > 0$, in Figure 4.10.

Since we have assumed no generation/recombination in the transition zone (**Starting assumption #5**) we can write:

$$J_p(-l_p) = J_p(l_n) \text{ and } J_n(-l_p) = J_n(l_n) \qquad (4.4.26)$$

The current at the boundaries of the space-charge region is entirely due to the minority carriers which have been injected. As a result, the total current in the device will be the sum of these two components, *i.e.* the sum of expressions 4.4.23 and 4.4.24 evaluated at l_n and $-l_p$, respectively.

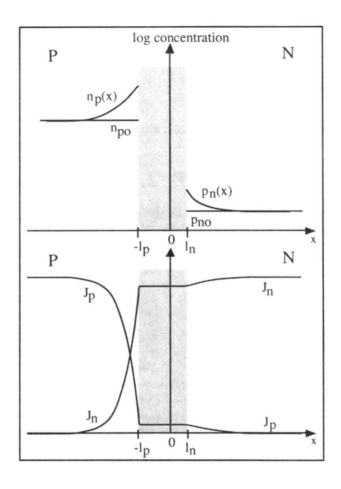

Figure 4.10: Minority carrier concentrations (top) and electron and hole current densities (bottom). [3]

Using the two latter Relationships we can write:

$$J_{TOTAL} = J = J_n(-l_p) + J_p(l_n)$$

$$\Downarrow$$

$$J = \frac{qD_n n_{po}}{L_n} exp\left(\frac{-l_p + l_p}{L_n}\right)\left[exp\left(\frac{qV_a}{kT}\right) - 1\right] + \frac{qD_p p_{no}}{L_p} exp\left(\frac{-(l_n - l_n)}{L_p}\right)\left[exp\left(\frac{qV_a}{kT}\right) - 1\right]$$

$$\Downarrow$$

$$J = \left\{\frac{qD_n n_{po}}{L_n} + \frac{qD_p p_{no}}{L_p}\right\}\left[exp\left(\frac{qV_a}{kT}\right) - 1\right]$$

$$\Downarrow$$

Current Density in the ideal PN junction

$$J = J_S \left[exp\left(\frac{qV_a}{kT}\right) - 1 \right] \qquad (4.4.27)$$

where J_S is called the "saturation current density" and is equal to:

Saturation Current Density

$$J_S = \frac{qD_n n_{po}}{L_n} + \frac{qD_p p_{no}}{L_p} = q\, n_i^2 \left(\frac{1}{N_a} \sqrt{\frac{D_n}{\tau_n}} + \frac{1}{N_d} \sqrt{\frac{D_p}{\tau_p}} \right) \qquad (4.4.28)$$

It is worthwhile noting that the magnitude of the current flowing in a reverse-biased PN junction ($V_a<0$) is equal to J_S. J_S is independent of the applied bias and of the magnitude of the electric field in the structure. It is, however, quite dependent on temperature.

The current in the device can readily be obtained by multiplying the current density, J, of expression 4.4.27 by the cross-sectional area of the junction, A such that $I = AJ$ (amperes). The current expression obtained in Relationship 4.4.27 is in good agreement with experimental current-voltage characteristics, since Expression 4.4.27 is equivalent to Expression 4.1.1, where $I_S = A\, J_S$. Note that the reverse-bias current of the diode, $-I_S$, is sometimes called a "leakage current".

4.4.2. Generation/recombination current

We have so far calculated the current-voltage characteristics of an "ideal diode" and neglected generation/recombination mechanisms in the transition region. Actual diodes are, unfortunately, non-ideal and the effects of generation/recombination have to be taken into account to accurately model experimental device characteristics.

When an external bias, V_a, is applied, the *pn* product in the transition region is different from its equilibrium value, n_i^2, since excess carriers are injected into ($V_a>0$) or extracted from ($V_a<0$) the transition region. As a result the Fermi level splits into two quasi Fermi levels (E_{Fn} for electrons and E_{Fp} for holes). The difference between the two quasi Fermi levels is the applied voltage, V_a. According to Expression 2.7.6 the *pn* product in the transition region is equal to:

$$pn = n_i^2\, exp\left[\frac{E_{Fn} - E_{Fp}}{kT}\right] = n_i^2\, exp\left[\frac{qV_a}{kT}\right] \qquad (4.4.29)$$

Therefore, the SRH generation/recombination rate is equal to (Expression 3.5.14):

$$U = \frac{pn - n_i^2}{\tau_o\left(p + n + 2n_i \cosh\left[\frac{E_t - E_i}{kT}\right]\right)} = \frac{n_i^2\left[\exp\left(\frac{qV_a}{kT}\right) - 1\right]}{\tau_o\left(p + n + 2n_i \cosh\left[\frac{E_t - E_i}{kT}\right]\right)} \quad (4.4.30)$$

or, considering that the recombination centers are located at midgap ($E_t = E_i$), where recombination is the most effective (see Expression 3.5.14):

$$U = \frac{n_i^2\left[\exp\left(\frac{qV_a}{kT}\right) - 1\right]}{\tau_o\left(p + n + 2n_i\right)} \quad (4.4.31)$$

Using the continuity equations 2.6.7a and 2.6.7b in steady state, which assumes $U_p = U_n \equiv U$, we can write:

$$\frac{dJ_n}{dx} = qU = -\frac{dJ_p}{dx} \quad (4.4.32)$$

and, integrating over the transition region one obtains:

$$J_n(l_n) = J_n(-l_p) + q \int_{-l_p}^{l_n} U(x)\,dx \quad (4.4.33)$$

The net current density is given by:

$$J = J_p(l_n) + J_n(l_n) = J_p(l_n) + J_n(-l_p) + q \int_{-l_p}^{l_n} U(x)\,dx$$

$$= J_s\left[\exp\left(\frac{qV_a}{kT}\right) - 1\right] + q \int_{-l_p}^{l_n} U(x)\,dx$$

which can be rewritten:

$$J = J_s\left[\exp\left(\frac{qV_a}{kT}\right) - 1\right] + J_{rg} \quad (4.4.34)$$

For a given forward bias, V_a, the generation/recombination rate will have a maximum value at that location in the transition region where the sum of the electron and hole concentration, $p+n$, is at a minimum value, based on Expression 4.4.31.[4] Since the product of the electron and hole concentrations, pn, is a constant, the conditions $d(p+n)=0$ and $d(pn)=0$ lead to:

$$dp = -dn \quad and \quad ndp + pdn = 0 \quad \Rightarrow \quad p = n$$

This condition exists at a location within the transition region where the intrinsic Fermi level, E_i, is half-way between the quasi-Fermi level for electrons, E_{Fn}, and for holes, E_{Fp}. There, the carrier concentrations are given by 4.4.29:

$$n = p = n_i \, exp(qV_a/2kT)$$

and, the recombination rate, U, can be found using Expression 4.4.31:

$$U = \frac{n_i^2 \left[exp\left(\frac{qV_a}{kT}\right) - 1 \right]}{\tau_o \left(p + n + 2n_i \right)} = \frac{n_i^2 \left[exp\left(\frac{qV_a}{kT}\right) - 1 \right]}{2n_i \left[exp\left(\frac{qV_a}{2kT}\right) + 1 \right] \tau_o} = \frac{n_i \left[exp\left(\frac{qV_a}{2kT}\right) - 1 \right]}{2\tau_o}$$

Assuming the latter expression is valid (*i.e.* generation/recombination is maximum) over the entire transition region Equation 4.4.33 can be solved analytically. The assumption of maximum generation/recombination over the entire transition region will slightly overestimate the current *Jrg*, but it accurately reproduces its exponential dependence on $qV_a/2kT$. Using Relationship 4.3.4 for calculating the width of the transition region we find:

$$J_{rg} = q \int_{-l_p}^{l_n} U(x) \, dx = q \, U \, (l_n + l_p)$$

$$= \frac{qn_i}{2\tau_o} \sqrt{\frac{2\varepsilon_s}{q} (\Phi_o - V_a) \frac{N_a + N_d}{N_a N_d}} \left[exp\left(\frac{qV_a}{2kT}\right) - 1 \right] \quad (4.4.35)$$

For a silicon diode the generation/recombination current is larger than the diffusion current for small forward bias and adds to the reverse current when $V_a < 0$. At small forward biases, therefore, the current dependence on the applied voltage follows an $exp\left(\frac{qV_a}{2kT}\right)$ law, which is characteristic of a recombination-dominated current. At higher bias values, however, the $exp\left(\frac{qV_a}{kT}\right)$ variation due to the diffusion current takes over (Figure 4.11) and completely overshadows the recombination current.

Generation current can be observed in the reverse-bias current-voltage characteristics. The physical origin of that current is the following: when the junction is reverse biased ($V_a < 0$), the *pn* product, given by Equation 4.4.29, is smaller than n_i^2. Therefore, the SRH generation mechanism forces an increase in the *pn* product towards its equilibrium value. The generated carriers are separated by the electric field in the transition region. The generated holes are swept into the P-type quasi-neutral region, and the generated electrons into the N-type region. The motion of these carriers constitutes the generation current.

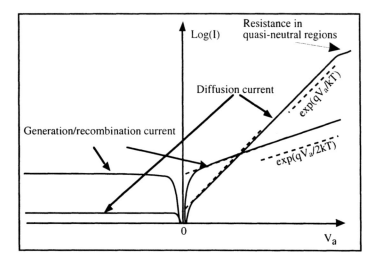

Figure 4.11: Diffusion, generation and recombination currents in the junction.

Often diffusion and generation/recombination currents are regrouped into a single current expression:

$$I = I_s \left[exp\left(\frac{qV_a}{nkT}\right) - 1 \right]$$ (4.4.36)

where n is called the "ideality factor". The ideality factor ranges between 1 and 2. It is equal to 1 in a diode where the current is completely dominated by diffusion mechanisms (ideal diode), and it is equal to 2 when the current is completely dominated by generation/recombination mechanisms.

Another divergence from ideality exists. At high forward-bias current levels the resistance in the quasi-neutral regions can no longer be neglected. If R is the sum of the resistances in the P and N neutral regions, then the potential difference at the edges of the transition region is not V_a, but rather $V_a - IR$. This causes a reduction of the current with applied voltage at high current levels (Figure 4.11).

4.4.3. Junction breakdown

When a PN junction is strongly reversed biased the electric field near the metallurgical junction can reach high values. The value of that field is given by Expression 4.2.10, where l_{no} and l_{po} are replaced by l_n and l_p, respectively. Carriers accelerated in that field can accumulate enough kinetic energy that they can, through a collision process, generate electron-hole pairs through impact ionization (see end of Section 3.3). The generated carriers can in turn be accelerated, and again through impact ionization, generate additional carriers. This carrier multiplication effect is a positive-feedback mechanism called avalanche multiplication and is characterized by a multiplication factor, M, which is defined as:

$$M = \frac{I_{ii}}{I_o} \qquad (4.4.37)$$

where I_o is the current that would flow in the absence of the impact ionization mechanism, and I_{ii} is the current measured when impact ionization is present. The multiplication factor can be related to the applied voltage using the following relationship:

$$M = \frac{1}{1 - \left(\frac{V_a}{BV}\right)^n} \qquad (4.4.38)$$

where BV is the junction breakdown voltage and V_a is the applied voltage. The multiplication factor tends to infinity as $V_a \rightarrow BV$. The value of n ranges between 4 and 6, depending on the impurity concentration profile.

When breakdown occurs in a reverse-biased junction a sudden increase of current is observed (Figure 4.12). The term "breakdown" does not necessarily imply that the device is "broken"; it is simply the term used for a device operating in the breakdown regime. If no current-limiting circuitry is provided, however, the junction can by destroyed by thermal effects.

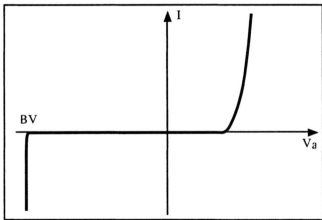

Figure 4.12: Breakdown of a reversed-biased PN junction. [5, 6]

There exists another breakdown mechanism in reverse-biased PN junctions, called "Zener breakdown". This effect takes place in diodes where both the N-type and P-type regions are heavily doped. As a result the width of the transition region is small and electrons can directly tunnel from the P-type valence band into the N-type conduction band.

This is a quantum-mechanical effect described in Section 14.1.1. In such diodes, called "Zener diodes" the breakdown voltage can be accurately controlled by means of adjusting doping concentrations. Zener diodes, are therefore, often used as voltage references.

4.4.4. Short-base diode

In Section 4.4.1 we assumed the PN junction was a "long-base diode", which implied that the length of the quasi-neutral regions was much larger than the diffusion length of minority carriers in those regions. In this section we will calculate the current in a diode where one of the quasi-neutral regions is shorter than the diffusion length of the minority carriers. The short-base diode is an essential element for the operation of the bipolar transistor, and in fact, it is used for the base of that device, hence its name.

Consider Figure 4.13. The P-type region is a "long base" having a length $x_L \gg L_n$. This region is identical to the P-type neutral region treated in Section 4.4.1 and shall be considered accordingly.

In Section 4.4.1 the N-type region was also considered long such that $x_{R2} \gg L_n$. Here, we will reduce the length of the N-type region to a value $x_R \ll L_n$ to see the implication of this base length reduction.

The continuity equation for holes in the N-type quasi-neutral region is, in steady-state:

$$0 = \frac{dp}{dt} = -\frac{1}{q}\frac{dJ_p}{dx} + (G_p - U_p) \qquad (4.4.39)$$

If the width of the base, x_R, is small enough, minority carriers in transit do not have time to recombine. To simplify the problem, we will assume that their lifetime, relative to the dimension of the base, is infinite:

$$\tau_p \rightarrow \infty \quad \Rightarrow \quad U_p = \frac{p-p_o}{\tau_p} \cong 0 \qquad (4.4.40)$$

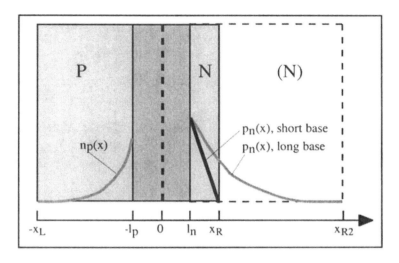

Figure 4.13: Geometry of the short-base diode and minority carrier concentration profiles.

Assuming no generation from an external source ($G_p = 0$) and using Expressions 4.4.39 and 4.4.40, we find, in the N-type region:

$$J_p(x) = \text{constant} \qquad (4.4.41)$$

Using the drift-diffusion equation for holes:

$$J_p = q\,\mu_p\,n\,\mathcal{E} - qD_p\frac{dp}{dx} \qquad (4.4.42)$$

and assuming, as in the case of the long-base diode, that $\mathcal{E}=0$ in the neutral N-type region, we have:

$$J_p = -qD_p\frac{dp}{dx} \quad \Rightarrow \quad p(x) = -\frac{J_p\,x}{qD_p} + B \qquad (4.4.43)$$

where B is an integration constant. Using the Boltzmann relationships as a boundary condition at the edge of the transition region:

$$p(x=l_n) = p_{no}\,exp(qV_a/kT) \qquad (4.4.44)$$

we find, from Relationship 4.4.43:

$$B = p_{no}\,exp(qV_a/kT) + \frac{J_p\,l_n}{qD_p} \qquad (4.4.45)$$

We will assume that the n-type region is connected to a metal at $x=x_R$. Such a contact usually brings about an infinite surface recombination velocity which implies that p is equal to p_{no} at $x=x_R$ (see Section 3.6). Using this as a boundary condition we have:

$$p(x=x_R) = p_{no} = -\frac{J_p \, x_R}{qD_p} + B = -\frac{J_p \, x_R}{qD_p} + p_{no} \, exp(qV_a/kT) + \frac{J_p \, l_n}{qD_p}$$

$$= \frac{J_p \, (l_n - x_R)}{qD_p} + p_{no} \, exp(qV_a/kT) \qquad (4.4.46)$$

Solving the latter equation for J_p we find:

$$J_p = \frac{q \, D_p \, p_{no} \, (exp(qV_a/kT)-1)}{(x_R - l_n)}$$

$$= qD_p \, \frac{n_i^2/N_d}{(x_R - l_n)} \, (exp(qV_A/kT)-1) \qquad (4.4.47)$$

From Equation 4.4.43 we know that the hole distribution is a linear function of x. The hole concentrations at $x=l_n$ and $x=x_R$ have been calculated using the boundary conditions, and are equal to $p_{no} \, exp(qV_a/kT)$ and p_{no}, respectively. Therefore, the hole concentration profile can easily be plotted in Figure 4.13. Note that the slope of the straight line is higher than the slope of the profile for the long-base case at $x=l_n$. Since the magnitude of the current is directly proportional to the slope of the minority carrier concentration, the short base will have a higher diffusion current flow $I = A(J_n + J_p)$ than a long-base diode.

4.5. PN junction capacitance

So far we have only considered the steady-state characteristics of the PN junction. Transient effects resulting from varying the applied voltage will now be considered. As we have seen earlier the application of a bias $V_a \neq 0$ gives rise to distribution of charges in the transition region and in the quasi-neutral regions which is different from the case $V_a = 0$. Some of these charges are located in the transition region, and their variation with the applied bias gives rise to a "transition capacitance", also called "depletion capacitance". In addition, under forward bias conditions, other charges are present, due to the injection of excess minority carriers in the quasi-neutral regions. These give rise to a capacitive component called "diffusion capacitance".

4.5.1. Transition capacitance

The width of the space-charge regions at the junction is given by Expressions 4.3.2 and 4.3.3:

$$l_p = \sqrt{\frac{2\varepsilon_s}{q} \, \frac{(\Phi_o - V_a) \, N_d}{N_a \, (N_a+N_d)}} \quad \text{and} \quad l_n = \sqrt{\frac{2\varepsilon_s}{q} \, \frac{(\Phi_o - V_a) \, N_a}{N_d \, (N_a+N_d)}} \qquad (4.5.1)$$

The charge of the fixed, ionized doping impurities in each depletion zone is, in absolute value:

$$Q_j \text{ (coulombs)} = AqN_d l_n = AqN_a l_p \qquad (4.5.2)$$

where A is the cross-sectional area of the junction.

The variation of that charge with applied bias (Figure 4.14) is due to the movement of majority carriers in and out of the depletion zones, and is therefore, a very fast process that takes on the order of a picosecond. Time constants associated with this charge variations, can therefore, be neglected, and the associated capacitances can be considered frequency-independent.

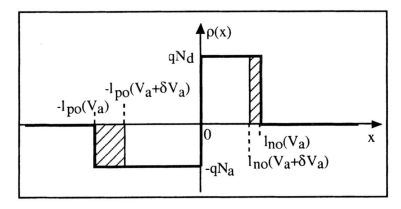

Figure 4.14: Variation of the depletion charge with applied bias.

The capacitance associated with the variation of the depletion charge is given by:

$$C_T = \left| \frac{dQ_j}{dV_a} \right| = AqN_d \left| \frac{dl_n}{dV_a} \right| = A \sqrt{\frac{q\varepsilon_s}{2} \frac{N_a N_d}{(N_a + N_d)}} \frac{1}{\sqrt{\Phi_o - V_a}} \qquad (4.5.3a)$$

or, using Relationship 4.3.4:

$$C_T = A \frac{\varepsilon_s}{l_n + l_p} \qquad (4.5.3b)$$

This expression corresponds to the capacitance of a classical parallel-plate capacitor, where the plates, separated by a distance $l_n + l_p$, have an area A, and where the dielectric material between them has a permittivity ε_s.

4.5.2. Diffusion capacitance

The diffusion capacitance is due to the variation of the charge of minority carriers into the quasi-neutral regions, with applied bias variation. The hole concentration in the N-type quasi-neutral region is given by Expression 4.4.22:

$$p_n(x) = p_{no} + p_{no}\left[exp\left(\frac{qV_a}{kT}\right) - 1\right]exp\left(\frac{-(x-l_n)}{L_p}\right) \qquad (4.5.4)$$

The *excess* hole concentration, is therefore, equal to (Figure 4.15):

$$p'_n(x) = p_n(x) - p_{no} = p_{no}\left[exp\left(\frac{qV_a}{kT}\right) - 1\right]exp\left(\frac{-(x-l_n)}{L_p}\right) \qquad (4.5.5)$$

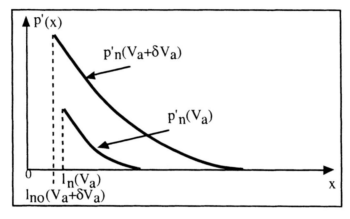

Figure 4.15: Distribution of excess holes injected in the N-type quasi-neutral region for two applied bias values. [7]

The charge per unit area carried by these excess minority carriers is given by:

$$Q_p = q\int_{l_n}^{\infty} p'_n(x)\,dx = q\,p_{no}\left[exp\left(\frac{qV_a}{kT}\right) - 1\right]\int_{l_n}^{\infty}exp\left(\frac{-(x-l_n)}{L_p}\right)dx$$

$$= qL_p p_{no}\left[exp\left(\frac{qV_a}{kT}\right) - 1\right] \qquad (4.5.6)$$

Under forward bias conditions for which $exp\left(\frac{qV_a}{kT}\right) >> 1$ the diffusion capacitance created by the presence of holes injected in the N-type quasi-neutral region, C_{Dp}, is equal to:

$$C_{Dp} = \frac{dQ_p}{dV_a} = \frac{q^2 L_p p_{no}}{kT}exp\left(\frac{qV_a}{kT}\right) \qquad (4.5.7)$$

Using Relationship 4.4.23 which gives: $J_p(l_n) = \dfrac{qD_p p_{no}}{L_p}\left[\exp\left(\dfrac{qV_a}{kT}\right)-1\right]$,

Expression 4.5.6 can be rewritten as:

$$C_{Dp}\ (\text{coulombs/cm}^2) = \frac{q}{kT}\frac{L_p^2}{D_p}J_p(l_n) = \frac{q}{kT}\tau_p\,J_p(l_n) \qquad (4.5.8)$$

A similar expression can be derived for electrons injected into the P-type region and yields the diffusion capacitance created by the presence of electrons injected in the P-type quasi-neutral region, C_{Dn}. The total diffusion capacitance is obtained by adding C_{Dp} and C_{Dn}:

$$C_D = \frac{dQ_p}{dV_a} + \frac{dQ_n}{dV_a} = C_{Dp} + C_{Dn} = \frac{q}{kT}\left(\tau_p\,J_p(l_n) + \tau_n\,J_n(-l_p)\right) \qquad (4.5.9)$$

4.5.3. Charge storage and switching time

Let us apply a constant forward step current to a P^+-N junction, such that $I = 0$ for $t < 0$ and $I = I_F$ for $t \geq 0$. Initially the excess hole concentration increases from zero to $p'_n(x)$ (Equation 4.5.5).

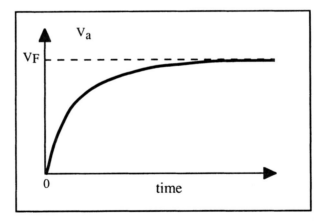

Figure 4.16: Diode voltage as a function of time

The build-up of a minority carrier charge in the N-type quasi-neutral region is called the charge storage. Some of the current initially injected in the junction is "used" to build up the charge Q_p described by Equation 4.5.6 followed by the forward bias the of the device. As a result, a negative exponential rise of the junction bias, from zero volt to $V_F = \dfrac{kT}{q}$ $ln\left(1 + \dfrac{I_F}{I_S}\right)$ is observed (Figure 4.16). Similarly, the excess minority carriers in a forward-biased PN junction must be removed when the device is turned off. Let the applied bias be switched from a positive value V_F to a negative value $-V_R$ at $t = 0$ (Figure 4.18A). Figure 4.17 shows the evolution of the excess minority carrier profile at different times after the switch at $t=0$.

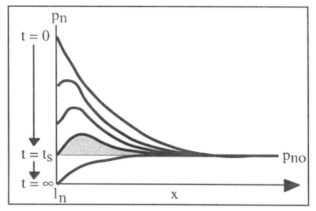

Figure 4.17: Evolution of the excess minority carrier profile at different times following the application of a negative bias ($-V_R$).

When the applied bias is switched from V_F to $-V_R$ the current caused by the excess minority carriers instantly changes direction, but its value, $-I_R$, is much larger than that of the saturation current, I_s (Figure 4.18B). The magnitude of the initial reverse current, $-I_R$, is a function of the stored charge. Current $-I_R$ remains constant until the excess minority carrier concentration at the edge of the transition region ($x=l_n$) drops to zero. During that time interval `the voltage drop across the transition region remains equal to V_F. The time elapsed for the removal of the excess minority carrier concentration at the edge of the transition region will be noted t_s (Figure 4.17).

For $t > t_s$ the stored charge is no longer sufficient to support the constant current $-I_R$, and the current decays exponentially to its equilibrium value, $-I_s$. The time necessary for the reverse current to reach a value equal to 10% of $-I_R$ is called the "reverse recovery time" and noted $t_s + t_f$, as shown in Figure 4.18B, where t_f is called the fall time. Between t_s and t_f the voltage drop across the transition region gradually evolves from V_F to $-V_R$ (Figure 4.18C). It is worthwhile noting that the time required to turn off the diode is typically larger than the time needed to turn it on. To improve the switching speed, metallic impurities are sometimes introduced in the semiconductor (*e.g.* gold in silicon). These impurities increase the SRH recombination rate. Thus they aid in the decrease of minority carrier lifetime, and hence, reduce minority carrier charge storage effects.

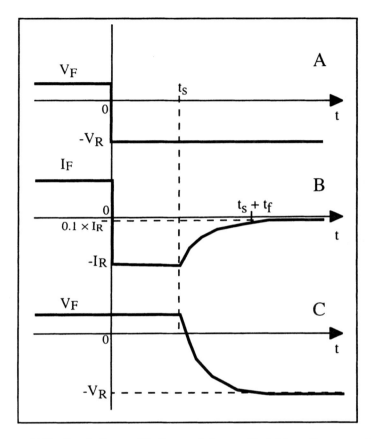

Figure 4.18: Evolution with time of A: applied bias; B: current; C: voltage drop across the transition region.[8,9]

4.6. Models for the PN junction

As we have seen earlier, the static current-voltage characteristics of the PN diode is described by a simple exponential equation:

$$I = I_s \left[exp\left(\frac{qV_a}{kT}\right) - 1 \right] \qquad (4.6.1)$$

The fact that this equation is non-linear can pose serious numerical problems regarding its use in a circuit simulator. As a result several linear models have been developed which can be used for the diode.

4.6.1. Quasi-static, large-signal model

The quasi-static, large-signal model for the diode stems from a linear approximation of Equation 4.6.1. This model is valid for a wide range of applied biases and does not account for transient or capacitive effects of any kind.

As illustrated in Figure 4.19, the characteristics of an actual diode (case A) can be approximated by:

B: an idealized diode having the following characteristics: $I=0$ when $V<0$ and $V=0$ when $I>0$

C: an idealized diode in series with a voltage source having the following properties: $I=0$ when $V<V_j$ and $V=V_j$ when $I>0$. V_j is approximately equal to 0.7 V in a silicon diode and 0.35 V in a germanium diode.

D: an idealized diode in series with a voltage source and a resistor having a conductance equal to $G = 1/R$. The current-voltage characteristics of this model are: $I=0$ when $V<V_j$ and $V=V_j+ I/G$ when $I>0$.

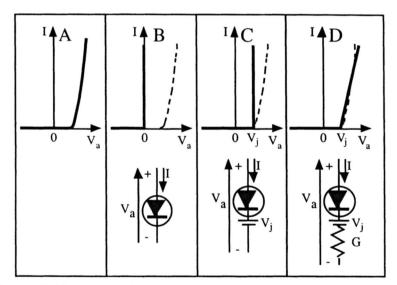

Figure 4.19: A: actual I-V characteristics, B: idealized diode, C: idealized diode + voltage source, D: idealized diode + voltage source + conductance.

4.6.2. Small-signal, low-frequency model

The quasi-static, small-signal model for the diode stems from a linear approximation of Equation 4.6.1. This model is valid for small signal variations and does not account for transient or capacitive effects.

Consider the case where the applied bias, $v(t)$, is composed of the superposition of a large continuous dc bias, V_o, and a small, low-frequency ac signal, $v_1(t)$:

$$v(t) = V_o + v_1(t) \qquad (4.6.2)$$

The corresponding current, $i(t)$, will encompass both a dc current component, I_o, and a small-signal ac component, $i_1(t)$ (Figure 4.20):

$$i(t) = I_o + i_1(t) \qquad (4.6.3)$$

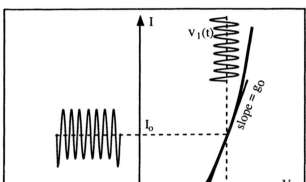

Figure 4.20: Response of the diode to a small ac signal.

The dynamic conductance, g_o, is defined by $g_o = \dfrac{i_1(t)}{v_1(t)}$ and is equal to:

$$g_o = \frac{i_1(t)}{v_1(t)} = \frac{di(t)}{dv(t)}\bigg|_{v=V_o} = \frac{d\left(I_s\left[\exp\left(\frac{qv(t)}{kT}\right) - 1\right]\right)}{dv(t)}\bigg|_{v=V_o} = \frac{q}{kT}I_s \exp\left[\frac{qV_o}{kT}\right]$$

$$= \frac{q}{kT}I_s\left(\exp\left[\frac{qV_o}{kT}\right] - 1\right) + I_s = \frac{q}{kT}(I_o + I_s) \qquad (4.6.4)$$

When the diode is forward biased the saturation current, I_s, is much smaller than I_o, and the dynamic conductance can be approximated as:

$$g_o = \frac{q}{kT}I_o \qquad (4.6.5)$$

The corresponding dynamic resistance is simply equal to:

$$r_o = \frac{1}{g_o} = \frac{kT}{q\,I_o} \tag{4.6.6}$$

As shown in Figure 4.21, the response of a diode to a small-signal, low-frequency signal can be modeled by a simple resistor, the value of which is inversely proportional to the dc bias current flowing through the diode.

Figure 4.21: Small-signal, low-frequency equivalent circuit.

4.6.3. Small-signal, high-frequency model

The small-signal, high-frequency, equivalent circuit of a PN junction is shown in Figure 4.22. It consists of the parallel association of the dynamic resistance 4.6.6, the transition capacitance 4.5.3b and the diffusion capacitance 4.5.8 (Figure 4.22).

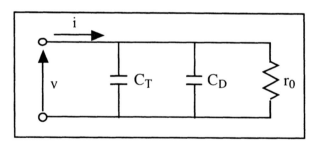

Figure 4.22: Small-signal, high-frequency, equivalent circuit for a PN junction. [10]

4.7. Solar cell

A solar cell is a PN junction in which the generation of carriers by an external source of energy, usually sunlight, is utilized to generate electrical power. In other words a solar cell directly converts solar energy into electrical power. The design of most solar cells is quite elaborate, such that the efficiency of energy conversion is maximized. In this Section, however, we will exemplify the operation of a solar cell using a simple PN junction structure. Solar cell operation is based on the generation of electron-hole pairs in the transition region, and the separation of both types of carriers by the junction electric field. Let us take the example of the P^+N junction shown in Figure 4.23. We will assume that illumination by sunlight uniformly generates G electron-hole pairs per cubic centimeter and per second, at any location in the semiconductor material. Using the same notations as before, the transition region extends from $-l_p$ to l_n. The bias applied to the device is noted V_a.

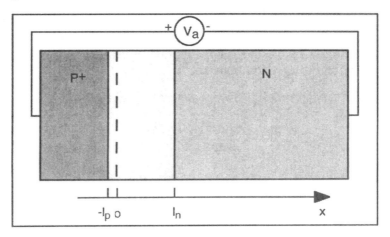

Figure 4.23: Geometry of a simple solar cell.

We will first calculate the diffusion current in the quasi-neutral regions. In the N-type material, far from the junction, we know from Expression 3.4.4 that:

$$p_n(\infty) = p_{no} + G\,\tau_p \qquad (4.7.1)$$

Assuming, as was the case for the simple PN junction, there is no electric field in the N-type quasi-neutral region, the current density for holes is:

$$J_p = -\,qD_p\frac{dp_n}{dx} \qquad (4.7.2)$$

Using the continuity equation for holes in the N-type quasi-neutral region one obtains, in the steady-state regime:

$$\frac{\partial p_n}{\partial t} = -\frac{1}{q}\frac{\partial J_p(x)}{\partial x} + (G - \frac{p_n\!-\!p_{no}}{\tau_p}) = 0 \qquad (4.7.3)$$

Combining the two latter equations we obtain:

$$D_p\frac{d^2p_n}{dx^2} = \frac{p_n\!-\!p_{no}}{\tau_p} - G \qquad (4.7.4)$$

The solution of Equation 4.7.4 is in the form:

$$p_n = p_{no} + \tau_p G + A\,exp(x/L_p) + B\,exp(-x/L_p) \qquad (4.7.5)$$

where A and B are integration constants and where $L_p = \sqrt{D_p \tau_p}$.

Using Expression 4.7.1 as a boundary condition for $x=\to\infty$ we find that $A=0$ in Relationship 4.7.5. Assuming low-level injection conditions the excess hole concentration at the edge of the transition region, on the N-type side, is given by Expression 4.4.10:

$$p'_n(l_n) = p_n(l_n) - p_{no} = p_{no}\left[\exp\left(\frac{qV_a}{kT}\right) - 1\right] \qquad (4.7.6)$$

Using Equation 4.7.6 as the second boundary condition for Expression 4.7.5 we find the integration constant B:

$$p_n(l_n) = p_{no}\ exp(qV_A/kT) = p_{no} + \tau_p G + B\ exp(-l_n/L_p)$$

$$\Downarrow$$

$$B = \frac{p_{no}\ (exp(qV_A/kT)-1) - \tau_p G}{exp(-l_n/L_p)}$$

Introducing A and B into 4.7.5 yields the minority carrier (hole) concentration as a function of x in the N-type quasi-neutral region:

$$p_n = p_{no} + \tau_p G + \frac{p_{no}\ (exp(qV_A/kT)-1) - \tau_p G}{exp(-l_n/L_p)}\ exp(-x/L_p) \qquad (4.7.7)$$

A similar calculation made for electrons in the P-type quasi-neutral region would yield (note that in this case $x<0$):

$$n_p = n_{po} + \tau_n G + \frac{n_{po}\ (exp(qV_A/kT)-1) - \tau_n G}{exp(-l_p/L_n)}\ exp(x/L_n) \qquad (4.7.8)$$

The total current density is given by:

$$J = J_n(-l_p) + J_p(l_n)$$

where the hole current density at $x=l_n$ is equal to:

$$J_p = -qD_p\ \frac{dp_n}{dx}\bigg|_{x=l_n} = \frac{qD_p}{L_p}\ \frac{p_{no}\ (exp(qV_A/kT)-1) - \tau_p G}{exp(-l_n/L_p)}\ exp(-l_n/L_p) \qquad (4.7.9)$$

and the electron current at $x=-l_p$ is given by:

$$J_n = qD_n\ \frac{dn}{dx}\bigg|_{x=-l_p} = \frac{qD_n}{L_n}\ \frac{n_{po}\ (exp(qV_A/kT)-1) - \tau_n G}{exp(-l_p/L_n)}\ exp(-l_p/L_n) \qquad (4.7.10)$$

The net current in the diode is finally obtained by adding Expressions 4.7.9 and 4.7.10, and by multiplying the result by the area of the diode, A:

$$I = A(J_n + J_p)$$

$$\Downarrow$$

$$I = Aq \left\{ \frac{D_p}{L_p} \left[p_{no}(exp(qV_A/kT)-1))-\tau_p G \right] + \frac{D_n}{L_n} \left[n_{po}(exp(qV_A/kT)-1) -\tau_n G \right] \right\}$$

$$= Aq \left[\frac{D_n n_{po}}{L_n} + \frac{D_p p_{no}}{L_p} \right] (exp(qV_A/kT)-1) - AqG(L_n+L_p) \qquad (4.7.11)$$

We now have to calculate the current produced by the photogeneration of electron-hole pairs within the transition region. Once such pairs are generated the electrons and holes are separated by the electric field in the transition region. The electrons are accelerated towards the N-type quasi-neutral region and the holes are accelerated towards the P-type quasi-neutral region. The resulting current is obtained by solving the continuity equation for holes and electrons. Under steady-state conditions, and neglecting recombination in the transition region one can write:

$$\frac{dJ_n}{dx} = - qG = - \frac{dJ_p}{dx} \qquad (4.7.12)$$

Integrating over the transition region one obtains:

$$J_n(l_n) = J_n(-l_p) - q \int_{-l_p}^{l_n} G \, dx = J_n(-l_p) - qG(l_n + l_p) \qquad (4.7.13)$$

The current density due to photogeneration in the transition region is thus equal to:

$$I_{pt} = - qAG(l_n + l_p) \qquad (4.7.14)$$

Adding this current to Expression (4.7.11) we obtain the current-voltage characteristics of the solar cell:

$$I = Aq \left[\frac{D_n n_{po}}{L_n} + \frac{D_p p_{no}}{L_p} \right] (exp(qV_A/kT)-1) - AqG(l_n + l_p + L_n+L_p) \qquad (4.7.15)$$

Comparing the latter Expression with the diode current "in the dark" (Relationship 4.4.27), we conclude that the current-voltage characteristics of the solar cell under illumination are the ideal characteristics in the dark shifted by a current amount equal to $- AqG(l_n + l_p + L_n+L_p)$ due to generation:

$$I_{light}(V) = I_{dark}(V) - AqG(l_n + l_p + L_n+L_p) \qquad (4.7.16)$$

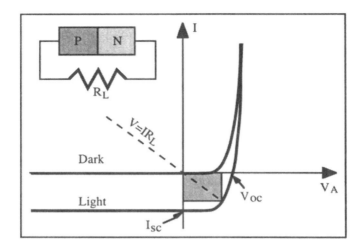

Figure 4.24: Current-voltage characteristics of a solar cell in the dark and under illumination. The area gray rectangle represents the power supplied by the cell to a load having a resistance R_L.

Figure 4.24 shows the current-voltage characteristics of a solar cell in the dark and under illumination. The insert shows a simple circuit where the solar cell under illumination delivers electrical power to a load resistor, R_L. The operation point of the circuit is given by the intersection of the *I-V* characteristics of the illuminated cell with the load line $V=IR_L$. The area of the gray rectangle represents the power supplied by the solar cell to the load. Optimization of solar cell performance involves the use of anti-reflection coatings, which increases light absorption, and therefore, the generation rate, G. The use of high-quality semiconductor material with a high minority carrier lifetime improves the diffusion lengths, L_n and L_p, which increases current generation. It is also desirable to used low doping concentrations to maximize the width of the transition region, l_n+l_p, and therefore maximize the amount of current generated in it. Further optimization can be achieved by the choice of a load resistance value, R_L, that maximizes the power transferred to the load (*i.e.*: which maximizes the area of the gray rectangle in Figure 4.24).

The short-circuit current and the open-circuit voltage of an illuminated solar cell are noted I_{sc} and V_{oc}, respectively. Assume the gray-colored rectangle in Figure 4.23 is the largest possible rectangle that can be inscribed between the axes and the I-V characteristics of the device, *i.e.* a rectangle that represents the maximum power that the solar cell can deliver for a given level of illumination. Let its area be noted S. One can then define a "fill factor", *FF*, by the following relationship:

$$FF = \frac{S}{V_{oc} I_{sc}} \qquad (4.7.13)$$

The fill factor depends on the design and the fabrication parameters of a solar cell and is optimized to increase the energy conversion efficiency of the device.

4.8. PiN diode

The structure of a PiN diode is shown in Figure 4.25. It consists of a PN junction with a wide intrinsic region sandwiched between the N and the P region. In practice, the intrinsic region is very lightly doped, either P-type (called π-type) or N-type (called ν-type).

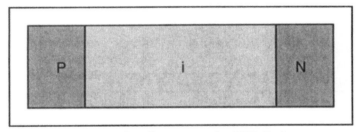

Figure 4.25: Structure of a PiN diode.

The lightly doped (intrinsic) region is basically completely depleted in every mode of operation. In the forward mode, holes injected from the P-type diffusion into the intrinsic region recombine with electrons injected from the N-type diffusion, such that the current density in the device is given by:

$$J = \int_{intrinsic\ region} q\, U\, dx \qquad (4.8.1)$$

If the electron and hole lifetimes are assumed to be equal in the intrinsic region, $\tau_n = \tau_p = \tau_0$, Expression 4.8.1 can be rewritten:

$$J = \frac{q\, n'\, W}{\tau_0} \qquad (4.8.2)$$

where n' is the average injected excess electron concentration, and W is the width of the intrinsic region. [11]

Because of their large depletion zone, reverse-biased PiN diodes have a large photon collection volume and are commonly used as photodetectors, including X-ray detectors.

Important Equations

Junction Potential

$$\Phi_o = \Phi_{no} - \Phi_{po} = \frac{qN_a}{2\varepsilon_s} l_{po}^{\;2} + \frac{qN_d}{2\varepsilon_s} l_{no}^{\;2} = \frac{kT}{q} \ln\left(\frac{N_aN_d}{n_i^2}\right) \qquad (4.2.9)$$

Maximum electric field

$$\mathcal{E}_{max} = -\frac{qN_a}{\varepsilon_s} l_{po} = -\frac{qN_d}{\varepsilon_s} l_{no} \qquad (4.2.10)$$

Width of Depletion Regions

$$l_{po} = \sqrt{\frac{2\varepsilon_s}{q} \frac{\Phi_o N_d}{N_a (N_a+N_d)}} \qquad (4.2.11a)$$

and

$$l_{no} = \sqrt{\frac{2\varepsilon_s}{q} \frac{\Phi_o N_a}{N_d (N_a+N_d)}} \qquad (4.2.11b)$$

Current Density in the ideal PN junction

$$J = J_s \left[exp\left(\frac{qV_a}{kT}\right) - 1 \right] \qquad (4.4.27)$$

Saturation Current Density

$$J_s = \frac{qD_n n_{po}}{L_n} + \frac{qD_p p_{no}}{L_p} = q\, n_i^2 \left(\frac{1}{N_a} \sqrt{\frac{D_n}{\tau_n}} + \frac{1}{N_d} \sqrt{\frac{D_p}{\tau_p}} \right) \qquad (4.4.28)$$

Problems

Problem 4.1

Consider a silicon PN junction in which the doping profile varies linearly as shown in Problem Figure 4.1. Such a junction is called a "gradual junction". The metallurgical junction is located at x=0, where the dopant type changes polarity. We have $N_d-N_a=ax$, where *a* is a constant.

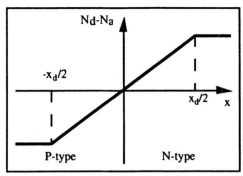

Problem Figure 4.1

1: The transition region is given and extends from $-x_d/2$ to $+ x_d/2$. We will assume $\Phi(x=0) = 0$. Find an analytical expression for the built-in junction potential (Φ_o) using the depletion approximation.

2: Noting that the doping concentrations at $x=-x_d/2$ and $x=x_d/2$ are $N_a=ax_d/2$ and $N_d=ax_d/2$, respectively, the junction potential can be calculated using Relationship

4.2.1: $\Phi_o = \dfrac{kT}{q} \ \ln\left(\dfrac{(ax_d/2)^2}{n_i^2}\right)$. Find the value of $x_d/2$ and Φ_o using an iteration technique with Matlab (*i.e.* solve $x_d = F(x_d)$ iteratively) using the following data: $T=300K$, $a = 10^{20}$ cm^{-4} and $V_a = 0$ V.

3: Plot $\rho(x)$, $\mathcal{E}(x)$ and $\Phi(x)$ for $-x_d < x < x_d$ using the depletion approximation. Also plot $n(x)$ and $p(x)$. For the y-axis of each curve, choose either a linear or a logarithmic scale, whenever most appropriate. N_d-N_a is constant for $x>x_d/2$ and $x<-x_d/2$.

Problem 4.2
Consider comparable PN junctions made in Si, Ge and GaAs. The junction area is 1 mm^2, $N_d = 10^{18}$ cm^{-3}, $N_a = 10^{15}$ cm^{-3} and the minority carrier lifetime is 1 μs.

To simplify the problem we will assume that the presence of doping impurities does not degrade carrier mobility and that the mobility does not vary with temperature.

Question: For each semiconductor calculate the current flowing through the junction when the applied bias is $V_a = -5V$, at the following temperatures: 20°C and 200°C.

The following data are given:

Symbol	Meaning	Value	Unit
$E_{g,GaAs}$	GaAs bandgap energy	1.42	eV
$E_{g,Ge}$	Ge bandgap energy	0.67	eV
$E_{g,Si}$	Si bandgap energy	1.124	eV
κ_{GaAs}	GaAs dielectric constant	13.1	-
κ_{Ge}	Ge dielectric constant	16	-
ε_0	vacuum permittivity	8.854×10^{-14}	F cm^{-1}
κ_{Si}	Si dielectric constant	11.7	-
h	Planck constant	6.63×10^{-34}	J s
k	Boltzmann constant	1.38×10^{-23}	J K^{-1}
kT/q	thermal voltage (@ 300K)	0.02586	V
$N_{c,GaAs}$	Effect. Dens. Cond. Band (GaAs)	4.7×10^{17}	cm^{-3}
$N_{c,Ge}$	Effect. Dens. Cond. Band (Ge)	1.04×10^{19}	cm^{-3}
$N_{c,Si}$	Effect. Dens. Cond. Band (Si)	2.8×10^{19}	cm^{-3}
$N_{v,GaAs}$	Effect. Dens. Valence Band (GaAs)	7×10^{18}	cm^{-3}
$N_{v,Ge}$	Effect. Dens. Valence Band (Ge)	6×10^{18}	cm^{-3}
$N_{v,Si}$	Effect. Dens. Valence Band (Si)	1.04×10^{19}	cm^{-3}
q	electron charge	1.6×10^{-19}	C
$\mu_{n,GaAs}$	electr. mobility (intrinsic GaAs)	8800	cm^2 V^{-1} s^{-1}
$\mu_{n,Ge}$	electr. mobility (intrinsic Ge)	3900	cm^2 V^{-1} s^{-1}
$\mu_{n,Si}$	electr. mobility (intrinsic Si)	1417	cm^2 V^{-1} s^{-1}
$\mu_{p,GaAs}$	hole mobility (intrinsic GaAs)	400	cm^2 V^{-1} s^{-1}
$\mu_{p,Ge}$	hole mobility (intrinsic Ge)	1900	cm^2 V^{-1} s^{-1}
$\mu_{p,Si}$	hole mobility (intrinsic Si)	471	cm^2 V^{-1} s^{-1}

<u>Problem 4.3</u>
Consider a silicon PN junction. Its area, A, is equal to 1 cm^2. The impurity concentrations are: $N_a = 10^{15}$ cm^{-3} and $N_d = 10^{17}$ cm^{-3}. The diode is reverse biased with an applied voltage, $V_a = -5$ volts. The following data are given :

$$T = 300K, \quad n_i = 1.45 \times 10^{10} \text{ cm}^{-3}, \quad \frac{kT}{q} = 0.0256 \text{ V}$$

$$\varepsilon_{si} = 11.7 \times 8.854 \times 10^{-14} \text{ F/cm}$$

μ_n (in the P region) = 1300 cm^{2}V^{-1}s^{-1} and μ_p (in the N region) = 300 cm^{2}V^{-1}s^{-1}
$$\tau_n = 1 \text{ } \mu s \quad \text{and} \quad \tau_p = 1 \text{ } \mu s$$

1) Calculate the current flowing through the diode, neglecting recombination in the transition region (U=0 in the transition region).
2) Calculate the current flowing through the diode for the same applied bias, when the diode is illuminated with light in such a way that 10^{14} electron-hole pairs are created per cm^3 and per second in the transition region. Neglect recombination phenomena in the transition region (U=0 in the transition region).

<u>Problem 4.4</u>:

Consider a silicon PN^+ junction with $N_a=10^{15}$ cm^{-3} and $N_d=10^{20}$ cm^{-3}. Plot the transition capacitance as a function of temperature from 0 to 400°C with $V_a = 0$ V.

<u>Problem 4.5</u>:

A silicon PN junction has the following parameters:

```
Area=0.01;      %junction area (cm2)
q=1.6e-19;      %Electron charge (C)
es=11.7*8.854e-14;  % Permittivity of silicon (F/cm)
kTq=0.0256;     %kT/q (V)
Na=1e16;        % Doping concentration, P-type region (cm-3)
Nd=1e19;        % Doping concentration, N-type region (cm-3)
ni=1.45e10;     % Intrinsic carrier concentration (cm-3)
mun=800;mup=400;  % Electron and hole mobility (cm2 V-1 s-1)
taun=5e-9;% Lifetime of electrons in the P-type neutral region (s)
taup=5e-10;  % Lifetime of holes in the N-type neutral region (s)
tau0=1e-6;   % Lifetime of carriers in the transition region (s)
```

Plot the following two current components as a function of the applied voltage for $-1V<V_a<0.7V$

1) The diffusion current *vs.* V_a, and

2) The total current (diffusion + generation/recombination current) *vs.* V_a.

The two curves must be on the same graph. The y-axis minimum and maximum is 1 pA and 1 A.

<u>Problem 4.6</u>:

Plot the electron and hole current density as a function of x (Figure 4.10) in a silicon PN junction using the following parameters:

```
q=1.6e-19;      %Electron charge (C)
esi=11.7*8.854e-14;  % Permittivity of silicon (F/cm)
kTq=0.0256;     %kT/q (V)
Na=1e16;    % Doping concentration, P-type region (cm-3)
Nd=2e16;    % Doping concentration, N-type region (cm-3)
ni=1.45e10;    % Intrinsic carrier concentration (cm-3)
mun=600;mup=300;    % Electron and hole mobility (cm2 V-1 s-1)
taun=5e-11; % Electron lifetime in the P-type neutral region (s)
taup=5e-11; % Hole lifetime in the N-type neutral region (s)
V = 0.3;    % Applied voltage (V)
```

<u>Problem 4.7</u>:

Plot $\rho(x)$, $\mathcal{E}(x)$, $\Phi(x)$, $n(x)$ and $p(x)$ for $-2\,l_p < x < 2\,l_n$ in a silicon PN gradual junction (see problem 4.1) using the depletion approximation and the numerical technique described in Problem 2.4. Plot n(x) and p(x) on the same graph. For the y-axis of each curve, choose either a linear or a logarithmic scale, when most appropriate. $N_a = 5 \times 10^{15}$ cm^{-3} and $N_d = 8 \times 10^{15}$ cm^{-3}, T=300K and $V_a = 0$ V. Use 120 mesh points.

Problem 4.8

Solve Problem 4.1 using a numerical technique.

Plot $\rho(x)$, $\mathcal{E}(x)$, $\Phi(x)$, n(x) and p(x) for -2 $x_d/2 < x < 2$ $x_d/2$ using the depletion approximation and the numerical technique described in Problems 2.4 and 4.7. Plot n(x) and p(x) on the same graph. For the y-axis of each curve, choose either a linear or a logarithmic scale, when most appropriate. N_d-$N_a(x \geq x_d/2) = N_d$-$N_a(x=x_d/2)$ and N_d-$N_a(x \leq -x_d/2) = N_d$-$N_a(x= -x_d/2)$, T=300K , a = 10^{20} cm^{-4} and $V_a = 0$ V. Use 120 mesh points. Plot $\rho(x)$, $\mathcal{E}(x)$, $\Phi(x)$, n(x) and p(x) derived analytically from Problem 4.1 on the same graphs. Comment on the results.

References

1   S.M. Sze, *Physics of Semiconductor Devices*, J. Wiley & Sons, p. 74, 1981
2   A.S. Grove, *Physics and Technology of Semiconductor Devices*, J. Wiley & Sons, p. 56, 1967
3   S.M. Sze, *Physics of Semiconductor Devices*, J. Wiley & Sons, p. 88, 1981
4   A.S. Grove, *Physics and Technology of Semiconductor Devices*, J. Wiley & Sons, p. 186, 1967
5   A.S. Grove, *Physics and Technology of Semiconductor Devices*, J. Wiley & Sons, p. 191, 1967
6   S.M. Sze, *Physics of Semiconductor Devices*, J. Wiley & Sons, p. 107, 1981
7   R.S. Muller and T.I. Kamins, *Device Electronics for Integrated Circuits*, J. Wiley and Sons, p. 250, 1986
8   M.S. Tyagi, *Introduction to Semiconductor Materials and Devices*, J. Wiley & Sons, p. 293, 1968
9   S.K. Ghandhi, *The Theory and Practice of Microelectronics*, J. Wiley & Sons, p. 119, 1981
10   A.B. Glaser and G.E. Subak-Sharpe, *Integrated Circuit Engineering*, Addison-Wesley, p. 24, 1979
11   S.M. Sze, *Physics of Semiconductor Devices*, 2nd edition, J. Wiley & Sons, p. 119, 1981

Chapter 5

METAL-SEMICONDUCTOR CONTACTS

This chapter analyzes the electrical characteristics of a metal-semiconductor contact. Two different types of contacts can be produced: a contact with non-linear, rectifying current voltage characteristics called a Schottky contact, and a linear, non-rectifying contact called an ohmic contact.

5.1. Schottky diode

A Schottky contact or Schottky diode is formed when a rectifying contact is formed between a metal and a semiconductor. The rectifying properties of the contact are similar to those of a PN junction diode. The first semiconductor devices, dating back to the end of the nineteenth century were rectifying, metal-semiconductor, "point-contact" diodes. The rectifying effect in metal-semiconductor contact diodes was discovered in 1874 by F. Braun and was explained by Schottky and Mott in 1938. A typical semiconductor material used at that time was galena, a naturally occurring lead sulfide crystalline mineral.

5.1.1. Energy band diagram

Consider an N-type semiconductor crystal and a metal. The energy band diagrams of these two materials are shown in Figure 5.1. We know because of the photoelectric effect (A. Einstein Nobel Prize, 1921), that electrons can be extracted from a metal in a vacuum, when light with a proper wavelength is shone onto the metal. In order to observe this effect the wavelength of the incident light must have a higher energy than a given critical value. In other words, the photons must carry enough energy to extract electrons from the metal and eject them into the vacuum. This energy $E = h\nu$ must be at least equal to the "work function" of the metal, noted $q\Phi_m$. The work function is, therefore, defined as the energy that must be supplied to an electron with an energy E_{Fm} (the metal Fermi level) in order for the electron to be ejected from the metal. Similarly, the work function of the semiconductor is the energy required to extract an electron located at its Fermi level, E_{Fsc}.

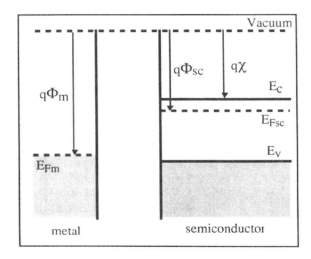

Figure 5.1: Energy bands in a metal and a semiconductor.

We know that in a semiconductor some electrons have an energy higher than E_{Fsc}. These can be found in the conduction band, and their energy is approximately equal to E_C. The energy needed to extract an electron from the conduction band into a vacuum is called the "electron affinity", and noted $q\chi$. In this Section we will consider an N-type semiconductor and a metal such that $E_{Fm} < E_{Fsc}$.

When the metal is contacted with the semiconductor the Fermi levels align and thermodynamic equilibrium is established through the transfer of electrons from the semiconductor conduction band into the metal, since $E_C > E_{Fm}$. These electrons "leave behind" positively charged donor impurity atoms in the semiconductor. A space-charge region corresponding to the zone depleted of electrons, is, therefore, formed in the semiconductor near the interface with the metal. The width of this depletion region is noted W_o. The metal is considered as a perfect conductor. An electron charge, equal in magnitude to the depletion charge, appears in the metal at the metal-semiconductor interface. For all practical purposes this charge can be considered infinitely thin. Such a charge distribution is often called a "charge sheet". Because of the alignment of the Fermi levels and the presence of a depletion region the band curvature in the semiconductor is equal to:

$$qV_i = q(\Phi_m - \Phi_{sc})\qquad\qquad(5.1.1)$$

This curvature corresponds to a potential barrier, V_i, which prevents further electrons from migrating into the metal. Electrons in the metal, on the other hand, see a potential barrier, Φ_b, having an amplitude equal to (Figure 5.2):

$$q\Phi_b = q(\Phi_m - \chi) = qV_i + (E_c - E_F) \tag{5.1.2}$$

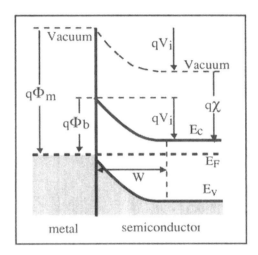

Figure 5.2: Energy band diagram of the Schottky contact.

At room temperature these potential barriers are significantly larger than kT/q and only a few electrons possess sufficient energy to overcome them. The current resulting from electrons from the semiconductor overcoming the barrier and migrating into the metal is noted $I_{m \to s}$. This notation is due to the fact that electrons carry a negative charge. Therefore, electrons migrating from the semiconductor into the metal corresponds to a "positive" current flow from the metal into the semiconductor.

At thermodynamic equilibrium and in the absence of any external bias the current $I_{m \to s}$ is exactly balanced by a current of electrons flowing from the metal into the semiconductor, noted $I_{s \to m}$. Thus, at equilibrium, we have: $I_{s \to m} = -I_{m \to s}$.

If a forward bias $V_a > 0$ is applied to the structure (+ on the metal side, and - on the semiconductor side) the potential barrier on the semiconductor side is decreased from V_i to $V_i - V_a$ (Figure 5.3A). A greater number of electrons can, therefore, flow from the semiconductor into the metal. On the other hand, the flow of electrons from the metal into the semiconductor, $I_{s \to m}$, remains constant because the potential barrier seen from the metal side, Φ_b, is unchanged. As a result, a net electron current flow from the semiconductor into the metal is observed.

If a reverse bias, $V_a<0$, is applied to the structure (+ on the semiconductor side, and - on the metal side) the potential barrier in the semiconductor is increased from V_i to V_i-V_a (Figure 5.3B). As a result the electron flow from the semiconductor into the metal, $I_{m \to s}$, is reduced while $I_{s \to m}$ remains unchanged. As a result a small reverse current of electrons flowing from the metal into the semiconductor, $I_{s \to m}$ - $I_{m \to s}$, is measured. The asymmetry between the forward and reverse current flow mechanisms create non-linear current-voltage characteristics similar to the PN junction.

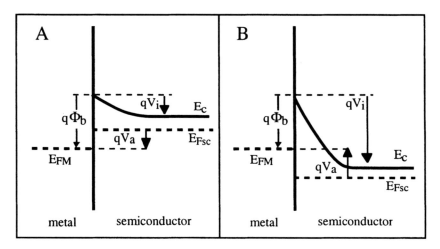

Figure 5.3: Energy band diagram under A: forward bias and B: reverse bias.

5.1.2. Extension of the depletion region

The width of the depletion zone in a Schottky diode can be calculated using the Poisson equation and the depletion approximation:

$$\frac{d^2\Phi(x)}{dx^2} = -\frac{\rho}{\varepsilon_{sc}} = -\frac{qN_d}{\varepsilon_{sc}} \qquad (5.1.3)$$

$$\Downarrow$$

$$\frac{d\Phi(x)}{dx} = \frac{qN_d}{\varepsilon_{sc}} (W\text{-}x) \qquad (5.1.4)$$

where W is the depth of the depletion region under an applied voltage V_a.

Note that the boundary conditions at $x=W$ are $\Phi(W) = 0$ and $\dfrac{d\Phi(W)}{dx} = 0$ since the potential, Φ, and the electric field ,$\mathcal{E}$, are equal to zero in the quasi-neutral part of the semiconductor. Integrating Expression 5.1.4 and applying the aforementioned boundary conditions we obtain:

$$\Phi(x) = -\frac{qN_d}{2\varepsilon_{sc}}(W-x)^2 \qquad (5.1.5)$$

The potential at $x=0$ is equal to the potential barrier on the semiconductor side, *i.e.* $V_i - V_a$ where V_a is the applied voltage taken as positive when the diode is forward biased. Substituting $V_i - V_a$ for $\Phi(x=0)$ in 5.1.5 gives the width of the depletion region:

Width of the depletion region

$$\boxed{W(V_a) = \sqrt{\frac{2\varepsilon_{sc}}{qN_d}(V_i - V_a)} \qquad (5.1.6a)}$$

The electric field at $x=0$ is $\mathcal{E}(0) = -qN_dW/\varepsilon_{sc}$, or, using Expression 5.1.4a:

$$\mathcal{E}(0) = -\sqrt{\frac{2qN_d}{\varepsilon_{sc}}(V_i - V_a)} \qquad (5.1.6b)$$

5.1.3. Schottky effect

The height of the potential barrier on the metal side, Φ_b, is not exactly constant and is slightly affected by the applied voltage. An actual lowering of Φ_b is observed. It is due to a mirror charge produced in the metal by electrons in the semiconductor. Electrostatics tells us that when a charge is near a "perfect" conductor (metal) a mirror charge of same magnitude but opposite sign is created inside the conductor, at a depth equal to the distance between the initial charge and the conductor surface (Figure 5.4). As a consequence, the charge is attracted by the metal, and in the case of the metal-semiconductor contact, the potential barrier is lowered.

The attraction exerted by the metal on an electron can be calculated as follows. Assuming the distance between the electron and the metal surface is x, the mirror charge bearing a charge $+q$ is located at a distance $-x$ inside the metal. Therefore, the Coulomb attraction force between the two charges is equal to $\dfrac{-q^2}{16\pi\varepsilon_{sc}x^2}$.

The force is equivalent to that exerted on an electron by an electric field $\mathcal{E}_m(x)$ obeying the relationship:

$$-q\mathcal{E}_m(x) = \frac{-q^2}{16\pi\varepsilon_{sc}x^2}$$

The resulting potential energy of the electron is equal to:

$$P(x) = -qV(x) = \int\limits_{x}^{\infty} -\frac{q^2}{16\pi\varepsilon_{sc}x^2}\,dx = \frac{-q^2}{16\pi\varepsilon_{sc}x}$$

the reference potential being $P(x=\infty) = 0$.

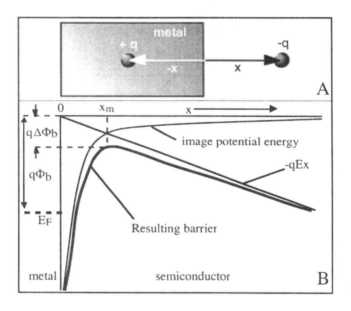

Figure 5.4: A: Mirror charge in a metal; B: the resulting lowering of the potential barrier. [1]

To find the total energy of the electron this potential energy must be added to the potential energy of the electron inside the semiconductor. In the depletion region the electric field is equal to $\dfrac{-qN_d}{\varepsilon_{sc}}$ $(W-x)$. This field gives the electron in the conduction band a potential energy which is equal to $-\dfrac{q^2N_d}{2\varepsilon_{sc}}$ $(W-x)^2 + E_c$. To simplify the problem we will assume that the electric field in the depletion region is constant . That field is noted $\mathcal{E}$ and gives the electron a potential energy $-q\mathcal{E}x$. The sum of the two potential energies (from the mirror charge and from the depletion region) yields the total potential energy $PE(x)$ of the electron:

$$PE(x) = -\frac{q^2}{16\pi\varepsilon_{sc}x} - q\,\mathcal{E}x + E_c \qquad (5.1.7)$$

The maximum potential energy can be found by writing $dPE(x)/dx=0$, which yields the maximum at $x = x_m = \sqrt{\dfrac{q}{16\pi\varepsilon_{sc}\mathcal{E}}}$. The potential energy at $x=x_m$ is equal to $PE(x_m) = -q\sqrt{\dfrac{q\mathcal{E}}{4\pi\varepsilon_{sc}}} < 0$, which corresponds to an effective lowering of the potential barrier $\Delta\Phi_b$ equal to:

$$\Delta\Phi_b = \sqrt{\frac{q\mathcal{E}}{4\pi\varepsilon_{sc}}} \qquad (5.1.8)$$

Using the value of the electric field at the semiconductor surface, given by equation 5.1.6b:

$$\mathcal{E}(0) = -\sqrt{\frac{2qN_d}{\varepsilon_{sc}}(V_i - V_a)} \qquad (5.1.9)$$

we find the magnitude of the potential barrier lowering, which constitutes the Schottky effect:

$$\Delta\Phi_b = \sqrt[4]{\frac{q^3 N_d}{8\pi^2 \varepsilon_{sc}{}^3}(V_i - V_a)} \qquad (5.1.10)$$

The resulting potential barrier height is equal to:

$$\Phi'_b = \Phi_b - \Delta\Phi_b \qquad (5.1.11)$$

5.1.4. Current-voltage characteristics

Electrons overcome the potential barrier between the metal and the semiconductor through a quantum-mechanical process called "thermionic emission". This process is activated by the thermal energy of the electrons. Although the potential barrier is clearly larger than kT/q at room temperature there exists a non-zero probability that some electrons gather enough energy to overcome the barrier. When a forward bias V_a is applied to the device the potential barrier that the electrons have to overcome to transit from the semiconductor into the metal is equal to Φ_b' - V_a. The resulting thermionic emission current is given by:

$$I_{m\to s} = A\,R^*\,T^2 \exp\left[-\frac{q(\Phi'_b - V_a)}{kT}\right] \qquad (5.1.12)$$

where R^* is called the "Richardson constant" and is equal to $\dfrac{4\pi\, m_e\, q\, k^2}{h^3}$

and A is the diode area.

Using the fact that $I_{m\rightarrow s} = -I_{s\rightarrow m}$ when $V_a = 0$, and that $I_{s\rightarrow m}$ is constant and independent of the applied voltage one can write:

$$I_{s\rightarrow m} = - A\, R^*\, T^2\, exp\left[\frac{-q\Phi'_b}{kT}\right] \qquad (5.1.13)$$

Since the net current in the diode is equal $I_{m\rightarrow s} + I_{s\rightarrow m}$, the expression of the current as a function of the applied voltage is:

$$I = A\, R^*\, T^2\, exp\left[\frac{-q\Phi'_b}{kT}\right]\left[exp\left(\frac{qV_a}{kT}\right) - 1\right] \qquad (5.1.14)$$

This equation describes a current-voltage characteristics similar to that of a PN junction. In addition the current depends on both the temperature and the height of the potential barrier between the metal and the semiconductor.

5.1.5. Influence of interface states

The equations derived previously describe the properties of a Schottky diode having an "ideal" metal-semiconductor interface, which means that the properties of the semiconductor are not affected by the presence of a metal. In an actual device the periodic nature of the semiconductor crystal is disturbed at the interface, which gives rise to a large number of permitted states in the bandgap of the semiconductor near the interface. These states are called "interface states" or "interface traps". They have energy values ranging from E_v to E_c and are occupied by electrons if they are below the Fermi level.

Consider the semiconductor before contact is made with the metal. Electrons trapped in the interface states originate from the semiconductor crystal. They form a negative surface charge, which creates a depletion zone in the semiconductor (Figure 5.5.A). Note the presence of an energy band curvature $q\Phi_o = E_F - E_v$ at the semiconductor surface.

Let us now bring the metal in contact with the semiconductor crystal. We recall from the previous sections that in absence of interface states, the alignment of the Fermi levels was achieved by a transfer of electrons from the semiconductor to the metal, resulting in the formation of a depletion zone and an upward curvature of the semiconductor energy bands. If we consider a *very large* interface state density an infinitesimal upwards increase of the band curvature, δE, will move a large number of interface states (all those with an energy between $E_F - \delta E$ and E_F) above the Fermi level. These states will lose trapped electrons, and as a consequence the alignment of the Fermi levels will be accomplished by the transfer of electrons from the traps into the metal, instead of from the semiconductor into the metal. The band curvature variation resulting from the alignment of the Fermi levels will, therefore, be negligible, and the height of the potential barrier will be: $q\Phi_b = E_g - q\Phi_o$.

In actual devices the interface state density is moderate, such that the height of the potential barrier is somewhere between $q\Phi_b = E_g - q\Phi_o$ and $q\Phi_b = q\Phi_m - q\chi$.

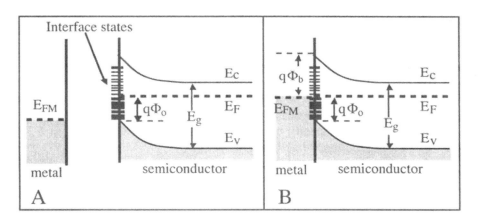

Figure 5.5: Energy bands in the presence of interface states; A: before contact with the metal; B: after contact. E_F shifts down relative to E_c at the interface and previously filled traps are emptied.

A more detailed analysis of the Schottky diode would show the existence of generation/recombination currents originating in the volume of the depletion zone. Because of the dependence of the potential barrier height on the applied bias and because of generation/recombination in the depletion zone the forward current takes the following form:

$$I = A R^* T^2 exp\left[\frac{-q\Phi_b'}{kT}\right]\left[exp\left(\frac{qV_a}{nkT}\right) - 1\right]$$ (5.1.15)

where $1 \le n \le 2$ is called the "ideality factor" of the diode (the diode is "ideal" when $n=1$).

5.1.6. Comparison with the PN junction

The current-voltage equations for the PN and Schottky diodes are the following:

Current-voltage characteristics (PN junction and Schottky diodes)

PN junction diode:

$$I = A\, q\, n_i^2 \left(\frac{D_p}{N_d L_p} + \frac{D_n}{N_a L_n} \right) \left(exp\left(\frac{qV_a}{nkT} \right) - 1 \right) \qquad (5.1.16)$$

or:

$$I = I_s \left(exp\left(\frac{qV_a}{nkT} - 1 \right) \right) \qquad (5.1.17)$$

Schottky diode:

$$I = A\, R^* \, T^2 \, exp\left[\frac{-q\Phi'_b}{kT} \right] \left[exp\left(\frac{qV_a}{nkT} \right) - 1 \right] \qquad (5.1.18)$$

or:

$$I = I_s \left[exp\left(\frac{qV_a}{nkT} \right) - 1 \right] \qquad (5.1.19)$$

Introducing adequate numerical values into these equations one observes that the reverse saturation current of a Schottky diode is 100 to 1000 times larger than that of a PN junction which accounts for a larger leakage current. In the forward mode, the I-V characteristics of a silicon Schottky diode shows strong conduction at 0.2-0.3 V, compared to 0.7 V in a silicon PN junction diode (Figure 5.6).

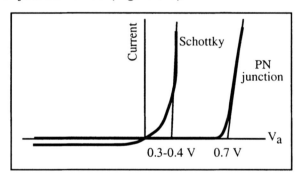

Figure 5.6: Current-voltage characteristics of a Schottky and a PN junction diode.

Schottky diodes are capable of very fast switching because their operation is based on majority carriers (unlike PN junction diodes where device operation is slowed down by storage and recombination of excess minority carriers). Majority carriers have a relaxation time on the order of ten picoseconds, which allows for operation at frequencies up to tens of gigahertz. The frequency performance of a Schottky diode can be appreciated by its cutoff frequency, which is given by $f_{co} = \dfrac{1}{2\pi RC}$ where

$R = \left. \dfrac{dV_a}{dI} \right|_{V_a=0}$ is the diode dynamic resistance, and where the depletion capacitance is equal to $C = A \, \mathcal{E}_{sc}/W(V_a)$. In a P$^+$N junction diode the cutoff frequency is given by $f_{co} = \dfrac{1}{2\pi R(C_D + C_T)}$ where C_D is the diffusion capacitance (Expression 4.5.9) and C_T is the transition capacitance (Expression 4.5.3b). The diffusion capacitance is proportional to the lifetime of minority carriers, ranging from 100 psec to several μs, which limits the frequency response of PN junction diodes.

5.2. Ohmic contact

An ohmic contact is a non-rectifying contact. The current-voltage characteristics of the contact should obey Ohm's law $V=IR$ and the resistance of the contact should be as low as possible. Consider the contact between the metal and the semiconductor shown in Figure 5.7. In this particular example $E_{FM} > E_F$ such that the energy bands of the N-type semiconductor are bent downwards near the contact. The magnitude of the band bending and its extension into the semiconductor are very small. As a result there is virtually no potential barrier between the metal and the semiconductor and electrons can flow freely through the contact. Such a contact is ohmic.

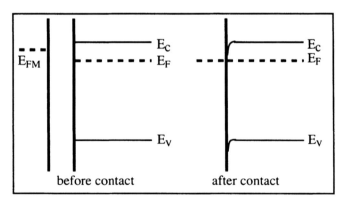

Figure 5.7: Energy bands of an ohmic contact.

It is also possible to obtain an ohmic contact between a metal and a semiconductor that would *a priori* form a Schottky diode, such as a metal where $E_{FM} < E_F$ in Figure 5.8. In practice a Schottky contact behaves as an ohmic contact if the impurity concentration in the semiconductor is high enough (e.g. $N_d = 10^{20}$ cm^{-3}). The width of the depletion region in the semiconductor is given by Expression 5.1.6a:

$$W(V_a) = \sqrt{\frac{2\mathcal{E}_{sc}}{qN_d}} \, (V_i - V_a)$$

where V_i is the built-in potential barrier height and V_a is the applied bias. If, for instance, $N_d=10^{20}$ cm^{-3}, $V_i=0.5$V and $V_a=0$ the thickness of the depletion zone is only 2.5 nm. Electrons can easily tunnel through such a thin potential barrier, which yields a low-resistance ohmic contact between the metal and the semiconductor. In metal-to-silicon contacts, current flow by tunnel effect becomes larger than current flow by thermionic emission when the doping concentration is larger than 10^{17} cm^{-3}. In practice, ohmic contacts between a metal and the terminals of semiconductor devices are always made on heavily doped areas.

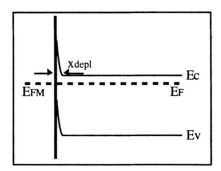

Figure 5.8: Energy bands in a contact between a metal and heavily doped silicon of an ohmic contact.

Important Equations

Width of the depletion region (Schottky diode)

$$W(V_a) = \sqrt{\frac{2\varepsilon_{sc}}{qN_d}} \, (V_i - V_a) \qquad (5.1.6a)$$

Current-voltage characteristics (PN junction and Schottky diodes)

PN junction diode:

$$I = A \, q \, n_i^2 \left(\frac{D_p}{N_d L_p} + \frac{D_n}{N_a L_n} \right) \left(exp \left(\frac{qV_a}{nkT} \right) - 1 \right) \qquad (5.1.16)$$

or:
$$I = I_s \left(exp \left(\frac{qV_a}{nkT} - 1 \right) \right) \qquad (5.1.17)$$

Schottky diode:

$$I = A \, R^* \, T^2 \, exp \left[\frac{-q\Phi'_b}{kT} \right] \left[exp \left(\frac{qV_a}{nkT} \right) - 1 \right] \qquad (5.1.18)$$

or:
$$I = I_s \left[exp \left(\frac{qV_a}{nkT} \right) - 1 \right] \qquad (5.1.19)$$

Problems

 Problem 5.1:

A Schottky diode is fabricated by depositing a layer of platinum on N-type silicon.

1) Plot the current in the diode as a function of applied voltage for $0 < Va < 0.5$ V, neglecting the potential barrier lowering effect (Schottky effect).
2) On the same graph, plot the current in the diode as a function of applied voltage for $0 < Va < 0.5$ V, taking the potential barrier lowering effect (Schottky effect) into account.
Use the following data:

```
epsil=8.854e-14;      % Permittivity of vacuum (F/cm)
esi=epsil*11.7;       % Permittivity of silicon (F/cm)
ni=1.45e10;           % Intrinsic carrier concentration
                      % in silicon at room temperature (cm-3)
Eg=1.12;              % Bandgap of silicon (eV)
kTq=0.0259;           % Thermal voltage (V)
Area=0.01;            % Diode area (cm-2)
T=300;                % Temperature (K)
FiB=0.8;              % Potential barrier (V)
Nd=1e16;              % Doping concentration (cm-3)
R=120;                % Richardson constant (A cm-2 K-2)
```

References

1   R.S. Muller and T.I. Kamins, *Device electronics for integrated circuits*, J. Wiley and Sons, p. 139, 1986

Chapter 6

JFET AND MESFET

6.1. The JFET

The Junction Field-Effect Transistor, or in short, JFET, is composed of a piece of semiconductor of one type (N-type, for example) and two diffusions with opposite doping polarity (P^+-type, in this case). Figure 6.1 represents such a device.

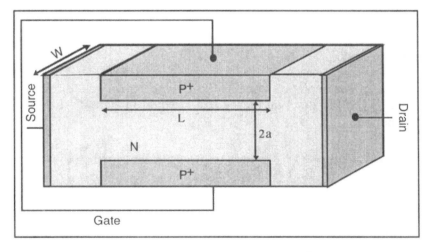

Figure 6.1: N-channel JFET.[1]

Two contacts are made to the N-type semiconductor and are labeled "source" and "drain". If the source voltage is taken as a reference ($V_S = 0$), the drain is biased with a positive voltage ($V_D > 0$). The two P^+-type regions are connected together and biased with a negative voltage and thus are reverse-biased with respect to the n-type region. These junctions form what is called the "gate" of the device. The N-type region connecting the source to the drain between the P$^+$-type regions is called the "channel". Because the source, drain and channel are all N-type an electron current can flow between source and drain. Conveniently the different parts of the device have been named after equivalent notions in fluid mechanics, such that the electron current originates at the source, flows in the channel, and ends up in the drain. A JFET in which current flow is due to the motion of electrons is called an N-channel JFET. In a P-channel JFET the semiconductor is P-type and the gate consists of two N$^+$-type diffusions.

If the drain is biased at a small positive value $\delta V_D > 0$ while the gate voltage, V_G, is equal to zero, the current of electrons flowing from source to drain is simply given by the expression of the current in a resistive bar of semiconductor having a length L and cross-section $(2a-2x_{depl})W$. The distance x_{depl} is the extension of the P$^+$N junction depletion zone in the N-channel. The current of electrons flowing from source to drain, called the drain current, is due to a drift mechanism and is equal to:

$$I_D = q\,\mu_n\,N_d\,\frac{2\,(a-x_{depl})\,W}{L}\,\delta V \qquad (6.1.1)$$

where μ_n is the electron mobility, N_d is the doping concentration in the N-type material, q is the electron charge, and $2(a-x_{depl})W$ is the cross-sectional area of the device.

The width of the depletion zone in the N-type semiconductor at equilibrium is given by the PN junction theory and is equal to Relationship 4.2.13 for a P$^+$N junction:

$$x_{depl} = \sqrt{\frac{2\varepsilon_{si}}{qN_d}}\,\Phi_o \quad \text{with } \Phi_o = \frac{kT}{q}\,ln\left(\frac{N_aN_d}{n_i^2}\right) \qquad (6.1.2)$$

where N_a is the doping concentration in the P$^+$ regions.

If we now apply a negative bias to the gate the width of the depletion regions in the N-type semiconductor will increase according to Relationship 4.3.3:

$$x_{depl} = \sqrt{\frac{2\varepsilon_{si}}{qN_d}(\Phi_o - V_G)} \quad \text{with } V_G < 0 \qquad (6.1.3)$$

When a negative gate voltage is applied the cross-sectional area of the channel through which electrons flow shrinks, which increases the resistance of the channel and decreases the drain current (Figure 6.2). The resistance of the channel can thus be modulated by the application of a gate bias. There exists a gate voltage for which the depletion zones from the two junctions come in contact, in which case we have $x_{depl} = a$. When the two depletion regions meet no current can flow between source and drain, since the depletion zones are emptied of carriers. The gate voltage for which the depletion zones meet is called the "threshold voltage" because it defines a threshold between conduction and non-conduction in the channel. Using the condition $x_{depl} = a$ and Expression 6.1.3 we find the threshold voltage:

Threshold voltage

$$V_{TH} = \Phi_o - \frac{q\,N_d\,a^2}{2\,\varepsilon_{si}} \qquad (6.1.4)$$

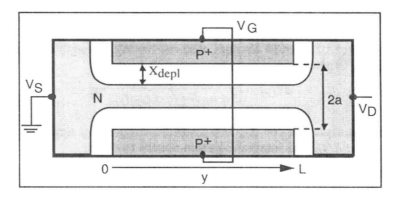

Figure 6.2: Cross-section view of the JFET for $V_D \cong 0V$. [2]

Let us now apply a larger drain voltage, the gate voltage being more positive than the threshold voltage such that current can flow between source and drain. Since the channel basically behaves as a resistor the current flow from source to drain gives rise to a progressive potential drop along the channel. The potential in the channel, noted $V(y)$, varies from $V(y=0) = V_S = 0$ at the source to $V(y=L) = V_D$ at the drain. Along the y-axis (source to drain) the reverse bias across the PN junctions is equal to $V_G - V(y)$.

As a consequence the width of the depletion zone varies as a function of y in such a way that:

Saturation drain current

$$I_{Dsat} = g_o \left\{ \frac{q N_d a^2}{6 \, \varepsilon_{si}} - (\Phi_o - V_G) + \frac{2}{3} \sqrt{\frac{2\varepsilon_{si}}{qN_d a^2}} (\Phi_o - V_G)^{3/2} \right\} \quad (6.1.11)$$

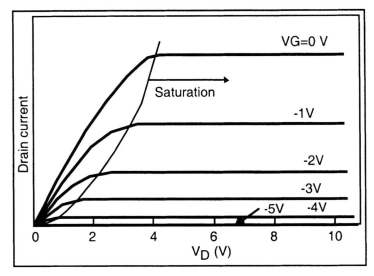

Figure 6.5: Output characteristics of an N-channel JFET: drain current as a function of drain voltage for different values of the gate voltage. [3]

It is worthwhile noting that the gate current of a JFET is equal to zero, with the exception of the small leakage current of the reverse-biased PN junctions. Therefore, JFETs have a very high input impedance which makes them useful in the fabrication of the input stage of high-sensitivity amplifiers and electrometers.

Among the important parameters of a JFET are its output conductance and transconductance. The output conductance is defined by:

$$g_D = \left. \frac{dI_D}{dV_D} \right|_{V_G=constant} \quad (6.1.12)$$

According to the simple model developed above the output conductance is equal to zero when the device is operating in saturation, which is not the case in practice when second-order effects are taken into consideration. Among these is the influence of the source and drain resistance (Figure 6.6).

One can easily make a correction to the model such that the influence of these resistances are taken into account. This can be done by replacing the output conductance, g_D, by an effective output conductance, g'_D, which is given by:

$$\frac{1}{g'_D} = \frac{1}{g_D} + R_S + R_D \qquad (6.1.13)$$

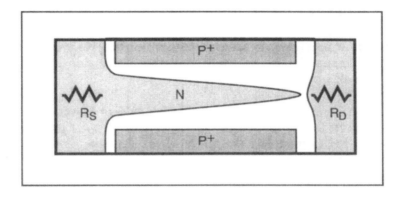

Figure 6.6: Source and drain resistances.

The transconductance, g_m, is defined as:

$$g_m = \frac{dI_D}{dV_G}\bigg|_{V_D=constant} \qquad (6.1.14)$$

When the transistor is saturated, its transconductance is equal to:

$$g_m = g_o \left\{ 1 - \sqrt{\frac{2\varepsilon_{si}}{qN_d\,a^2}} \sqrt{\Phi_o - V_G} \right\} \qquad (6.1.14)$$

Transconductance and output conductance are the most important parameters affecting the amplification gain that can be obtained from a JFET.

6.2. The MESFET

The acronym MESFET stands for "**ME**tal-**S**emiconductor **F**ield-**E**ffect **T**ransistor". It is widely used in gallium arsenide technology since it does not require the growth of a quality oxide nor the tailoring of complex diffusion patterns which are fabrication techniques that are much harder to achieve in

GaAs than in silicon. MESFETs can be operated at very high frequencies (> 100 GHz) because they are based on high-mobility semiconductor materials and on fast-recovery Schottky diodes.

The MESFET is basically a JFET in which the width of the depletion region that pinches the channel is due to the presence of a Schottky diode instead of PN junction. A typical MESFET is realized in a thin semiconductor having a thickness *a* (Figure 6.7) and a doping concentration N_d. This layer is sitting on top of a lightly doped, high-resistivity semiconductor substrate. The substrate resistivity is so high that it is often referred to as a "semi-insulating" material. The substrate plays no active role in the device and simply acts as a mechanical substrate.

As in the case of the n-channel JFET the gate is biased with a negative voltage with respect to the source, which we will consider grounded. The gate voltage is used to modulate the width of the depletion zone, and therefore, the conductivity of the channel (Figure 6.7). The drain voltage is positive and higher than that of the source. The metal gate forms a reverse-biased Schottky diode with the N-type semiconductor, such that there is no gate current, except for a small leakage current.

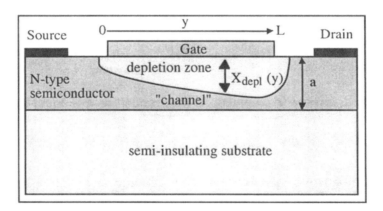

Figure 6.7: Cross-section view of a MESFET. [4]

If the drain voltage is small the width of the depletion zone can be obtained from the Schottky diode theory (Expression 5.1.6a):

$$x_{depl} = \sqrt{\frac{2\varepsilon_{sc}}{qN_d}(V_i - V_G)} \qquad (6.2.1)$$

where V_i is the Schottky diode potential barrier on the semiconductor side and ε_{sc} is the permittivity of the semiconductor. The threshold voltage is the gate voltage for which $x_{depl}=a$. Using 6.2.1 we find:

Threshold voltage

$$V_{TH} = V_i - \frac{qN_d\, a^2}{2\varepsilon_{sc}} \tag{6.2.2}$$

The threshold voltage can be either positive or negative, depending on the thickness of the N-type layer, the doping concentration N_d, and the metal used to form the Schottky gate. If the threshold voltage is negative the MESFET is a depletion-mode device; if it is positive, it is an enhancement-mode MESFET.

The current in the MESFET can be calculated as a function of gate and drain voltage using a technique similar to that which was used for the JFET. Since the channel basically behaves as a resistor the current flow from source to drain gives rise to a progressive potential drop along the channel. The potential in the channel, noted $V(y)$, varies from $V(y=0) = V_S = 0$ at the source to $V(y=L) = V_D$ at the drain. In each vertical section located at a position y the reverse bias across the Schottky junction is, therefore, equal to $V_G - V(y)$. As a consequence the width of the depletion zone varies as a function of y in such a way that:

$$x_{depl}(y) = \sqrt{\frac{2\varepsilon_{sc}}{qN_d}\,(V_i - V_G + V(y))} \tag{6.2.3}$$

The expression for the current is obtained by integrating Ohm's law from source to drain:

$$dV = I_D\, dR = \frac{I_D\, dy}{q\,\mu\, N_d\, W\, (a - x_{depl}(y))} \tag{6.2.4}$$

where W is the device width.

Replacing $x_{depl}(y)$ by its value from 6.2.3 we obtain:

$$dV = \frac{I_D\, dy}{q\,\mu\, N_d\, W\left(a - \sqrt{\dfrac{2\varepsilon_{sc}}{qN_d}\,(V(y) + V_i - V_G)} \right)} \tag{6.2.5}$$

which can be re-written as:

$$q \, \mu \, N_d \, W \int_0^{V_D} \left(a - \sqrt{\frac{2\varepsilon_{sc}}{qN_d} \, (V(y) + V_i - V_G)} \right) dV = I_D \int_0^L dy \quad (6.2.6)$$

Performing the integration we obtain:

Drain current

$$I_D = g_o \left(V_D - \frac{2\left((V_D + V_i - V_G)^{3/2} - (V_i - V_G)^{3/2}\right)}{3\sqrt{V_p}} \right) \quad (6.2.7)$$

where $\qquad g_o = \dfrac{q \, \mu \, N_d \, W \, a}{L} \quad$ and $\quad V_p = \dfrac{q \, N_d \, a^2}{2 \, \varepsilon_{sc}} \qquad (6.2.8)$

These equations are valid if the channel is not pinched-off, *i.e.* if the device is not in saturation. Pinch-off occurs when $x_{depl}(L) = a$, at which point $V_D \equiv V_{Dsat} = V_p - V_i + V_G$.

Saturation drain voltage

$$V_{Dsat} = V_p - V_i + V_G = V_G - V_{TH} \qquad (6.2.9)$$

The drain saturation current is obtained by replacing V_D by the saturation drain voltage, V_{Dsat}, in Expression 6.2.7, which yields:

Saturation drain current

$$I_{Dsat} = g_o \left(\frac{V_p}{3} + \frac{2 \, (V_i - V_G)^{3/2}}{3\sqrt{V_p}} - V_i + V_G \right) \qquad (6.2.10)$$

The transconductance in saturation is given by:

$$g_m = \left. \frac{\partial I_D}{\partial V_G} \right|_{V_D = constant} = g_o \left(1 - \sqrt{\frac{V_i - V_G}{V_p}} \right) \qquad (6.2.11)$$

Important Equations

Threshold voltage (JFET)

$$V_{TH} = \Phi_o - \frac{q\,N_d\,a^2}{2\,\varepsilon_{si}} \qquad (6.1.4)$$

Drain current (JFET)

$$I_D = g_o \left\{ V_D - \frac{2}{3} \sqrt{\frac{2\varepsilon_{si}}{qN_d\,a^2}} \left[(V_D + \Phi_o - V_G)^{3/2} - (\Phi_o - V_G)^{3/2} \right] \right\} \qquad (6.1.8)$$

with

$$g_o = \frac{2\,q\,\mu_n\,N_d\,a\,W}{L} \qquad (6.1.9)$$

Saturation drain voltage (JFET)

$$V_{Dsat} = \frac{q\,N_d\,a^2}{2\,\varepsilon_{si}} - (\Phi_o - V_G) = V_G - V_{TH} \qquad (6.1.10)$$

Saturation drain current (JFET)

$$I_{Dsat} = g_o \left\{ \frac{q\,N_d\,a^2}{6\,\varepsilon_{si}} - (\Phi_o - V_G) + \frac{2}{3} \sqrt{\frac{2\varepsilon_{si}}{qN_d\,a^2}} (\Phi_o - V_G)^{3/2} \right\} \qquad (6.1.11)$$

Threshold voltage (MESFET)

$$V_{TH} = V_i - \frac{qN_d\,a^2}{2\varepsilon_{sc}} \qquad (6.2.2)$$

Drain current (MESFET)

$$I_D = g_o \left(V_D - \frac{2\left((V_D + V_i - V_G)^{3/2} - (V_i - V_G)^{3/2} \right)}{3\sqrt{V_p}} \right) \qquad (6.2.7)$$

where

$$g_o = \frac{q\,\mu\,N_d\,W\,a}{L} \quad and \quad V_p = \frac{q\,N_d\,a^2}{2\,\varepsilon_{sc}} \qquad (6.2.8)$$

Saturation drain voltage (MESFET)

$$V_{Dsat} = V_p - V_i + V_G = V_G - V_{TH} \qquad (6.2.9)$$

Saturation drain current (MESFET)

$$I_{Dsat} = g_o \left(\frac{V_p}{3} + \frac{2\,(V_i - V_G)^{3/2}}{3\sqrt{V_p}} - V_i + V_G \right) \qquad (6.2.10)$$

References

1   S.M. Sze, *Physics of semiconductor devices*, J. Wiley & Sons, p. 314, 1981

2   R.S. Muller and T.I. Kamins, *Device electronics for integrated circuits*, J. Wiley and Sons, pp. 203-212, 1986

3   R.S. Muller and T.I. Kamins, *Device electronics for integrated circuits*, J. Wiley and Sons, pp. 209, 1986

4   S.M. Sze, *Physics of semiconductor devices*, J. Wiley & Sons, p. 336, 1981

Chapter 7

THE MOS TRANSISTOR

7.1. Introduction and basic principles

The MOS transistor, also called MOSFET (Metal-Oxide-Semiconductor Field-Effect Transistor) or IGFET (Insulated-Gate Field-Effect Transistor) is the most widely used semiconductor device and is at the heart of every digital circuit. Without the MOSFET there would be no computer industry, no digital telecommunication systems, no video games, no pocket calculators and no digital wristwatches. MOS transistors are also increasingly used in analog applications such as switched-capacitor circuits, analog-to-digital converters, and filters.

The exponential progress of MOS technology is best illustrated by the evolution of the number of MOS transistors integrated in a single memory chip or single microprocessor, as a function of calendar year. Each memory cell of a dynamic random-access memory (DRAM) contains a MOS transistor and a capacitor. It can be observed from Figure 7.1 that there is a four-fold increase in the number of transistors in a DRAM every three years. This exponential growth of integration density with time is known as Moore's law.[1]

The integration density of memory circuits is about 5 to 10 times higher than that of logic circuits such as microprocessors because of the more repetitive layout of transistors in memory chips. The increase in integration density is essentially due to the reduction of transistor size. The first experimental 1-gigabit DRAMs were reported in 1995 [2] where 1-gigabit DRAM contains over a billion MOSFETs. About 400 of these chips can be fabricated on a single silicon wafer, 40 centimeters in diameter. Such a wafer, therefore, contains over 400,000,000,000 transistors. This number is equal to the number of stars in our galaxy... More MOSFETs have been fabricated during the last ten years than grains of rice have been harvested by humans since the dawn of mankind.

The first description of a device called IGFET dates back to the 1930's in patents by Lilienfeld and Heil.[3,4] Because of technological limitations the IGFET could not be successfully fabricated at that time. The first working MOS transistor was realized in 1960 by Kahng and Attala.[5] A few years later, the integrated circuit industry took off to reach incredible proportions and has become one of the leading industries worldwide.

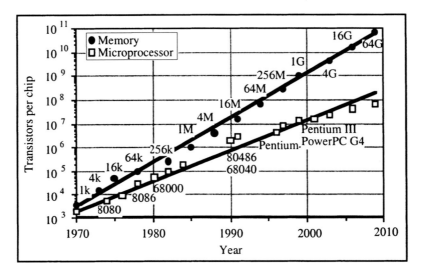

Figure 7.1: Actual and predicted evolution of circuit complexity in DRAMs and microprocessors.

There are two types of MOS transistors: the n-channel MOSFET, in which current flow is due to electron transport, and the p-channel MOSFET in which holes are responsible for current flow. A circuit containing only n-channel devices is produced by an nMOS process. Similarly, a pMOS process fabricates circuits that contain only p-channel transistors. Today the most commonly used technology is CMOS (Complementary MOS) in which both n-channel and p-channel transistors are fabricated. Here we will limit our analysis to n-channel devices. The current-voltage expressions describing a p-channel device can readily be derived from the n-channel equations, provided the appropriate changes of sign are made.

An n-channel MOS transistor is fabricated in a P-type semiconductor substrate, usually silicon. Two N-type diffusions are made in the substrate and the current flow will take place between these two diffusions. The diffusion with the lowest applied potential is called the "source" and the diffusion with the highest applied potential is called the "drain". Above the substrate, and between the source and the drain lies a thin insulating layer, usually silicon dioxide, and a metal electrode called "gate" (Figure 7.2). An electron-rich layer referred to as the "channel" can be created between the source and the drain underneath the gate insulator when a positive bias is applied to the gate. With appropriate voltages applied at the source and drain electrons can then flow from the source into the drain, through the channel. In a p-channel transistor an N-type substrate is used. The P-type drain is at a lower potential than the P-type source and the application of a negative bias to the gate enables the formation of a hole-enriched channel between source and drain. The metal-insulator-semiconductor structure is often referred to as a "MIS" structure, where the "I" stands for the insulator. When the insulator is an oxide, it is called a "MOS" structure.

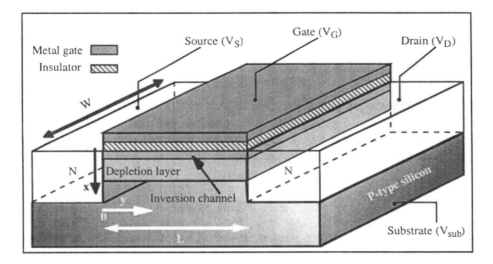

Figure 7.2: N-channel MOS transistor.

The basic operation of the n-channel MOSFET is the following. We will first consider the case where the gate voltage is equal to zero while the P-type substrate and the source are grounded ($V_{sub} = V_S = 0$). The drain is connected to a positive voltage source ($V_D = 5$ volts, for instance). Since the source and the substrate are at the same potential there is no current flow in the source-substrate junction. The drain-substrate junction is reverse biased and except for a small negligible reverse leakage current no current flows in that junction either. Under these conditions there is no channel formation, and therefore, no current flow from source to drain.

In the second case a constant positive bias is applied to the gate. There is no gate current since the metal electrode is dielectrically insulated from the silicon. Because it is positively biased the gate electrode does, however, attract electrons from the semiconductor, and a thin, electron-rich layer forms under the gate insulator. These electrons are supplied by the source and the drain which, being N-type, are large reservoirs of electrons. The electron-rich layer underneath the gate is called "channel". The N-type source and the N-type drain are connected by the electron-rich channel, and current is now free to flow between source and drain. The effect of the gate voltage controlling the concentration of electrons in the semiconductor through the gate oxide is called "field effect". The bias on the gate creates an electric field which can either induce or inhibit the formation of an electron-rich region at the surface of the semiconductor. The terms "source", "drain", "channel" and "gate" come to mind quite naturally since the electrons originate at the source, flow through the channel and are finally collected by the drain, the whole process being controlled by the bias on the gate.

The current in the channel, from source to drain can, to aa first approximation, be estimated using Ohm's law. Using $V=IR$ in a small channel element having a length dy and a width W we obtain:

$$dV(y) = I \, dR(y) \qquad (7.1.1)$$

The channel resistance as a function of y is obtained from Equation 2.3.3 where the electron concentration in the channel per unit area (unit: cm^{-2}) results from integrating the electron concentration per unit volume (unit: cm^{-3}) over the thickness of the device:

$$dR(y) = \frac{dy}{q \, \mu_n \, W \int_0^\infty n(x,y) \, dx} \qquad (7.1.2)$$

where x is the depth in the silicon ($x = 0$ at the silicon/SiO_2 interface). Note that the electron charge per unit area in the channel element can be written as:

$$Q_n(y) = q \int_0^\infty n(x,y) \, dx \quad (C \, cm^{-2}) \qquad (7.1.3)$$

The formation of a channel occurs when the gate voltage is positive and sufficiently high. In practice, the channel forms if the gate voltage is larger than a given value called the "threshold voltage", noted V_{TH}. Considering that the Metal-Oxide-Semiconductor structure forms a parallel-plate capacitor, we can write:

$$Q_n(y) = C_{ox} \, (V_G - V_{TH} - V(y)) \qquad (7.1.4)$$

where C_{ox} is the capacitance of the gate oxide per unit area and $V(y)$ is the local potential in the channel element, which varies from $V(y=0) = V_S = 0$ near the source to $V(y=L) = V_D$ near the drain.

Introducing Equations 7.1.2 and 7.1.4 into Expression 7.1.1 we obtain:

$$I \, dy = \mu_n \, W \, C_{ox} \, (V_G - V_{TH} - V(y)) \, dV \qquad (7.1.5)$$

Since $V_S = 0V$ and since the current I is constant from source to drain, the integration of Equation 7.1.5 yields:

$$I \int_0^L dy = \mu_n \, W \, C_{ox} \int_0^{V_D} (V_G - V_{TH} - V(y)) \, dV \qquad (7.1.6)$$

$$\Downarrow$$

$$I = \mu_n \, C_{ox} \, \frac{W}{L} \left((V_G - V_{TH}) \, V_D - \frac{V_D^2}{2} \right) \qquad (7.1.7)$$

If the local potential between source and drain, $V(y)$, becomes equal to or larger than $V_G - V_{TH}$ the formation of a channel can locally no longer be supported near the drain and the channel exists only between $y=0$ and a location y where $V(y) = V_G - V_{TH}$. In practice, that location is very close to L, and the current is obtained by replacing V_D by $V_G - V_{TH}$ in Expression 7.1.7. The current is then called the "saturation current" and noted I_{sat}. Saturation takes place when $V_D \geq V_G - V_{TH}$, and replacing V_D by $V_G - V_{TH}$ in Equation 7.1.7 we obtain:

$$I_{sat} = \mu_n \, C_{ox} \, \frac{W}{L} \, \frac{(V_G - V_{TH})^2}{2} \qquad (7.1.8)$$

Note that the current in saturation is no longer a function of the drain voltage and that the potential drop in the y-direction in the channel is fixed at a value equal to $V_G - V_{TH}$ in saturation.

In a p-channel MOSFET the source is at the highest potential and supplies holes to the channel. The holes are finally collected by the drain, which is at a lower potential than the source. In this case a negative bias relative to the substrate must be applied to the gate to create a hole-rich p-type channel.

A study of the metal-insulator-semiconductor structure, called the "MOS capacitor", will aid in the understanding of the detailed operation of the MOS transistor.

7.2. The MOS capacitor

The MOS capacitor is comprised of a metal gate, an insulating oxide layer, and a semiconductor. The thickness of the oxide typically varies between 5 to 50 nanometers. The semiconductor chosen for the example of Figure 7.3 is P-type silicon, which corresponds to the substrate of an n-channel device (nMOS).

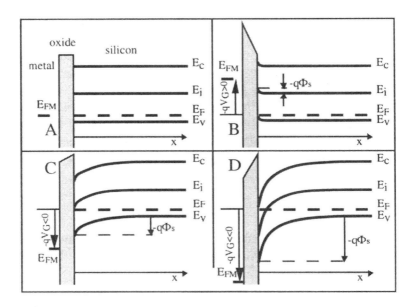

Figure 7.3: Energy bands in the MOS structure. [6] A: Flat energy bands, B: Accumulation, C: Depletion, D: Inversion

We will first consider the case of an hypothetical metal that has the same Fermi level as the silicon. When the structure is fabricated the Fermi level of the system is unique, and since the metal has the same Fermi level as the silicon, the band structure is that shown in Figure 7.3A. This condition is referred to as flat band for obvious reasons.

7.2.1. Accumulation:

If a negative bias is applied to the metal gate while the silicon substrate is grounded the structure behaves like a parallel-plate capacitor where the two electrodes are the silicon and the metal, and the oxide is the insulator between them. The application of the bias gives rise to a negative charge on the gate. This is a surface charge in the metal, located at the metal-oxide interface. An equal charge of opposite sign appears at the surface of the silicon, at the silicon-oxide interface (Figure 7.3B). The charge in the silicon can also be considered a surface charge, as we will demonstrate next. Its thickness is approximately 10 nanometers. This thin, hole-rich layer is called an accumulation layer. The capacitance of the MOS structure in accumulation is that of a parallel-plate capacitor between the metal gate and the accumulation layer. Its value (in Farads per unit area) is equal to:

$$C = \frac{\varepsilon_{ox}}{t_{ox}} \equiv C_{ox} \quad \text{(unit: F cm}^{-2}\text{)} \tag{7.2.1}$$

where ε_{ox} is the permittivity of silicon dioxide and t_{ox} is the thickness of the gate oxide. C_{ox} is called the gate oxide capacitance. The permittivity of SiO_2, ε_{ox}, is equal to $\kappa_{SiO2} \times \varepsilon_o$ where ε_o is the permittivity of vacuum, equal to 8.854×10^{-14} F/cm, and κ_{SiO2} is the dielectric constant of SiO_2, equal to 3.9.

Thickness of the accumulation layer
A derivation of the accumulation layer thickness as a function of substrate doping concentration will show that the layer is very small and hence can be considered as a surface charge.[7] The distribution of the charge as a function of depth, x, can be found using Poisson's equation:

$$\frac{d^2\Phi(x)}{dx^2} = -\frac{\rho}{\varepsilon_{si}} = -\frac{q}{\varepsilon_{si}} \, (p - n + N_d - N_a) \tag{7.2.2}$$

with:

$$p(x) = p_{po} \, exp\left(-\frac{q\Phi(x)}{kT}\right) = N_a \, exp\left(-\frac{q\Phi(x)}{kT}\right) \tag{7.2.3}$$

and:

$$n(x) = n_{po} \, exp\left(\frac{q\Phi(x)}{kT}\right) = \frac{n_i^2}{N_a} \, exp\left(\frac{q\Phi(x)}{kT}\right) \tag{7.2.4}$$

where p_{po} is the equilibrium hole concentration in the P-type material, n_{po} is the equilibrium electron concentration in the same material, and $\Phi(x)$ is the potential in the silicon as a function of depth. Far from the surface of the silicon the potential is equal to zero: $\Phi(x=\infty)=0$, which will be used as a boundary condition for Equation 7.2.2.

In the hole accumulation layer formed in P-type material one can assume that $n<<p$ and that $N_d<<N_a$, thus Equation 7.2.2 can be rewritten as:

$$\frac{d^2\Phi(x)}{dx^2} = -\frac{\rho}{\varepsilon_{si}} = -\frac{q}{\varepsilon_{si}} \, N_a \left[exp\left(-\frac{q\Phi(x)}{kT}\right) - 1\right] \tag{7.2.5}$$

where ε_{si}, the permittivity of silicon is equal to $\kappa_{Si} \times \varepsilon_o$ where κ_{Si} is the dielectric constant of silicon ($\kappa_{Si} = 11.7$). In the accumulation layer the hole concentration is greater than the hole concentration due to doping concentration, and therefore, $p \gg N_a$ in the accumulation layer. The following approximation can thus be used:

$$\frac{d^2\Phi(x)}{dx^2} \cong -\frac{qN_a}{\varepsilon_{si}} \exp\left(-\frac{q\Phi(x)}{kT}\right) \tag{7.2.6}$$

To integrate this equation we must first multiply both terms of the equation by $2\dfrac{d\Phi(x)}{dx}$, which yields:

$$2\frac{d\Phi(x)}{dx}\frac{d^2\Phi}{dx^2} = -2\frac{qN_a}{\varepsilon_{si}} \exp\left(-\frac{q\Phi(x)}{kT}\right)\frac{d\Phi(x)}{dx}$$

or:

$$\frac{d}{dx}\left(\frac{d\Phi}{dx}\right)^2 = 2\frac{N_akT}{\varepsilon_{si}}\frac{d}{dx}\left(\exp\left(-\frac{q\Phi(x)}{kT}\right)\right) \tag{7.2.7}$$

which can be rewritten:

$$\frac{d\mathcal{E}^2}{dx} = 2\frac{N_akT}{\varepsilon_{si}}\frac{d}{dx}\left(\exp\left(-\frac{q\Phi(x)}{kT}\right)\right) \tag{7.2.8}$$

Integrating from x to x_{acc}, where x_{acc} is the thickness of the accumulation layer, and noting that $\mathcal{E}(x=x_{acc}) = \Phi(x=x_{acc}) = 0$, since the silicon underneath the accumulation layer is neutral, one obtains:

$$\mathcal{E}^2(x) - 0 = \frac{2kTN_a}{\varepsilon_{si}}\left[\exp\left(-\frac{q\Phi(x)}{kT}\right) - 1\right]$$

$$= \frac{2}{L_D^2}\left(\frac{kT}{q}\right)^2\left[\exp\left(-\frac{q\Phi(x)}{kT}\right) - 1\right] \tag{7.2.9}$$

with:

$$L_D = \sqrt{\frac{\varepsilon_{si}kT}{q^2N_a}} \tag{7.2.10}$$

L_D is called the "Debye length". For example, L_D has a value of 40, 18 and 13 nanometers for doping impurity concentrations of 10^{16}, 5×10^{16} and 10^{17} cm^{-3}, respectively. Noting that $d\Phi(x)/dx > 0$, Equation 7.2.9 can be rewritten as follows:

$$\frac{d\Phi(x)}{\dfrac{kT}{q}\sqrt{\exp\left(-\dfrac{q\Phi(x)}{kT}\right) - 1}} = \frac{\sqrt{2}\,dx}{L_D} \tag{7.2.11}$$

The latter expression can be integrated using the following boundary conditions: $\Phi(x=x_{acc}) = 0$ and $\Phi(x=0) = \Phi_s$ where Φ_s is the potential at the semiconductor surface and is called the "surface potential". Equation 7.2.11 can be rewritten as:

$$\frac{d(-\frac{q\Phi(x)}{2kT})}{\sqrt{exp\left(-\frac{q\Phi(x)}{kT}\right)-1}} = -\frac{dx}{\sqrt{2}\,L_D} \qquad (7.2.12)$$

Numerator and denominator of the left-hand term are then multiplied by $exp\left(\frac{-q\Phi(x)}{2kT}\right)$:

$$\frac{exp\left(\frac{-q\Phi(x)}{2kT}\right)d(-\frac{q\Phi(x)}{2kT})}{exp\left(\frac{-q\Phi(x)}{2kT}\right)\sqrt{exp\left(-\frac{q\Phi(x)}{kT}\right)-1}} = -\frac{dx}{\sqrt{2}\,L_D} \qquad (7.2.13)$$

Changing variables and writing $u = exp\left(\frac{-q\Phi(x)}{2kT}\right)$, one obtains:

$$\frac{du}{u\sqrt{u^2-1}} = -\frac{dx}{\sqrt{2}\,L_D} \qquad (7.2.14)$$

Posing $u = \frac{1}{cos(\theta)} = sec\ (\theta)$, the latter equation becomes:

$$\frac{du}{u\sqrt{u^2-1}} = d\theta = -\frac{dx}{\sqrt{2}\,L_D} \implies \theta = C - \frac{x}{\sqrt{2}\,L_D} \qquad (7.2.15)$$

where C is an integration constant. We can conclude that:

$$exp\left(\frac{-q\Phi(x)}{2kT}\right) = u = sec\ (C - \frac{x}{\sqrt{2}\,L_D}\,) \qquad (7.2.16)$$

and, therefore:

$$\Phi(x) = -\frac{kT}{q}\ ln\left\{sec^2\left[C - \frac{x}{\sqrt{2}\,L_D}\right]\right\} \qquad (7.2.17)$$

The integration constant, C, can be related to the surface potential, Φ_S, by the following relationship:

$$\Phi(x=0) = \Phi_S \implies C = cos^{-1}\left(exp(\frac{q\Phi_S}{2kT})\right)$$

Finally we find that the thickness of the accumulation layer, x_{acc}, can be found using the condition that $\Phi(x=x_{acc}) = 0$:

$$x_{acc} = \sqrt{2}\,L_D\ cos^{-1}\left(exp(\frac{q\Phi_S}{2kT})\right)$$

The thickness of the accumulation layer, x_{acc}, can thus vary between 0 and $\frac{\sqrt{2}}{2} \pi L_D$, depending on the accumulation charge (Figure 7.4).

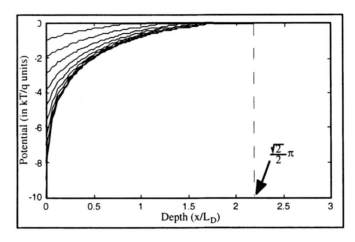

Figure 7.4: Potential (normalized to kT/q) in an accumulation layer (holes in P-type silicon) as a function of depth (normalized to L_D) for different values of surface potential.

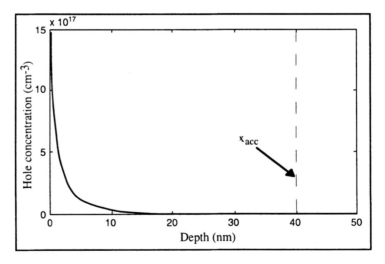

Figure 7.5: Hole concentration profile in an accumulation layer. The substrate doping concentration, N_a, is equal to 10^{16} cm^{-3} and the surface potential, Φ_S, is equal to -5kT/q

The hole concentration is an exponential function of the potential. Therefore, the charge density increases very rapidly close to the surface and most of the accumulation charge is concentrated within a depth much smaller than x_{acc} (Figure 7.5). Therefore, the charge in the accumulation layer can be considered as a surface charge. One can also consider that the surface potential, Φ_S, is very small. It actually is slightly negative and in practice reaches only a few $-kT/q$ (kT/q is equal to 25.9 mV at room temperature).

The application of a negative bias V_G on the gate gives rise to a negative surface charge Q_G in the metal at the metal-oxide interface. The accumulation charge in the semiconductor, Q_{acc}, is equal to Q_G, with opposite sign ($Q_{acc} = -Q_G$). Integrating Poisson's equation (Expression 7.2.2) from $x = 0$ to $x = +\infty$ we obtain:

$$Q_{acc} = \int_0^\infty \rho(x)\, dx = = q \int_0^\infty (p - n + N_d - N_a)\, dx = -\mathcal{E}_{si}\, \mathcal{E}(0)$$

Within the accumulation layer ($0 < x < x_{acc}$) we assumed that $n \ll p$, $N_d \ll N_a$ and $p \gg N_a$ while the silicon underneath the accumulation layer is neutral ($\rho = 0$ for $x > x_{acc}$). The charge in the semiconductor is, therefore, equal to the accumulation charge Q_{acc}. Using Equation 7.2.9 evaluated for $\mathcal{E}(x=0)$ the expression for the accumulation charge is:

$$Q_{acc} = -\, \mathcal{E}_{si} \frac{\sqrt{2}}{L_D} \frac{kT}{q} \sqrt{\exp\left(-\frac{q\Phi(x=0)}{kT}\right) - 1}$$

The exact value of the surface potential $\Phi_S \equiv \Phi(x=0)$ is related to the applied gate voltage V_G in the following way. V_G is equal to the potential drop across the oxide, V_{ox}, added to the potential drop Φ_S within the semiconductor:

$$V_G = \Phi_S + V_{ox} = \Phi_S - \frac{Q_{acc}}{C_{ox}}$$

or

$$V_G = \Phi_S - \frac{\mathcal{E}_{si}}{C_{ox}} \frac{\sqrt{2}}{L_D} \frac{kT}{q} \sqrt{\exp\left(-\frac{q\Phi_S}{kT}\right) - 1} \qquad (7.2.18)$$

The magnitude of the surface potential, Φ_S, is very small (only a few $\frac{-kT}{q}$), even for large applied negative gate voltage values. Since the accumulation charge Q_{acc} has a negligible thickness it can be considered as a surface charge and the approximation previously given for the capacitance of the MOS structure:

$$C = C_{ox} \equiv \frac{\mathcal{E}_{ox}}{t_{ox}} \qquad (7.2.19)$$

holds for the MOS structure in accumulation.

7.2.2. Depletion:

If a small positive bias is applied to the gate (Figure 7.3C) holes near the silicon surface are repelled by the gate. Because the acceptor doping atoms cannot move in the silicon lattice a negative charge appears underneath the gate oxide. Similarly a positive charge of equal magnitude can be found in the gate electrode, at the metal-oxide interface. The gate charge is a surface charge, but the charge in the silicon is not. It is a depletion charge which extends to a non-negligible depth into the silicon. The potential in the depletion region can be found integrating by Poisson's equation. Using $n<<p$ and $N_d<<N_a$ one can write:

$$\frac{d^2\Phi(x)}{dx^2} = -\frac{\rho}{\varepsilon_{si}} = -\frac{q}{\varepsilon_{si}}(p - N_a) = -\frac{q}{\varepsilon_{si}}N_a\left[exp\left(-\frac{q\Phi(x)}{kT}\right)-1\right] \quad (7.2.20)$$

The potential in the depletion region near the oxide/silicon interface is positive. Therefore, the exponent term of Equation 7.2.20 is small and can be neglected, which implies $p<<N_a$:

$$\Phi(x)>0 \Rightarrow exp\left(-\frac{q\Phi(x)}{kT}\right) \cong 0 \quad (7.2.21)$$

Using this approximation Equation 7.2.20 becomes:

$$\frac{d^2\Phi(x)}{dx^2} = -\frac{\rho}{\varepsilon_{si}} = \frac{qN_a}{\varepsilon_{si}} \quad (7.2.22)$$

This result is the *depletion approximation* which assumes that the charge density is constant and equal to $-qN_a$ in the depletion region. The depth up to which holes are repelled is called the depletion depth (or width) and noted x_d. Outside the depletion region the silicon is assumed to be neutral, such that $\rho(x)$, $\mathcal{E}(x)$ and $\Phi(x)$ are equal to zero for $x \geq x_d$. The potential in the silicon can be found by integrating the Poisson equation 7.2.22 with the following boundary conditions:

$$\Phi(x_d) = 0 \quad \text{and} \quad \frac{d\Phi(x_d)}{dx} = 0 \quad (7.2.23)$$

which yields:

$$\Phi(x) = \frac{qN_a}{2\varepsilon_{si}}(x - x_d)^2 \quad (7.2.24)$$

The surface potential at the oxide/silicon interface where $x=0$ is equal to:

$$\Phi_S = \Phi(x=0) = \frac{qN_a}{2\varepsilon_{si}}x_d^2 \quad (7.2.25)$$

Equation 7.2.25 can be used to evaluate the depletion depth expressed as a function of the surface potential:

$$x_d = \sqrt{\frac{2\varepsilon_{si}\Phi_s}{qN_a}} \tag{7.2.26}$$

The charge per surface area in the region from $x=0$ to $x=x_d$, called "depletion charge" is equal to:

$$Q_d = -qN_a x_d = -\sqrt{2q\varepsilon_{si}N_a\Phi_s} \tag{7.2.27}$$

The gate voltage, V_G, is equal to the potential drop across the oxide added to the potential variation in the semiconductor:

$$V_G = \Phi_s + V_{ox} = \Phi_s - \frac{Q_d}{C_{ox}} \tag{7.2.28}$$

The capacitance of the structure can be calculated as follows:

$$C = \frac{dQ_G}{dV_G} = -\frac{dQ_d}{dV_G} = -\frac{dQ_d}{d(-\dfrac{Q_d}{C_{ox}}+\Phi_s)} = -\frac{dQ_d/d\Phi_s}{d(-\dfrac{Q_d}{C_{ox}}+\Phi_s)/d\Phi_s} = \frac{1}{\dfrac{1}{C_{ox}}+\dfrac{1}{C_D}} \tag{7.2.29}$$

where

$$C_D = -\frac{dQ_d}{d\Phi_S} = \frac{\varepsilon_{si}}{x_d} \tag{7.2.30}$$

The overall capacitance is thus the series association of the gate oxide capacitance and the depletion region capacitance, ε_{si}/x_d. The capacitance can also be expressed as a function of the gate voltage by rewriting expression 7.2.28 in the following way:

$$V_G = -\frac{Q_d}{C_{ox}} + \Phi_s = \frac{qN_a x_d}{C_{ox}} + \frac{qN_a}{2\varepsilon_{si}}x_d^2 \tag{7.2.31}$$

x_d can be expressed as a function of the gate voltage:

$$x_d = -\frac{\varepsilon_{si}}{C_{ox}} + \sqrt{\left(\frac{\varepsilon_{si}}{C_{ox}}\right)^2 + \frac{2\varepsilon_{si}}{qN_a}V_G} \tag{7.2.32}$$

Substituting x_d into Equation 7.2.29 we obtain the capacitance as a function of the gate voltage:

$$C = \frac{C_{ox}}{\sqrt{1 + \dfrac{2C_{ox}^2 V_G}{qN_a\varepsilon_{si}}}} \tag{7.2.33}$$

7.2.3. Inversion:

If a larger positive voltage is applied to the gate the surface potential will continue to increase. The hole concentration near the surface decreases while the electron concentration increases, according to the following relationships:

$$p(x=0) = N_a \exp\left(-\frac{q\Phi_S}{kT}\right) \tag{7.2.34}$$

and:

$$n(x=0) = \frac{n_i^2}{N_a} \exp\left(\frac{q\Phi_S}{kT}\right) \tag{7.2.35}$$

Since $n = n_i \exp\left(\frac{E_F-E_i}{kT}\right)$, $p = n_i \exp\left(\frac{E_i-E_F}{kT}\right)$ and $pn = n_i^2$, the electron surface concentration is equal to the hole surface concentration ($n(0) = p(0) = n_i$) when E_i coincides with E_F at $x=0$. This happens when $\Phi_S = \Phi_F \equiv \frac{kT}{q} \ln\left(\frac{N_a}{n_i}\right)$ (Figure 7.6).

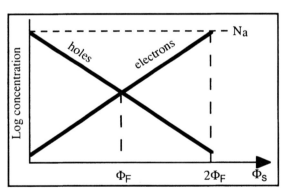

Figure 7.6: Hole and electron surface concentration as a function of Φ_S.

If the gate voltage is increased further the electron surface concentration increases up to a point where $n(x=0)$ becomes equal to $p_{po} = N_a$, which is the original hole concentration in the substrate. This happens when the band curvature at the surface ($x=0$) places E_i at an energy $q\Phi_F$ below E_F. In other words the band curvature is equal to $2(E_i-E_F)$ or:

$$\Phi_S = 2\,\Phi_F \tag{7.2.36}$$

When this condition is met, the semiconductor surface is said to be in "strong inversion". For $\Phi_F \leq \Phi_S < 2\,\Phi_F$ the electron concentration is larger than the hole concentration, and the surface is in weak inversion, while for $\Phi_S \geq 2\ \Phi_F$ it is in strong inversion (Figure 7.7).

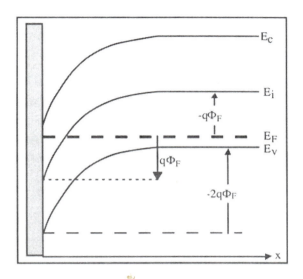

Figure 7.7: Energy band curvature under strong inversion at the surface.

Example

Calculate the electron concentration at the oxide/semiconductor interface when the surface potential is equal to 1) Φ_F and 2) $2\Phi_F$. The P-type doping concentration is N_a.

$$\Phi_F = \frac{kT}{q} \ln \left(\frac{N_a}{n_i}\right) \quad \text{and} \quad n_o = \frac{n_i^2}{N_a}$$

1) $n(x=0) = \dfrac{n_i^2}{N_a} \exp\left(\dfrac{q\Phi_F}{kT}\right) = \dfrac{n_i^2}{N_a} \exp\left(\dfrac{q\frac{kT}{q} \ln \left(\frac{N_a}{n_i}\right)}{kT}\right) = n_i$

2) $n(x=0) = \dfrac{n_i^2}{N_a} \exp\left(\dfrac{2q\Phi_F}{kT}\right) = \dfrac{n_i^2}{N_a} \exp\left(\dfrac{q\frac{kT}{q} \ln \left(\frac{N_a}{n_i}\right)^2}{kT}\right) = N_a$

The inversion layer is rich in electrons, and therefore, a good conductor. The MOS capacitor consists of two conducting electrodes (the metal gate and the inversion layer at the silicon surface). As in the case of accumulation, the capacitance of the MOS structure is once again equal to C_{ox}.

When an inversion layer is formed electrons are locally majority carriers at the surface. Any subsequent increase in gate voltage increases the electron concentration in the inversion layer, and a larger inversion charge, Q_{inv}, is produced. However, the thickness of the inversion layer remains very small. Its actual thickness is similar to that of an accumulation layer (derived in Section 7.2.1). The electron charge in an inversion layer can, therefore, be considered as a surface charge. As in the case of an accumulation layer the inversion charge depends exponentially on the surface potential ($Q_{inv} \propto exp\left(\dfrac{q\Phi_S}{kT}\right)$). When the gate voltage is increased beyond inversion formation the surface potential, Φ_S, increases only very slightly above $2\Phi_F$ and for all practical purposes one can assume that $\Phi_S = 2\Phi_F$ when an inversion layer is present, regardless of the gate voltage. Therefore, the depth of the depletion region is given by Equation 7.2.26 where $\Phi_S = 2\Phi_F$:

$$x_{dmax} = \sqrt{\frac{4\varepsilon_{si}\Phi_F}{qN_a}} \qquad (7.2.37)$$

Since the semiconductor is P-type one may wonder where the electrons in the inversion layer come from. They are produced by thermal generation, which is a rather slow process at room temperature. They can also be produced by external generation (if a light source is present, for example). If the semiconductor is in the dark and at cryogenic temperature the inversion layer may never form.

Figure 7.8 shows the capacitance of an MOS capacitor as a function of the applied gate bias. Such a curve is often called a capacitance curve, or C-V curve. Different types of measurements can be made, each of these probing a different aspect of the device properties.

In a first measurement the gate voltage is slowly ramped from negative to positive values, and a small ac signal is superimposed to this quasi-dc bias. The small signal is used to measure the value of the capacitance at the various dc gate biases. Different curves can be obtained for a given device depending on the frequency of the ac signal.

◊   Let us first consider the case of a low-frequency ac signal (quasi-static curve in Figure 7.8). When the gate voltage is negative an accumulation layer is present. As the gate voltage varies a corresponding variation of the accumulation charge occurs, and the capacitance of the structure is equal to C_{ox} (Expression 7.2.1). When the gate voltage is increased the silicon surface becomes depleted, and the variations of gate voltage induce variations of the depletion charge. The value of the capacitance is then given by the series combination of the gate and depletion region capacitances (Equation 7.2.33). As the gate voltage is further increased an inversion layer is formed and variations of gate voltage give rise to variations of inversion charge and thus the measure capacitance is again equal to C_{ox}.

◊   If we repeat the same measurement using a higher frequency for the small ac signal (1 MHz, typically), thermal generation cannot create minority carriers fast enough to support a variation of charge in the inversion layer. Therefore, while the portions of the curve in accumulation and depletion are identical to the previous experiment, the inversion part of the curve is not. The variation of charge due to the variation of the gate voltage is no longer supported by the inversion charge, but by a variation of the depletion charge (Figure 7.8). The depth of the depletion region is equal to $x_{dmax} \pm \Delta x_d$, where Δx_d is a small modulation of the depletion depth due to the application of the small ac gate bias. In this case the capacitance of the structure is given by the series association of the gate capacitance, C_{ox}, and the depletion capacitance, $\varepsilon_{si}/x_{dmax}$.

◊   If a fast gate voltage ramp is used there is no time for generation of minority carriers (electrons). Majority carriers are readily available to form an accumulation layer, so that the accumulation part of the curve remains unchanged. When the gate voltage is ramped up, a depletion layer is formed, but no inversion layer can be formed. Therefore, only a depletion charge can respond the gate voltage variation, and the depletion depth can be larger than x_{dmax}. Such operation is called the deep-depletion regime, and the value of the capacitance is given by Equation 7.2.33 where the surface potential is not clamped at $2\Phi_F$.

◊   If a very high-frequency ac signal is used, even majority carriers may not have time to react to the gate voltage variation. Frequencies of 1 GHz or higher must be used for this effect to appear. The higher the doping concentration, the higher the frequency. In this case the whole semiconductor sample behaves as a dielectric (dielectric mode of operation in Figure 7.8), such that the capacitance of the structure is given by the series association of C_{ox} and ε_{si}/d_{si}, where d_{si} is the thickness of the silicon wafer.[8]

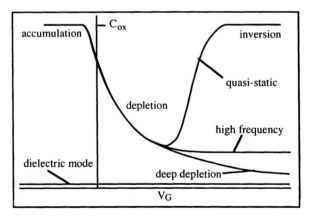

Figure 7.8: C-V curves of a MOS capacitor on a P-type substrate. [9]

In summary the following rules will be used to describe the relationships between the charge on the metal gate and the charge in the accumulation, depletion and inversion layers (Figure 7.9):

$$-Q_G = Q_{acc} \qquad \text{(accumulation)} \quad (7.2.38a)$$
$$-Q_G = Q_d \qquad \text{(depletion)} \quad (7.2.38b)$$
$$-Q_G = Q_d + Q_{inv} \qquad \text{(inversion)} \quad (7.2.38c)$$

$-Q_G$ = charge at the backside contact of the sample (dielectric mode) (7.2.38d)

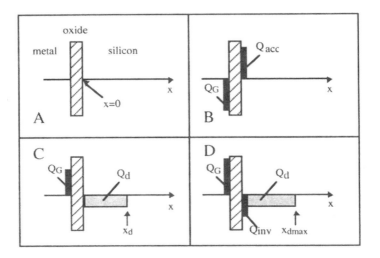

Figure 7.9: Charges in the MOS structure:
A: Flat band, B: Accumulation, C: Depletion, D: Inversion

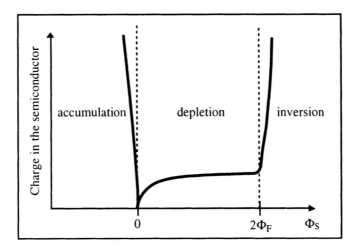

Figure 7.10: Absolute value of the charge in the semiconductor as a function of the surface potential.[10]

The total charge in the semiconductor can be plotted as a function of the surface potential (Figure 7.10). Accumulation and inversion charges are exponential functions of the surface potential, while the depletion charge varies as a square root.

7.3. Threshold voltage

The threshold voltage of a MOS transistor is the voltage that must be applied on the gate to form an inversion layer. It depends on several device parameters which will be described next.

7.3.1 Ideal threshold voltage

In a MOS transistor the gate voltage is equal to sum of the potential drops in the semiconductor and the oxide. If one assumes that the back of the semiconductor is grounded, one can write:

$$V_G = \Phi_S + \frac{Q_G}{C_{ox}} \tag{7.3.1}$$

where Q_G is equal to the positive charge on the gate electrode. An equal amount of negative charge exists in the semiconductor, comprised of ionized impurities in the depletion zone, and free electrons at the oxide/silicon interface a inversion. If we assume that the charge due to the free electrons is much smaller than the charge due to ionized impurities when the inversion layer starts being formed then Equation 7.3.1 can be written as:

$$V_G = 2\Phi_F - \frac{Q_d}{C_{ox}} = 2\Phi_F + \frac{\sqrt{4q\varepsilon_{si}N_a\Phi_F}}{C_{ox}} \equiv V'_{THo} \tag{7.3.2}$$

V'_{THo} is called the "ideal threshold voltage" and it is measured with respect to the source. In this definition of the threshold voltage both the source and the substrate are grounded.

Example

Calculate the depletion and inversion charges for $N_a = 10^{16}$ cm^{-3}, $t_{ox} = 30$ nm = 30×10^{-7} cm, $n_i = 1.45 \times 10^{10}$ cm^{-3} and $\Phi_S = 2\Phi_F$.

$$\Phi_F = \frac{kT}{q} \ln\left(\frac{N_a}{n_i}\right) = .0259 \ \ln\left(\frac{10^{16}}{1.45 \times 10^{10}}\right) = 0.348V$$

$$Q_d = -\sqrt{4q\varepsilon_{si}N_a\Phi_F}$$

$$= -\sqrt{4 \times 1.6 \times 10^{-19} \times 11.7 \times 8.854 \times 10^{-14} \times 10^{16} \times 0.348} = -5.5 \times 10^{-8} \text{ C cm}^{-2}$$

The free electron charge density at the oxide/silicon interface is equal to $n(\Phi_S = 2\Phi_F) = N_a$. Assuming the thickness of the inversion layer is equal to a

tenth of the Debye length $0.1 \times L_D = 0.1 \sqrt{\dfrac{\varepsilon_{si}kT}{q^2 N_a}} = 4$ nm (Equation 7.2.10) and assuming the electron concentration is constant as a function of depth in the inversion layer, the inversion charge can be approximated by $Q_n \cong q \, N_a \, L_D =$ -6.5×10^{-10} C cm^{-2}. The inversion charge at threshold is much smaller than the depletion charge.

7.3.2. Flat-band voltage

Equalization of the Fermi levels
We have so far assumed that the Fermi level of the metal gate was equal to that of the silicon. In practice this is not the case. In modern devices the gate material is not an actual metal, but heavily doped polycrystalline silicon, also called polysilicon. The doping concentration used for that material is so high ($\approx 10^{20}$ cm^{-3}) that it can be considered as a metal, for all practical purposes. Let us first consider the metal and the semiconductor separately. The energy which is necessary to extract an electron with an energy E_{FM} from the metal is called the "work function", $q\Phi_m$. Similarly, the work function in the semiconductor is noted $q\Phi_{sc}$.

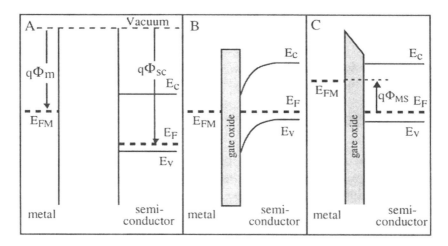

Figure 7.11: Energy bands in the MOS structure. A: The metal and the semiconductor are taken separately; B: Alignment of the Fermi levels with no applied bias, C: A bias equal to Φ_{ms} is applied to the gate.[11]

When the two materials are put together to form the MOS structure, the Fermi levels align, and the charge transfer resulting from this process curves the energy bands in the semiconductor, near the semiconductor-oxide interface (Figure 7.11). To recover to a flat-band condition a voltage must be applied to the gate. This voltage is equal to the difference of the work functions between the two materials, called the "work function difference", and is noted Φ_{ms}:

$$\Phi_{ms} = \Phi_m - \Phi_{sc} = \frac{E_F - E_{FM}}{q} \tag{7.3.3}$$

Example
Calculate Φ_{ms} for $N_a = 10^{16}$ cm^{-3}, $n_i = 1.45 \times 10^{10}$ cm^{-3} and $\Phi_S = 2\Phi_F$. The gate material is made N-type polysilicon with $N_d = 10^{20}$ cm^{-3}.

$$\Phi_{ms} = \frac{E_F - E_{FM}}{q} = -\frac{kT}{q}\left(ln\left(\frac{N_a}{n_i}\right) + ln\left(\frac{N_d}{n_i}\right)\right) = -0.94 \text{ V}$$

Charges in the oxide
Oxides grown on silicon contain positive charges due to the presence of contaminating metallic ions or imperfect Si-O bonds. These charges can either be fixed or mobile in the oxide. Mobile ions such as sodium and potassium can move in the presence of an electric field if the temperature is high enough. Here, we shall consider only the case of fixed charges.

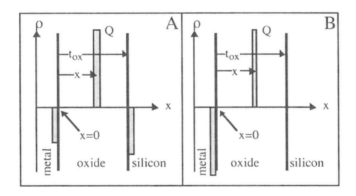

Figure 7.12: Single charge in an oxide. A: $V_G = 0$; B: a compensation gate voltage is applied.[12]

Let us consider an elementary areal positive charge Q (C cm^{-2}) at a depth x in the oxide, where $x=0$ is now defined at the metal/oxide interface. (Figure 7.12A). To insure charge neutrality negative charges will appear in the metal and the silicon. The sum of these three charges is equal to zero. The charge in the silicon can be removed if an appropriate negative voltage is applied to the gate. This voltage can be found by integrating Poisson's equation in the oxide between 0 and x. It is given by:

$$V_Q = -\frac{Qx}{\varepsilon_{ox}} = -\frac{x}{t_{ox}}\frac{Q}{C_{ox}} \tag{7.3.4}$$

If the charge is closer to the semiconductor a larger compensation bias on the gate is required to remove the charge in the semiconductor. In an actual device charges are distributed throughout the oxide according to an arbitrary concentration profile, $\rho(x)$ (cm^{-3}). The compensation voltage, V_Q, is obtained by integrating the contribution of every single charge throughout the oxide. The compensating voltage is thus equal to:

$$V_Q = -\frac{1}{C_{ox}} \int_0^{t_{ox}} \frac{x}{t_{ox}} \rho(x) \, dx \equiv -\frac{Q_{ox}}{C_{ox}} \quad \left(\frac{\text{C cm}^{-2}}{\text{F cm}^{-2}}\right) \qquad (7.3.5)$$

Example

Calculate V_Q for the following oxide charge distributions: 1) delta distribution at the metal/oxide interface $\rho(x) = q \, N_{ox} \, \delta(x=0)$; 2) delta distribution at the oxide/semiconductor interface $\rho(x) = q \, N_{ox} \, \delta(x=t_{ox})$; 3) a constant charge density throughout the oxide $\rho(x) = q \, N_{ox}/t_{ox}$. N_{ox} is measured in cm^{-2}. Note that the total charge in the oxide, $Q_{ox} = q \int_0^{t_{ox}} N_{ox} \, dx$, is the same for the three distributions.

Using Equation 7.3.5 we find 1) $V_Q = 0$; 2) $V_Q = -\dfrac{q \, N_{ox}}{C_{ox}}$; 3) $V_Q = -\dfrac{q \, N_{ox}}{2C_{ox}}$.

Interface traps

The presence of the silicon-SiO$_2$ interface at the silicon surface introduces an obvious perturbation to the periodic crystal structure of the semiconductor and causes some Si-Si bonds to be unfulfilled or "dangling". As a result there are energy states in the bandgap at the silicon surface. These states are called "interface states" or "interface traps". They can be charged positively or negatively, depending on their nature and their energy with respect to the Fermi level, and thus, will affect the surface potential.

If the interface density trap is noted N_{it} (unit: cm^{-2} V^{-1}) a charge $qN_{it}(\Phi_S)$ is present at the semiconductor surface. The charge is usually negative in n-channel transistors and is due to electrons trapped in the interface states. If the surface potential increases from Φ_S to $\Phi_S + \Delta\Phi_S$ the trapped charge increases by a amount equal to $-qN_{it}\Delta\Phi_S$. When an inversion layer is present the surface potential is equal to $2\Phi_F$. To compensate for these charges, a bias must be applied to the gate. Its value is:

$$V_{it} = \frac{2qN_{it}\Phi_F}{C_{ox}} \qquad (7.3.6)$$

Flat-band voltage

The "flat-band voltage" is the voltage that must be applied to the gate to bring the semiconductor energy bands to a flat level. Flatband is achieved by applying a gate voltage which compensates for 1) differences in work functions of the semiconductor and the gate electrode, 2) the presence of charges in the oxide, and 3) interface traps. The sum of all these effects is found by adding Expressions 7.3.3, 7.3.5, and 7.3.6:

$$V_{FB} = \Phi_{ms} + V_Q + V_{it} = \Phi_{ms} - \frac{Q_{ox}}{C_{ox}} + \frac{2qN_{it}\Phi_F}{C_{ox}} \qquad (7.3.7)$$

7.3.3. Threshold voltage

The flat-band voltage must be added to the expression for the threshold voltage calculated previously (Expression (7.3.2)), in order to accurately describe the actual, "non-ideal" threshold voltage:

$$V_{THo} = V_{FB} + 2\Phi_F - \frac{Q_d}{C_{ox}}$$

$$= \Phi_{ms} - \frac{Q_{ox}}{C_{ox}} + \frac{2qN_{it}\Phi_F}{C_{ox}} + 2\Phi_F + \frac{\sqrt{4q\varepsilon_{si}N_a\Phi_F}}{C_{ox}} \qquad (7.3.8)$$

The threshold voltage can be either positive or negative, depending on the doping concentration N_a, the material used to form the gate electrode, etc. If the threshold voltage is negative the n-channel MOSFET is a depletion-mode device; if V_{THo} is positive, the device is an enhancement-mode MOSFET. Depletion-mode devices will have an inversion layer when the gate voltage is equal to zero. These devices are sometimes referred to as "normally on". Enhancement-mode devices require an applied positive gate voltage to create the inversion layer. They are sometimes called "normally off". The value of the threshold voltage can be adjusted by introducing a controlled amount of doping impurities in the channel region during device fabrication (see Sections 11.3.1 and 11.9 and Problem 11.5).

7.4. Current in the MOS transistor

The current in the channel is due to the drift of electrons from source to drain. We can define a local potential at the surface of the silicon between source and drain. The value of this local potential, noted $V(y)$, ranges between V_S at $x = 0$ and V_D at $y = L$. To illustrate this notion of local surface potential, consider a case where the channel runs from source to drain. The channel can be viewed as a simple resistor through which current can flow between source and drain. In this representation the local potential, $V(y)$, can be viewed as the potential at any point y along the resistive channel. Both the source terminal and the substrate are grounded. The electric field in the channel is equal to

$$\mathcal{E}_y = -dV(y)/dy \qquad (7.4.1)$$

and the drift current is equal to:

$$J_{ny} = q\,\mu_n\,n\,\mathcal{E}_y \tag{7.4.2}$$

or

$$J_{ny} = -\,q\,\mu_n\,n\,\frac{dV(y)}{dy} \tag{7.4.3}$$

The total inversion charge in the channel is given by:

$$Q_{inv}(y) = -\,q \int_{channel} n(x,y)\;dx \tag{7.4.4a}$$

or

$$Q_{inv}(y) = -\,q \int_{x=0}^{\infty} n(x,y)\;dx \tag{7.4.4b}$$

Using 7.4.3 the current in the channel, which is also called the drain current, I_D, can be derived:

$$I_D = -\,W \times \int_{x=0}^{\infty} J_{ny}\;dx \qquad (\text{with } J_{ny} < 0)$$

$$= -\,W\,Q_{inv}(y)\,\mu_n\,\frac{dV(y)}{dy} \tag{7.4.6}$$

where W is the width of the channel (see Figure 7.1). The latter expression is simply Ohm's law applied to a small element of channel having a length dy and a resistance $dR(y)$:

$$dV(y) = I_D\,dR(y) \tag{7.4.7}$$

where

$$dR(y) = \frac{-\,dy}{W\mu_n Q_{inv}(y)} \qquad (\text{with } Q_{inv} < 0) \tag{7.4.8}$$

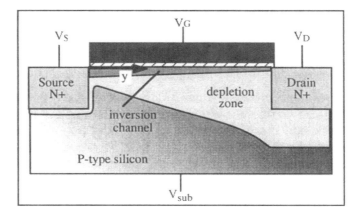

Figure 7.13: Cross section of a MOS transistor.

Since the source is grounded and the drain is at a positive bias, V_D, the potential in the channel will vary from $V(x=0) = V_S$ near the source to

$V(y=L)$ = V_D near the drain. The surface of the silicon is in strong inversion when the surface electron concentration is equal to the hole concentration in the quasi-neutral substrate $(n = N_a)$. This condition imposes the band curvature in the x-direction to be equal to $2\Phi_F$ near the source and $2\Phi_F + V_D$ near the drain. Generalizing to the entire channel we can write:

$$\Phi_s(y) = 2\Phi_F + V(y) \qquad (7.4.9)$$

What is the value of the inversion charge, $Q_{inv}(y)$? If we recall Expression 7.2.38c:

$$-Q_G(y) = Q_d(y) + Q_{inv}(y) \qquad (7.4.10)$$

and notice that the potential drop in the gate oxide above any location in the channel is given by:

$$V_{ox}(y) = V_G - [V_{FB} + 2\Phi_F + V(y)] = Q_G(y)/C_{ox} \qquad (7.4.11)$$

we can write:

$$Q_{inv}(y) = - C_{ox} \left\{ V_G - [V_{FB} + 2\Phi_F + V(y)] \right\} - Q_d(y) \qquad (7.4.12)$$

As mentioned earlier the channel can be considered as a simple N-type material resistor connecting source to drain. The potential at any point y along this resistor is equal to $V(y)$, which varies from V_S at $y = 0$ to V_D at $y = L$. In the x-direction the energy band curvature, from the substrate to the surface near the source, is equal to $2q\Phi_F + qV_S$. Similarly, the band curvature in the x-direction near the drain is equal to $2q\Phi_F + qV_D$. In other words, the electron-rich channel can be viewed as an actual N-type slab of semiconductor which forms a reverse-biased PN junction with the P-type substrate. Since the potential $V(y)$ along that N-type slab varies from V_S to V_D, the depth of the depletion region in the x-direction will vary and grow larger near the drain. The local width of the depletion layer is given by:

$$Q_d(y) = - \sqrt{2q\varepsilon_{si}N_a (2\Phi_F + V(y))} \qquad (7.4.13)$$

Integrating 7.4.6 from source to drain one obtains:

$$I_D \int_0^L dy = I_D L = - W \mu_n \int_{V_S}^{V_D} Q_{inv}(y) \, dV(y) \qquad (7.4.14)$$

Using 7.4.12 and 7.4.13, and since I_D is constant at any location along the channel :

$$I_D = \mu_n C_{ox} \frac{W}{L} \int_{V_S}^{V_D} \left(V_G - [V_{FB} + 2\Phi_F + V(y)] - \frac{\sqrt{2q\varepsilon_{si}N_a (2\Phi_F + V(y))}}{C_{ox}} \right) dV(y)$$

$$(7.4.15)$$

If we define:
$$\gamma = \frac{\sqrt{2q\varepsilon_{si}N_a}}{C_{ox}} \qquad (7.4.16)$$

and $V_{DS} \equiv V_D - V_S$ the integration of 7.4.15 becomes (See Problem 7.21):

$$I_D = \mu_n C_{ox} \frac{W}{L} \left\{ \left(V_G - V_{FB} - 2\Phi_F - V_S - \frac{V_{DS}}{2} \right) V_{DS} - \frac{2}{3}\gamma \left[(2\Phi_F + V_D)^{3/2} - (2\Phi_F + V_S)^{3/2} \right] \right\}$$
$$(7.4.17)$$

Expression 7.4.17 yields the drain current as a function of source, gate and drain voltage, with the substrate grounded. If this equation as a function of drain voltage is plotted a bell-shaped curve is obtained, as shown in Figure 7.14. The left half of the curve correctly depicts the behavior of the actual device, but the right half does not. Expression 7.4.17 reaches a maximum when V_D is equal to a value called "drain saturation voltage", V_{Dsat}. Setting $\frac{dI_D}{dV_D} = 0$, the drain saturation voltage is obtained:

$$V_G - V_{FB} - 2\Phi_F - V_{Dsat} = \gamma \sqrt{2\Phi_F + V_{Dsat}}$$

$$\Downarrow$$

$$V_{Dsat} = V_G - V_{FB} - 2\Phi_F + \frac{\gamma^2}{2} - \gamma \sqrt{V_G - V_{FB} + \frac{\gamma^2}{4}} \qquad (7.4.18)$$

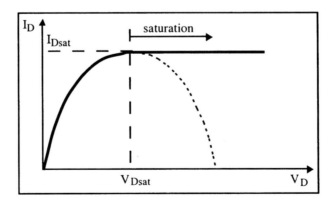

Figure 7.14: $I_D(V_{DS})$ for $V_G > V_{THo}$

Saturation appears when the gate voltage is no longer large enough (with respect to the local surface potential) to sustain the presence of an inversion layer near the drain junction. The current evaluated at V_{Dsat} is called the "drain saturation current", I_{Dsat}. It is obtained by replacing V_D by V_{Dsat} in Equation 7.4.17. Both V_{Dsat} and I_{Dsat} are functions of the gate voltage. It can easily be verified that the condition $\frac{dI_D}{dV_D} = 0$ for $V_D = V_{Dsat}$ is equivalent to writing $Q_{inv}(L) = 0$. When this happens the channel is said to be "pinched off" near the drain, and the transistor is said to be in saturation (Figure 7.15). For an ideal "long channel" device the lateral dimension of the pinch-off region is very small, even if the drain voltage is high. When the transistor is not in saturation, it is said to operate in the non-saturated or "triode" regime (Figure 7.14).

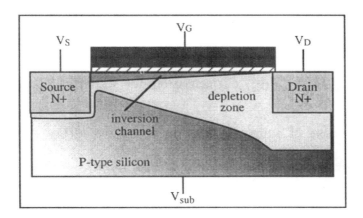

Figure 7.15: Cross section of a MOS transistor illustrating channel pinch-off.

Although the inversion channel is pinched at the drain, the device will still conduct current. The local potential at the channel pinch-off point is V_{Dsat} and the potential drop across the pinch-off region is equal to $V_D - V_{Dsat}$. Since the pinch-off lateral extension is very small, there is an intense electric field between the pinch-off point and the drain junction, which causes electrons to drift from the channel into the drain. The magnitude of this electron current is fixed by the potential drop across the channel, which is constant and equal to $V_{Dsat} - V_S$. As a result, when $V_D > V_{Dsat}$ the current remains constant and is independent of the drain voltage, as shown in Figure 7.14.

A complete set of $I_D(V_{DS})$ curves, called "output characteristics" of a MOSFET is shown in Figure 7.16. At the left of the dashed parabolic line the transistor operates in the non-saturated regime, also called the "triode regime"; past that line it is in saturation. The value of the voltage at the point where the triode and saturation regions meet is given by $V_D = V_{Dsat}$.

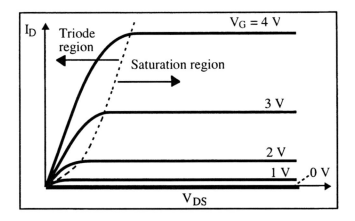

Figure 7.16: $I_D(V_{DS})$ characteristics for $V_G = 0,1,2,3$ and 4 V ($V_{THo} = 0.7$ V) [13]

7.4.1. Influence of substrate bias on threshold voltage

In the derivation of the threshold voltage we previously assumed that both the source and the substrate were grounded (Expression 7.3.8). However in many applications the source and substrate may be at a different potentials. Therefore we will now investigate the influence of a variation of source to substrate bias on the threshold voltage. Using relationships 7.3.8 and 7.4.16 the threshold voltage can be written as:

$$V_{THo} = V_{FB} + 2\Phi_F + \gamma\sqrt{2\Phi_F} \qquad (7.4.19)$$

with $V_S - V_{sub} = 0$, V_{sub} being the substrate voltage.

Let us now apply a negative bias to the substrate with the source grounded ($V_S=0$ and $V_{sub}<0$). All PN junctions in the device remain reverse biased. However, the reverse bias applied to all these junctions is larger than it was when the substrate was grounded. The bias across the source junction is V_{sub} and the bias across the drain junction is $-V_D + V_{sub}$. Hence, the energy band curvature between the inversion channel and the substrate is no longer equal to $2\Phi_F + V(y)$ but to $2\Phi_F + V(y) - V_{sub}$, the grounded source being taken as a reference. The depletion charge under the channel is obtained by introducing V_{sub} in 7.4.13:

$$Q_d(y) = -\sqrt{2q\varepsilon_{si}N_a\,(2\Phi_F + V(y) - Vsub)} \qquad (7.4.20)$$

When the gate voltage is equal to threshold voltage, we have:

$$V_G = V_{FB} + 2\Phi_F + V_S - \frac{Q_d(y=0)}{C_{ox}} \qquad (7.4.21)$$

or, using 7.4.20:

$$-C_{ox}[V_G - V_{FB} - 2\Phi_F - V_S] = -\sqrt{2\varepsilon_{si}\,qN_a}\,\sqrt{V_S + 2\Phi_F - V_{sub}} \qquad (7.4.22)$$

which yields the threshold voltage with the substrate effect included:

$$V_{TH} = V_{FB} + 2\Phi_F + V_S + \gamma\sqrt{V_S + 2\Phi_F - V_{sub}} \qquad (7.4.23)$$

Equation 7.4.23 is the general definition of threshold voltage which can be used for any source and substrate bias. It is, however, convenient to rewrite 7.4.23 for the particular cases where either the source or the substrate are grounded:

1) If the substrate is grounded and the source potential is positive ($V_{sub}=0$ and $V_S>0$), Expression 7.4.23 becomes:

$$V_{TH} = V_S + V_{FB} + 2\Phi_F + \gamma\sqrt{2\Phi_F + V_S} \qquad (7.4.24a)$$

where the threshold voltage is measured with respect to the substrate. V_{TH} is equal to V_{THo} when $V_S = 0$ (Expression 7.3.8).

2) If the source is grounded and the substrate bias is negative ($V_S=0$ and $V_{sub}<0$), Expression 7.4.23 becomes:

$$V_{TH} = V_{FB} + 2\Phi_F + \gamma\sqrt{2\Phi_F - V_{sub}} \qquad (7.4.24b)$$

where the threshold voltage is measured with respect to the source. V_{TH} is equal to V_{THo} when $V_{sub} = 0$ (Expression 7.3.8).

The threshold voltage increases as a function of the potential difference between source and substrate (Figure 7.17).

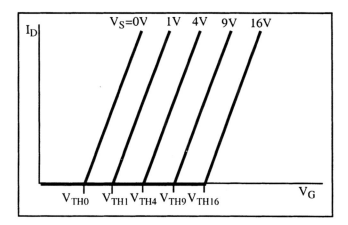

Figure 7.17: Drain current vs. gate voltage for different source bias. The substrate is grounded and V_{DS} is small. [14]

7.4.2. Simplified model

The model developed in equations 7.4.17 and 7.4.18 is often considered too cumbersome for practical use. It can be simplified by linearizing the maximum depth of the depletion region as a function of the local surface potential, $V(y)$. If the substrate is used as reference ($V_{sub} = 0$) the depletion charge can be linearized as follows:

$$- \frac{Q_d(y)}{C_{ox}} = \frac{\sqrt{2qN_a\varepsilon_{si}\,(2\Phi_F + V(y))}}{C_{ox}} \cong \gamma\sqrt{2\Phi_F} + \delta\,V(y) \quad (7.4.25)$$

where γ is defined in Equation 7.4.16 and δ is a constant that represents the linearized dependence of the depletion charge on $V(y)$. The threshold voltage is given by 7.4.23:

$$V_{TH} = V_{FB} + 2\Phi_F + V_S + \gamma\sqrt{V_S + 2\Phi_F}$$

When the source is grounded ($V_S = 0$) the threshold voltage is given by 7.3.8:

$$V_{THo} = V_{FB} + 2\Phi_F + \frac{\sqrt{4q\varepsilon_{si}N_a\Phi_F}}{C_{ox}}$$

Comparing these two equations we can write:

$$V_{TH} = V_{THo} + \gamma\left(\sqrt{(2\Phi_F + V_S)} - \sqrt{2\Phi_F}\right) \quad (7.4.26)$$

and the inversion charge (7.4.12) is given by:

$$Q_{inv}(y) = - C_{ox}\left\{V_G - [V_{FB} + 2\Phi_F + V(y)]\right\} - Q_d(y)$$

$$= - C_{ox}\left\{V_G - [V_{FB} + 2\Phi_F + V(y)]\right\} - \sqrt{2q\varepsilon_{si}N_a\,(2\Phi_F + V(y))}$$

or, using 7.4.25:

$$Q_{inv}(y) = - C_{ox}\left\{V_G - [V_{FB} + 2\Phi_F + \gamma\sqrt{2\Phi_F} + (1+\delta)V(y)]\right\}$$

$$= - C_{ox}\left\{V_G - [V_{THo} + (1+\delta)V(y)]\right\} \quad (7.4.27)$$

If we define the "body factor" (or "body effect coefficient"): $n \equiv 1 + \delta$, we can write:

$$Q_{inv}(y) = -C_{ox}\left\{V_G - [V_{THo} + nV(y)]\right\} \quad (7.4.28)$$

Integrating from source to drain one obtains:

$$\int_{o}^{L} I_D \, dy = - W\,\mu_n \int_{V_S}^{V_D} Q_{inv}(y)\, dV(y) \quad (7.4.29)$$

and, since I_D is constant at any position, y, in the channel:

$$I_D = \mu_n C_{ox} \frac{W}{L} \int_{V_S}^{V_D} \left(V_G - [V_{THo} + nV(y)] \right) dV(y) \qquad (7.4.30)$$

Writing the linearized dependence of the threshold voltage on source bias $V_{THS} = V_{THo} + nV_S$, we finally obtain:

Simplified Current-Voltage Relationship

$$I_D = \mu_n C_{ox} \frac{W}{L} \left\{ \left(V_G - V_{THS} \right) V_{DS} - \frac{1}{2} n V_{DS}^2 \right\} \qquad (7.4.31)$$

The latter equation describes a parabolic dependence of the drain current on drain voltage similar to the curve shown in Figure 7.14. The curve reaches a maximum when the drain voltage is equal to the drain saturation voltage, V_{Dsat}. V_{Dsat} is obtained by setting $dI_D/dV_D = 0$, which yields:

Saturation Drain Voltage

$$V_{Dsat} = \frac{(V_G - V_{THS})}{n} + V_S \qquad (7.4.32)$$

Replacing V_{DS} by V_{Dsat} in Equation 7.4.31 yields the drain saturation current:

Saturation Drain Current

$$I_{Dsat} = \mu_n C_{ox} \frac{W}{L} \frac{(V_G - V_{THS})^2}{2n} \qquad (7.4.33)$$

The transconductance of the transistor, defined as the variation of drain current with gate voltage is given by:

$$g_m \equiv \frac{dI_D}{dV_G} = \mu_n C_{ox} \frac{W}{L} V_{DS} \text{ in the non saturated regime, and}$$

$$g_{msat} \equiv \frac{dI_D}{dV_G} = \mu_n C_{ox} \frac{W}{L} \frac{(V_G - V_{THS})}{n} \quad \text{ in saturation.} \qquad (7.4.34)$$

In many instances an even more simplified model is used. Simplification is obtained by assuming that the maximum depth of the depletion region does not vary from source to drain. In mathematical terms this is equivalent to writing $\delta=0$ in Expression 7.4.25. As a result the body factor, n, is equal to 1, and Equations 7.4.31, 7.4.32, and 7.4.33 become:

$$I_D = \mu_n C_{ox} \frac{W}{L} \left\{ \left(V_G - V_{THS} \right) V_{DS} - \frac{1}{2} V_{DS}^2 \right\} \qquad (7.4.35)$$

$$V_{Dsat} - V_S = (V_G - V_{THS}) \qquad (7.4.36)$$

$$I_{Dsat} = \mu_n C_{ox} \frac{W}{L} \frac{(V_G - V_{THS})^2}{2} \qquad (7.4.37)$$

or, if both source and substrate are grounded:

$$I_D = \mu_n C_{ox} \frac{W}{L} \left\{ \left(V_G - V_{THo} \right) V_D - \frac{1}{2} V_D^2 \right\} \qquad (7.4.38)$$

$$V_{Dsat} = (V_G - V_{THo}) \qquad (7.4.39)$$

$$I_{Dsat} = \mu_n C_{ox} \frac{W}{L} \frac{(V_G - V_{THo})^2}{2} \qquad (7.4.40)$$

Note that Equations 7.4.38 and 7.4.40 are identical to Equations 7.1.7 and 7.1.8.

It is worthwhile noting that the body factor, n, is equal to $1 + \dfrac{C_D}{C_{ox}}$, where C_D is the depletion capacitance. Indeed, using Relationships 7.2.30 and 7.4.25 one can write:

$$\delta = \frac{-1}{C_{ox}} \frac{dQ_d(y)}{dV(y)} = \frac{C_D}{C_{ox}} \Rightarrow n = 1 + \delta = 1 + \frac{C_D}{C_{ox}} = 1 + \frac{\varepsilon_{si}/x_{dmax}}{C_{ox}} \qquad (7.4.41)$$

7.5. Surface mobility

The mobility, μ_n, used in the MOSFET model is not the mobility of electrons in the silicon crystal, called "bulk mobility". Rather, it is a "surface mobility". The surface mobility is lower than the bulk mobility because of increased scattering of the electrons at the silicon-oxide interface. The surface mobility depends on how much the electrons interact with the interface, and therefore, on the vertical electric field which "pushes" the electrons against the interface. We will note μ_{no} as the surface mobility in absence of such an electric field. The higher the electric field, the lower the surface mobility.

The current in the transistor is given by 7.4.14:

$$I_D = \frac{W}{L} \int_{V_S}^{V_D} \mu \left(-Q_{inv}(y) \right) dV(y)$$

In this expression the mobility is inside the integral because it is not constant (it depends on the vertical electric field in the channel, which varies from source to drain). Calculating this integral is a complex task. However, an "average" constant mobility value can be used instead of the electric-field dependent mobility. It will be called "effective mobility",

μ_{eff}. There exists an empirical relationship that describes the dependence of surface mobility on vertical electric field in the channel, $\mathcal{E}_x$:

$$\mu_n(y) = \frac{\mu_{no}}{1 + \Theta \mathcal{E}_x(y)} \qquad (7.5.1)$$

where Θ is called the "mobility reduction factor". The average electric field in the channel is:[15]

$$\mathcal{E}_x(y) = \frac{\mathcal{E}_{so}(y) + \mathcal{E}_{si}(y)}{2} \qquad (7.5.2)$$

where $\mathcal{E}_{so}$ is the electric field at the silicon-oxide interface and $\mathcal{E}_{si}$ is the vertical field at the boundary between the inversion layer and the depletion region. According to Gauss' law at the silicon-oxide interface, we can write:

$$\varepsilon_{si} \, \mathcal{E}_{so}(y) = \varepsilon_{ox} \, \mathcal{E}_{ox}(y)$$
$$\Downarrow$$
$$\varepsilon_{si} \, \mathcal{E}_{so}(y) = C_{ox} \, (V_G - V_{FB} - \Phi_S(y)) = - (Q_{inv}(y) + Q_d(y)) \qquad (7.5.3)$$

$\mathcal{E}_{si}$ can also be obtained using Gauss' law:

$$\varepsilon_{si} \, \mathcal{E}_{si}(y) = - Q_d(y) = \gamma \, C_{ox} \sqrt{\Phi_S(y)} \qquad (7.5.4)$$

Therefore from 7.5.2 we have:

$$\mathcal{E}_x(y) = \frac{C_{ox}}{2\varepsilon_{si}} \, (V_G - V_{FB} - \Phi_S(y) + \gamma \sqrt{\Phi_S(y)} \,) \qquad (7.5.5)$$

In order to calculate a simplified average effective mobility the drain current must satisfy the following condition:

I_D *calculated with* μ_n *depending on* $\mathcal{E}_x(y) = I_D$ *calculated with a constant* μ_{eff}

where:

$$\mu_n(y) = \frac{\mu_{no}}{1 + \Theta \dfrac{C_{ox}}{2\varepsilon_{si}} \, (V_G - V_{FB} - \Phi_S(y) + \gamma \sqrt{\Phi_S(y)} \,)} \qquad (7.5.6)$$

Let us consider a small element of length along the channel, dy. We can write the right-side of the condition as:

$$I_D \, dy = W \, \mu_{eff} \, (-Q_{inv}(y)) \, dV(y) \qquad (7.5.7)$$

and the left-side as:

$$I_D \, dy = \frac{W \, \mu_{no} \, (-Q_{inv}(y)) \, dV(y)}{1 + \Theta \dfrac{C_{ox}}{2\varepsilon_{si}} \, (V_G - V_{FB} - \Phi_S(y) + \gamma \sqrt{\Phi_S(y)} \,)} \qquad (7.5.8)$$

or upon rearranging:

$$I_D \left[1 + \Theta \frac{C_{ox}}{2\varepsilon_{si}} (V_G - V_{FB} - \Phi_{S(y)} + \gamma \sqrt{\Phi_{S(y)}}) \right] dy = W \mu_{no} (-Q_{inv}(y)) \, dV(y)$$

$$(7.5.9)$$

Substituting 7.5.9 into 7.5.7 and integrating from source to drain we obtain:

$$\int_0^L \left\{ 1 + \Theta \frac{C_{ox}}{2\varepsilon_{si}} \{ V_G - V_{FB} - \Phi_{S(y)} + \gamma \sqrt{\Phi_{S(y)}} \} \right\} dy = L \frac{\mu_{no}}{\mu_{eff}} \qquad (7.5.10)$$

Noting that $\Phi_S = 2\Phi_F + V(y)$ and that $\frac{dV}{dy} \cong \frac{V_D - V_S}{L}$ (linear approximation), we can write:

$$\int_0^L \left[V(y) - \gamma \sqrt{2\Phi_F + V(y)} \right] dy = \frac{L}{V_D - V_S} \int_{V_S}^{V_D} \left[V(y) - \gamma \sqrt{2\Phi_F + V(y)} \right] dV$$

$$= \frac{L}{V_D - V_S} \left[\frac{V_D^2}{2} - \frac{V_S^2}{2} - \frac{2}{3} \gamma \{ (2\Phi_F + V_D)^{3/2} - (2\Phi_F + V_S)^{3/2} \} \right] \quad (7.5.11)$$

Introducing this result in 7.5.10 we obtain:

$$\frac{\mu_{no}}{\mu_{eff}} = 1 + \frac{\Theta \, C_{ox}}{2\varepsilon_{si}} \left[V_G - (V_{FB} + 2\Phi_F) - \frac{V_S + V_D}{2} + \frac{2}{3} \gamma \frac{(2\Phi_F + V_D)^{3/2} - (2\Phi_F + V_S)^{3/2}}{V_{DS}} \right]$$

which can be rewritten:

$$\frac{\mu_{no}}{\mu_{eff}} = 1 + \frac{\Theta \, C_{ox}}{2\varepsilon_{si}} \left[V_G - V_{THo} + \gamma \sqrt{2\Phi_F} - \frac{V_S + V_D}{2} + \frac{2}{3} \gamma \frac{(2\Phi_F + V_D)^{3/2} - (2\Phi_F + V_S)^{3/2}}{V_{DS}} \right]$$

$$(7.5.12)$$

The latter expression yields μ_{eff} as a function of μ_{no} and the bias applied to the different terminals of the device. Equation 7.5.12 can be simplified if V_{DS} is small, in which case:

$$\frac{(2\Phi_F + V_D)^{3/2} - (2\Phi_F + V_S)^{3/2}}{V_{DS}} \cong \frac{3}{2} \sqrt{2\Phi_F + V_S}$$

Therefore, we obtain:

$$\mu_{eff} = \frac{\mu_{no}}{1 + \dfrac{\Theta \, C_{ox}}{2\varepsilon_{si}} \left(V_G - V_{THo} - V_S + \gamma \sqrt{2\Phi_F} + \gamma \sqrt{2\Phi_F + V_S} \right)} \qquad (7.5.13)$$

or $\quad \mu_{eff} = \dfrac{\mu_{no}}{1 + \dfrac{\Theta \, C_{ox}}{2\varepsilon_{si}} \left(V_G - V_{TH} - V_S + 2\gamma \sqrt{2\Phi_F + V_S} \right)} \qquad (7.5.14)$

Neglecting the influence of the depletion charge near the source, a series development of 7.5.14 yields a simpler relationship between μ_{eff} and μ_{no}, which is widely used in practice:[16]

$$\mu_{eff} = \frac{\mu_{no}}{1 + \frac{\Theta C_{ox}}{2\varepsilon_{si}} (V_G - V_{TH} - V_S)} \qquad (7.5.15)$$

Commonly $\theta = \dfrac{\Theta C_{ox}}{2\varepsilon_{si}}$ is used and the previous expression can be written as:

$$\mu_{eff} = \frac{\mu_{no}}{1 + \theta (V_G - V_{TH} - V_S)} \qquad (7.5.16)$$

The reduction of surface mobility increases as the gate voltage is increased, as illustrated by Figure 7.18.

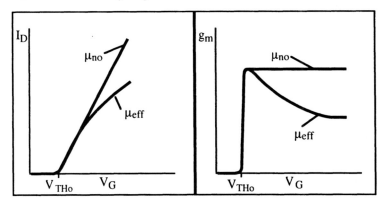

Figure 7.18: I_D and g_m with a constant mobility, μ_{no}, and with a gate-voltage dependent effective mobility, μ_{eff}. [17]

If the simplified model represented by Equations 7.4.31, 7.4.32 and 7.4.32 is used, then $V_G - V_{TH} - V_S$ in Expression 7.5.16 can be replaced by $(V_G-V_{TH})/n$, which yields:[18]

$$\mu_{eff} = \frac{\mu_{no}}{1 + \frac{\theta}{n} (V_G - V_{TH})} \qquad (7.5.17)$$

7.6. Carrier velocity saturation

All the expressions derived hitherto are based on the assumption of a linear dependence of the drift current on the lateral electric field:

$$q \mu_n n \, \mathcal{E}_y = - q v n \qquad (7.6.1)$$

where v is the carrier velocity in the inversion layer. In reality this linear dependence is observed for low electric field values only. At higher fields above a critical value, $\mathcal{E}_c$, the velocity of the carriers saturates. For electrons in silicon, the maximum velocity, v_{max}, is 10^7 cm/s (Figure 7.19). We will now assess the impact of this velocity saturation effect on the expression of the drain current of a MOSFET.

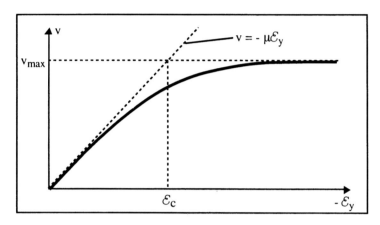

Figure 7.19: Carrier velocity vs. electric field.[19]

It is easy to show that the lateral electric field near the drain junction reaches high values when the drain voltage is equal or larger than the drain saturation voltage, V_{Dsat}. Consider the channel as a resistor connecting source to drain. The drain current is given by Expression 7.4.6:

$$I_D = - W \, Q_{inv}(y) \, \mu_n \frac{dV(y)}{dy} = W \, Q_{inv}(y) \, \mu_n \, \mathcal{E}_y$$

Since the current is constant along the channel, it is easy to observe that $\mathcal{E}_y \to \infty$ when $Q_{inv}(y) \to 0$, which is the case when V_D is equal to or greater than V_{Dsat}. Actually the lateral electric field does not become infinite, but reaches high values especially if the gate length is small.

The carrier velocity can be expressed as follows:

$$v = v_{max} \frac{\mathcal{E}_y/\mathcal{E}_c}{\mathcal{E}_y/\mathcal{E}_c - 1} \tag{7.6.2}$$

where $\mathcal{E}_c$ is the critical field defined by: $\mathcal{E}_c = v_{max}/\mu_n$ ($\mathcal{E}_c$ is taken positive while $\mathcal{E}_y < 0$ since $\mathcal{E}_y = -dV(y)/dy$ and $V(y)$ increases with y, which yields:

$$v(y) = v_{max} \frac{-\dfrac{1}{\mathcal{E}_c}\dfrac{dV(y)}{dy}}{-\dfrac{1}{\mathcal{E}_c}\dfrac{dV(y)}{dy} - 1} = \mu_n \frac{\dfrac{dV(y)}{dy}}{1 + \dfrac{1}{\mathcal{E}_c}\dfrac{dV(y)}{dy}} \tag{7.6.3}$$

The expression of the current corrected for the velocity saturation becomes:

$$I_D \left(1 + \frac{1}{\mathcal{E}_c} \frac{dV(y)}{dy} \right) = - W \mu_n Q_{inv}(y) \frac{dV(y)}{dy} \qquad (7.6.4)$$

which is equivalent to replacing μ_n by $\dfrac{\mu_n}{1 + \dfrac{1}{\mathcal{E}_c} \dfrac{dV(y)}{dy}}$ to take velocity

saturation effects into account. Integrating from source to drain we obtain (see 7.4.29 - 7.4.31):

$$I_D \left(L + \frac{V_{DS}}{\mathcal{E}_c} \right) = - \mu_n W \int_{V_S}^{V_D} Q_{inv}(y)\, dV(y) \qquad (7.6.5)$$

or:

$$I_D = - \frac{W}{L} \frac{\mu_n}{1 + \dfrac{V_{DS}}{L\mathcal{E}_c}} \int_{V_S}^{V_D} Q_{inv}(y)\, dV(y) \qquad (7.6.6)$$

which finally yields:

$$I_D = \mu_n C_{ox} \frac{W}{L(1 + \dfrac{V_{DS}}{L\mathcal{E}_c})} \left\{ (V_G - V_{THS})V_{DS} - \frac{1}{2} n V_{DS}^2 \right\} \qquad (7.6.7)$$

By imposing $dI_D/dV_{DS}=0$, the drain saturation voltage can be found. It is equal to:

$$V'_{Dsat} = L\mathcal{E}_c \left[\sqrt{1 + \frac{2(V_G - V_{THS})}{nL\mathcal{E}_c}} - 1 \right] \qquad (7.6.8)$$

Therefore, when velocity saturation is taken into account it is equivalent to making the channel longer (L is multiplied by $1 + \dfrac{V_{DS}}{L\mathcal{E}_c}$), and therefore, to reducing the drain saturation voltage and drain saturation current.

7.7. Subthreshold current - Subthreshold slope

We have so far assumed that the drain current is equal to zero when the gate voltage is smaller than the threshold voltage. There can actually be a significant amount of electrons near the semiconductor surface when the device operates below strong inversion. A brief look at Figure 7.6 reminds us that the electron surface concentration is larger than the hole surface concentration when $\Phi_F < \Phi_S < 2\Phi_F$. The actual dependence of the electron concentration at the surface is an exponential function of the surface potential.

It is experimentally observed that the drain current below threshold, called "subthreshold current", is independent of the drain voltage as long as V_{DS} is larger than a few kT/q. This suggests that the subthreshold current is caused by diffusion rather than by a drift mechanism. Based on this observation the electron current density from source to drain can be written:

$$J_{ny} = -q\, D_n \frac{dn}{dy} \;\;\Rightarrow\;\; I_D = q\, A\, D_n \frac{n(0) - n(L)}{L} \qquad (7.7.1)$$

where A is the cross-sectional area of a vertical section of the channel region through which the electrons flow, D_n is the diffusion coefficient for electrons, and $n(0)$ and $n(L)$ are the electron concentrations at the edge of the source and drain junction, respectively. The latter can be expressed as follows:

$$n(0) = n_{po}\, exp\left(\frac{q\Phi_S}{kT}\right) \; and \; n(L) = n_{po}\, exp\left(\frac{q(\Phi_S - V_D)}{kT}\right) \qquad (7.7.2)$$

where the source is considered at ground and $n_{po} = n_i^2/N_a$.

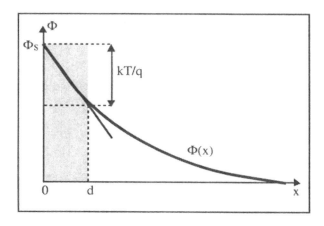

Figure 7.20: Calculation of the "channel depth", d. [20]

What is "the area of the vertical section of the channel region through which the electrons flow"? We know that the electron concentration varies as $exp(q\Phi(x)/kT)$. To simplify calculations we will approximate the exponential electron profile by a constant electron density extending to a depth d below the surface (gray area in Figure 7.20). The depth d is defined as the depth at which the potential has decreased by kT/q below the surface potential value. Therefore, one can write:

$$d = \frac{kT/q}{\mathcal{E}_s} \quad where \; \mathcal{E}_s = - \left. \frac{d\Phi(x)}{dx} \right|_{x=0} \tag{7.7.3}$$

The area of the section through which the electrons flow is thus equal to $A = W{\times}d$, where W is the transistor width. Using Equations 7.7.1-3 and Einstein's relationship $D_n = \frac{kT}{q} \mu_n$ we obtain:

Subthreshold Current

$$I_D = \mu_n \frac{W}{L} q \left(\frac{kT}{q} \right)^2 \frac{n_i^2}{N_a} \; [1 - exp\,(-qV_D/kT)] \; \frac{exp\,(q\Phi_S/kT)}{-\frac{d\Phi_S}{dx}} \tag{7.7.4}$$

where the electric field at the surface can be found using 7.2.24, 7.2.26 and 7.2.30:

$$- \left. \frac{d\Phi(x)}{dx} \right|_{x=0} = \mathcal{E}_s = \sqrt{\frac{2qN_a\Phi_S}{\varepsilon_{si}}} = \frac{qN_a}{C_D} \tag{7.7.5}$$

Relationship 7.7.4 shows that the subthreshold current is independent of the drain voltage, as long as V_{DS} is larger than a few kT/q. It also shows that the drain current is proportional to the electron concentration at the surface. Therefore, the subthreshold current increases exponentially with surface potential (Figure 7.21).

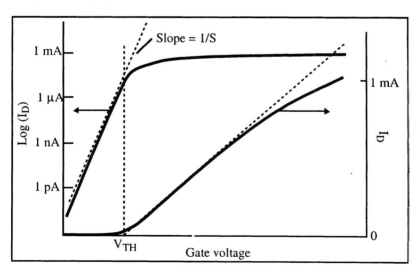

Figure 7.21: Drain current in both linear and logarithmic scales (V_{DS} is small). [21]

On a log plot such as Figure 7.21 the subthreshold current appears as a straight line. The inverse of the slope of that line is called "inverse subthreshold slope", "subthreshold swing", or more simply, "subthreshold slope". It is expressed in millivolts per decade, which means: "How many millivolts should the gate voltage be increased to increase the drain current by a factor 10". The lower the value of the subthreshold slope, S, the more efficient and rapid the switching of the device from the off state to the on state.

By definition the subthreshold slope is given by:

$$S = \frac{dV_G}{d\,log(I_D)} \qquad (7.7.6)$$

or, if we change the logarithm base to the natural logarithm base:

$$S = \frac{ln(10)}{\dfrac{d(ln\,(I_D))}{dV_G}} \qquad (7.7.8)$$

Since
$$\frac{d(ln(I_D))}{dV_G} = \frac{1}{I_D}\frac{d(ln(I_D))}{d\Phi_S}\frac{d\Phi_S}{dV_G} \qquad (7.7.9)$$

The log of the subthreshold current can be differentiated:

$$\frac{d(ln(I_D))}{dV_G} = \left[\frac{q}{kT} - \frac{\dfrac{d}{d\Phi_S}\left(-\dfrac{d\Phi_S}{dx}\right)}{-\dfrac{d\Phi_S}{dx}}\right]\frac{d\Phi_S}{dV_G} \qquad (7.7.10)$$

Using Equation 7.7.5 we can write:

$$\mathcal{E}_s = -\frac{d\Phi_S}{dx} = \frac{qN_a}{C_D} \qquad (7.7.11)$$

Note that:

$$\frac{d}{d\Phi_S}\left(-\frac{d\Phi_S}{dx}\right) = \frac{C_D}{\varepsilon_{si}} \qquad (7.7.12)$$

where C_D is the depletion capacitance, defined by $C_D = -\dfrac{dQ_d}{d\Phi_S} = \dfrac{\varepsilon_{si}}{x_d}$.

We also have:

$$\frac{\dfrac{d}{d\Phi_S}\left(-\dfrac{d\Phi_S}{dx}\right)}{-\dfrac{d\Phi_S}{dx}} = \frac{C_D^2}{qN_a\varepsilon_{si}} = \frac{1}{2\Phi_S} \qquad (7.7.13)$$

Since in weak inversion $\Phi_F < \Phi_S < 2\Phi_F$, $\dfrac{1}{2\Phi_S}$ is small compared to $\dfrac{q}{kT}$ and can be neglected in Equation 7.7.10. We thus obtain:

$$\frac{d(ln(I_D))}{dV_G} \cong \frac{q}{kT}\frac{d\Phi_S}{dV_G} \qquad (7.7.14)$$

An expression for $\frac{d\Phi_S}{dV_G}$ between gate voltage and surface potential is obtained by adding the flatband voltage to Equation 7.3.1 :

$$V_G = V_{FB} + \Phi_S - \frac{Q_d}{C_{ox}} \qquad (7.7.15)$$

from which we derive:

$$\frac{d\Phi_S}{dV_G} = \left(1 + \frac{C_D}{C_{ox}}\right)^{-1} \qquad (7.7.16)$$

Finally an expression for the subthreshold slope can be written as:

$$S = \frac{kT}{q} \ln(10) \left(1 + \frac{C_D}{C_{ox}}\right) = n \frac{kT}{q} \ln(10) \qquad (7.7.17)$$

where n is the body factor (Equation 7.4.41). The closer n is to unity, the sharper the transition between the transistor's off and on states.

Since $\frac{d(\ln(I_D))}{dV_G} = \frac{q}{kT} \left(1 + \frac{C_D}{C_{ox}}\right)^{-1}$ the subthreshold current varies exponentially as a function of gate voltage:

$$I_D \propto \exp\left(\frac{qV_G}{nkT}\right) \quad \text{where} \quad n = 1 + \frac{C_D}{C_{ox}} \qquad (7.7.18)$$

Influence of interface states

As mentioned in Expression 7.3.6 there are interface states, or interface traps in the silicon energy bandgap at the silicon-oxide interface. The density of these states is noted N_{it} (cm^{-2} V^{-1}). The charge trapped in those states depends on the value of the surface potential according to the relationship $Q_{it} = -qN_{it}\Phi_S$. A capacitance can be associated with these traps, which is simply given by: $C_{it} = -dQ_{it}/d\Phi_S = qN_{it}$. When the influence of the interface states is taken into account in the relationship between gate voltage and surface potential the following equations are obtained (see Equation 7.3.8):

$$V_G = \Phi_{ms} - \frac{Q_{ox}}{C_{ox}} + \frac{qN_{it}\Phi_S}{C_{ox}} + \Phi_S - \frac{Q_d}{C_{ox}}$$

$$= \Phi_{ms} - \frac{Q_{ox}}{C_{ox}} + \left(1 + \frac{qN_{it}}{C_{ox}}\right) \Phi_S - \frac{Q_d}{C_{ox}} \qquad (7.7.19)$$

and

$$\frac{dV_G}{d\Phi_S} = 1 + \frac{C_D + C_{it}}{C_{ox}} \qquad (7.7.20)$$

which yields the following expression for the subthreshold slope :

Subthreshold Slope

$$S = \frac{kT}{q} \, ln(10) \left(1 + \frac{C_D + C_{it}}{C_{ox}} \right)$$

$$= 60 \left(1 + \frac{C_D + C_{it}}{C_{ox}} \right) \text{mV/decade at room temperature (300K)} \quad (7.7.21)$$

7.8. Continuous model

The models developed in Sections 7.4 (for $V_G > V_{TH}$) and 7.7 (for $V_G < V_{TH}$) are based on the actual physics of semiconductors. Unfortunately, they do not connect well around $V_G = V_{TH}$ and a discontinuity in the equations appears when the gate voltage is close to the threshold voltage, as shown on Figure 7.22. This constitutes a problem for the design of analog MOS circuits, where gates are often biased with a voltage close to V_{TH}, and computing convergence problems arise when the previously mentioned models are used.

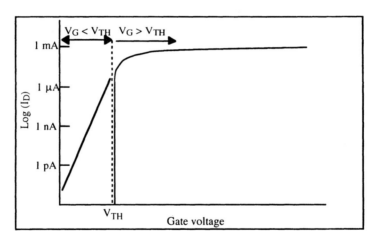

Figure 7.22: Discontinuity of the $I_D(V_G)$ characteristics at $V_G \cong V_{TH}$ using models developed in Sections 7.4 and 7.7.

A model which is valid both below and above threshold can, however, be derived. Such a model can be conveniently used in circuit simulators. Using Equation 7.4.29 we can write:

$$I_D = \mu_n \, C_{ox} \, \frac{W}{L} \int_{V_S}^{V_D} \frac{-Q_{inv}(y)}{C_{ox}} \, dV(y) \qquad (7.8.1)$$

which can be rewritten:

$$I_D = \mu_n\, C_{ox}\, \frac{W}{L} \left(\int_{V_S}^{V_P} \frac{-Q_{inv}(y)}{C_{ox}}\, dV(y) - \int_{V_D}^{V_P} \frac{-Q_{inv}(y)}{C_{ox}}\, dV(y) \right) = I_F - I_R \qquad (7.8.2)$$

where I_F and I_R are called "forward" and "reverse" currents. V_p is defined as:
$$V_P = \frac{(V_G - V_{TH})}{n}$$

Note that V_p is equal to the drain saturation voltage when the source and substrate are grounded and $V_G > V_{TH}$ (see Equation 7.4.32).

It is possible to find a mathematical function which describes the evolution of the current as a function of gate and drain voltage for all regimes of operation (depletion, weak and strong inversion). Following the work of Enz, Krummenacher and Vittoz (the so-called "EKV model"), one can rewrite Equation 7.8.2 as follows: [22,23]

EKV Model

$$I_D = I_F - I_R = 2n\, \mu_n\, C_{ox}\, \frac{W}{L} \left(\frac{kT}{q}\right)^2$$

$$\times \left[\left\{ ln \left[1 + exp\left(\frac{V_P - V_S}{2kT/q}\right) \right] \right\}^2 - \left\{ ln \left[1 + exp\left(\frac{V_P - V_D}{2kT/q}\right) \right] \right\}^2 \right] \qquad (7.8.3)$$

This expression is continuous for any bias applied to the transistor terminals, and so are its derivatives. Using this model and introducing the dependence of mobility on gate voltage given in Expression 7.5.16, a complete set of curves for the MOSFET can readily be obtained. Figure 7.23 shows such a set of curves.

Equation 7.8.3 looks very different from the current equations derived earlier. This is because it includes all the possible operation modes of the transistor. Depending on the applied bias, some terms in 7.8.3 become negligible with respect to others and the equation is reduced to expressions we have derived earlier. To illustrate this, let us consider the current for $V_S=0$, $V_D<V_P$ and $V_G>V_{TH}$ (i.e. the transistor is operating in the non-saturated regime). In that case the exponential terms are much larger than unity, and one can write:

$$I_D = 2n\, \mu_n\, C_{ox}\, \frac{W}{L} \left(\frac{kT}{q}\right)^2 \left[\left(\frac{V_P}{2kT/q}\right)^2 - \left(\frac{V_P - V_D}{2kT/q}\right)^2 \right]$$

$$= \frac{1}{2} n\, \mu_n\, C_{ox}\, \frac{W}{L} \left[2V_D V_P - V_D^2 \right]$$

$$= \frac{1}{2} n\, \mu_n\, C_{ox}\, \frac{W}{L} \left[2\frac{(V_G - V_{TH})\, V_D}{n} - V_D^2 \right]$$

$$= \mu_n C_{ox} \frac{W}{L}\left[\left(V_G - V_{TH} \right) V_D - \frac{1}{2} n\, V_D^2 \right]$$

which is identical to Equation 7.4.31 when $V_S = 0$.

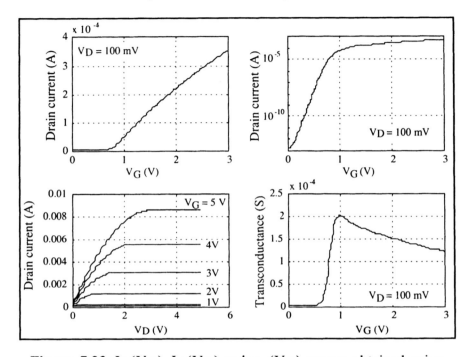

Figure 7.23: $I_D(V_G)$, $I_D(V_D)$ and $g_m(V_D)$ curves obtained using the EKV model; W=10 μm, L=1 μm, t_{ox}=10 nm, μ_0=600 cm²/Vs, θ=0.2 V⁻¹, V_{TH}=0.7 V, and n=1.4.

7.9. Channel length modulation

We have previously assumed that when $V_D > V_{Dsat}$ the drain current of a MOSFET is constant and equal to I_{Dsat} (Figure 7.16). This is because the magnitude of the current is fixed by the potential drop across the channel, which is equal to $V_{Dsat} - V_S$. Actually, when the drain voltage is increased beyond V_{Dsat} the depletion region and the local threshold voltage near the drain are increased. As a result the effective length of the channel shrinks and becomes equal to $L - \Delta L$ (Figure 7.24). This reduction in effective gate length increases the drain current, as will be demonstrated next.

To simplify calculations we will linearize the current variation as a function of drain voltage. If we define $I_{Dsat} \equiv I_D(V_D = V_{Dsat})$, the saturation current for $V_D > V_{Dsat}$, which we will call I'_{Dsat}, can be written as follows:

$$I'_{Dsat} = I_{Dsat}\left(1 + \frac{V_D - V_{Dsat}}{V_A + V_{Dsat}}\right) \qquad (7.9.1)$$

where V_A is a positive voltage value that can be obtained through direct measurement of the device output characteristics, as shown in Figure 7.25. V_A is often called the "Early voltage".

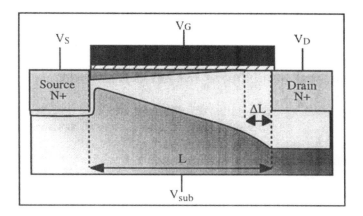

Figure 7.24: Channel length modulation.

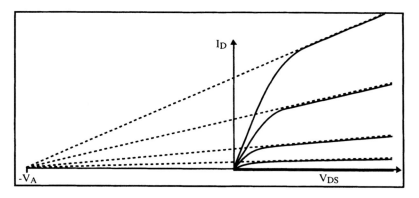

Figure 7.25: Influence of channel length modulation on the output characteristics of a MOSFET.[24]

Expressing I'_{Dsat} in terms of channel length modulation one finds:

$$\frac{I'_{Dsat}}{L} = \frac{I_{Dsat}}{L - \Delta L} = \frac{I_{Dsat}\left(1 + \dfrac{V_D - V_{Dsat}}{V_A + V_{Dsat}}\right)}{L} \qquad (7.9.2)$$

which yields:

$$\Delta L = L\left(1 - \cfrac{1}{1 + \cfrac{V_D - V_{Dsat}}{V_A + V_{Dsat}}}\right) \qquad (7.9.3)$$

The saturation output conductance, which was hitherto considered equal to zero, is now given by:

$$g_{Dsat} \equiv \frac{dI'_{Dsat}}{dV_D} = \frac{I_{Dsat}}{V_A + V_{Dsat}} \cong \frac{I_{Dsat}}{V_A} \qquad (7.9.4)$$

For $V_D > V_{Dsat}$ all lines tangent to the output characteristics in saturation intercept the V_{DS}-axis at $-V_A$, as shown in Figure 7.25.

7.10. Numerical modeling of the MOS transistor

The electrical characteristics of electron devices can be numerically simulated on a computer using finite-element techniques. These simulations are based on the discretization of the device into a series of nodes connected together by mesh elements. Figure 7.26 shows the cross section of a MOS transistor, and Figure 7.27 represents the mesh generated by a computer code which will be used for simulating the device.

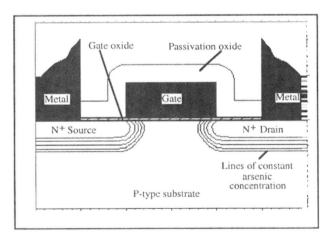

Figure 7.26: Cross section of the MOS transistor used in the simulation. The length of the polysilicon gate is 1 μm, and the effective channel length, L_{eff}, given by the distance between the source and drain junctions, is equal to 0.6 μm. [25]

Figure 7.26 was generated by a process simulator software code which emulated the device fabrication steps. The output file contains the topology of the device, the different materials used for fabrication, the doping type and concentration at every simulation node, as well as the distribution of charges in the oxide, etc. The doping concentration profile along a vertical cut passing through the drain junction is shown in Figure 7.28 as an example of the information contained in the file.

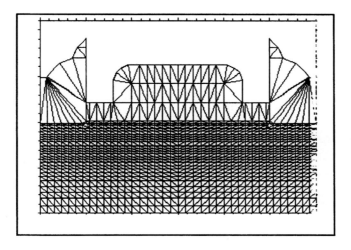

Figure 7.27: Mesh generated by the simulator.

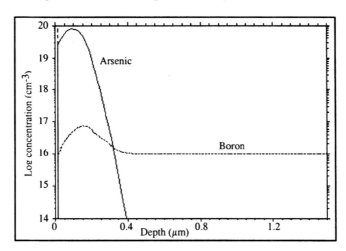

Figure 7.28: Impurity profile along a vertical line passing through the drain and into the substrate.

Once the device structure has been defined, another simulation code is used to solve the transport equations (Poisson, continuity and drift-diffusion) at each node in the semiconductor and the adjacent materials.

The drift-diffusion equations for holes and electrons are given by Expressions 2.6.1a and 2.6.1b:

$$J_p = q\,\mu_p\,p\,\mathcal{E} - qD_p\,grad\,p \quad \text{and} \quad J_n = q\,\mu_n\,n\,\mathcal{E} + qD_n\,grad\,n \quad (7.10.1)$$

The Poisson equation is given by Relationship 2.6.2:

$$\nabla^2 \Phi(x,y,z) = = -\frac{q}{\varepsilon_s}(p - n + N_d^+ - N_a^-) \quad (7.10.2)$$

The continuity equations are given by Expressions 2.6.7a and 2.6.7b:

$$\frac{\partial n}{\partial t} = \frac{1}{q}\,div\,J_n + (G_n - U_n) \quad (7.10.3a)$$

$$\frac{\partial p}{\partial t} = -\frac{1}{q}\,div\,J_p + (G_p - U_p) \quad (7.10.3b)$$

where the SRH recombination obeys Relationship 3.5.12:

$$U = \frac{pn - n_i^2}{\tau_{po}\left(n + n_i\,exp\left[\dfrac{E_t - E_i}{kT}\right]\right) + \tau_{no}\left(p + n_i\,exp\left[\dfrac{E_i - E_t}{kT}\right]\right)} \quad (7.10.4)$$

In addition to these basic semiconductor equations a whole series of effects can be introduced at will in the simulation. This allows one to refine the simulation and to better reproduce the behavior of the actual device. This is exemplified in Figures 7.29 and 7.30 where the current in a MOSFET has been simulated as a function of gate voltage, for V_{DS}=100 mV. In one curve the electron surface mobility is constant, and in the other curve a field-dependent model similar to that described by Equation 7.5.16 is used. The dramatic difference between the two curves can be seen immediately.

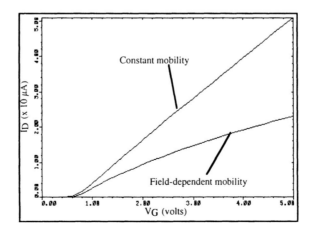

Figure 7.29: Drain current vs. gate voltage (V_{DS}=100 mV, L_{eff}=0.6 μm, W=1 μm).[26]

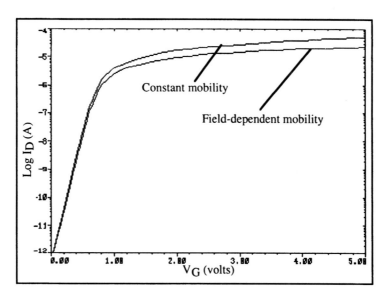

Figure 7.30: Drain current (log scale) vs. gate voltage (V_{DS}=100 mV, L_{eff}=0.6 μm, W=1 μm).

7.11. Short-channel effect

The evolution of semiconductor processing technology calls for constant reduction of device dimensions, especially gate length. The gate length of MOSFETs used in 256k DRAMs in 1984 was approximately 1.2 μm. Ten years later 64M DRAMs were routinely produced using 0.4 μm gates. Gate length is predicted to be reduced down to 100, 50 and 35 nm in years 2003, 2009 and 2012, respectively. Such an aggressive scaling trend results in the appearance of several undesirable effects. One of these is the so-called "short-channel effect" and will described next.

The threshold voltage of a MOSFET is given by Expression 7.3.8:

$$V_{THo} = V_{FB} + 2\Phi_F - \frac{Q_d}{C_{ox}}$$

The depletion charge, Q_d, used in the latter expression, can be represented by the trapezoid area shown in Figure 7.31, where the drain voltage is equal to zero. The trapezoid shape is due to the encroachment of the depletion regions from the source and drain reversed-bias junctions into the depletion zone created by the gate electrode.

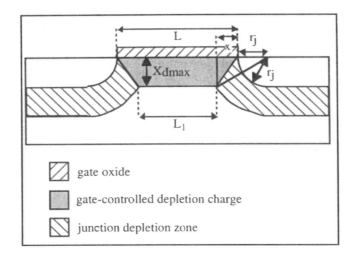

Figure 7.31: Parameters involved in the short-channel effect.[27]

Referring to Figure 7.31 the area of the trapezoid is equal to $(L+L_1)X_{dmax}/2$. If the channel is long, then $L \approx L_1$ and the area of the trapezoid is virtually equal to $L \times X_{dmax}$, which represents the area of a rectangle of width L and height X_{dmax}. In that case the device is referred to as a "long-channel transistor" and its threshold voltage is accurately described by the equations derived in Section 7.3. When the gate is short, on the other hand, $L_1 < L$, then the depletion charge due V_G under the gate electrode is reduced.

Consider a MOSFET at threshold ($V_G = V_{TH}$). The drain voltage is small ($V_{DS} \cong 0$). Based on geometrical considerations (Pythagorean theorem) and noting that the built-in potential of the source and drain junctions relative to the substrate, Φ_o, is approximately equal to the surface potential in the channel, $2\Phi_F$, such that the width of the depletion region around the source and drain is equal to X_{dmax} the following relationship is obtained:

$$X^2_{dmax} + (x + r_j)^2 = (r_j + X_{dmax})^2 \qquad (7.11.1)$$

where r_j is the junction radius of curvature, which is equal to the source and drain junction depth. The latter expression can be simplified, which yields:

$$x^2 + 2r_j x - 2\, r_j\, X_{dmax} = 0 \qquad (7.11.2)$$

From which the value of x can be extracted:

$$x = -r_j \pm \sqrt{r_j^2 + 2\, r_j X_{dmax}} \;=\; r_j\left(-1 \pm \sqrt{1 + \frac{2\, X_{dmax}}{r_j}}\,\right) \qquad (7.11.3)$$

Since x must have a positive value one obtains:

$$x = r_j \left(\sqrt{1 + \frac{2 X_{dmax}}{r_j}} - 1 \right) \tag{7.11.4}$$

Using this result L_1 can be calculated:

$$L_1 = L - 2x = L - 2r_j \left(\sqrt{1 + \frac{2 X_{dmax}}{r_j}} - 1 \right) \tag{7.11.5}$$

The depletion charge controlled by the gate voltage is then equal to:

$$\frac{Q_d}{Q_{do}} = \frac{q X_{dmax} N_a \left(\frac{L + L_1}{2} \right)}{q X_{dmax} N_a L} = \frac{1}{2} \left(1 + \frac{L_1}{L} \right)$$

$$= \frac{1}{2} \left(1 + 1 - \frac{2 r_j}{L} \left(\sqrt{1 + \frac{2 X_{dmax}}{r_j}} - 1 \right) \right) = 1 - \frac{r_j}{L} \left(\sqrt{1 + \frac{2 X_{dmax}}{r_j}} - 1 \right) \tag{7.11.6}$$

where Q_{do} is the depletion charge that would be found underneath the gate if the depletion region was rectangular instead of trapezoidal, as in a long-channel device. Using Equation 7.3.8 the threshold voltage can now be expressed as a function of gate length:

$$V_{THo} = V_{FB} + 2\Phi_F - \frac{Q_d}{C_{ox}}$$

$$= V_{FB} + 2\Phi_F + \frac{q X_{dmax} N_a}{C_{ox}} \left(1 - \frac{r_j}{L} \left(\sqrt{1 + \frac{2 X_{dmax}}{r_j}} - 1 \right) \right) \tag{7.11.7}$$

The short-channel effect is illustrated in Figure 7.32. The problem associated with the short-channel effect is not that devices with different channel lengths have different threshold voltages, since circuit designers typically use only one channel length (the minimum length allowed by processing parameters). Rather, the problem is that in short devices small statistical variations in gate length give rise to large statistical variations of threshold voltage, which poses a clear reproducibility problem in integrated circuit manufacturing. The short-channel effect, however, can be reduced by using shallower junctions and higher substrate doping concentrations, which reduces the extension of the source and drain depletion regions in the channel.

This model is valid as long as $L_1 > 0$. If the gate length is small and the drain voltage is high enough, the source and drain depletion regions can touch one another. In such a case the potential in the channel region is no longer controlled by the gate and a large, undesirable current flows between source and drain. This phenomenon is called "punchthrough".

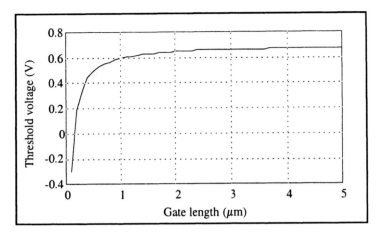

Figure 7.32: Short-channel effect in MOSFETs with t_{ox} = 25 nm, V_{FB} = -1 V, N_a = 6×10^{16} cm^{-3} and r_j = 300 nm.

7.12. Hot-carrier degradation

7.12.1. Scaling rules

The constant reduction of transistor dimensions has given rise to reliability issues not seen in long-channel devices. Although smaller dimensions were achieved in every new generation of devices a constant supply voltage (5V) was used for many years. This has led to increasingly intense electric fields inside the MOSFETs, causing device degradation problems.

Let us define a dimensionless scaling factor, λ, which is characteristic of the reduction of device dimensions from generation to generation. Taking $\lambda > 1$ and dividing the device dimensions by λ results in scaling, as illustrated by Figure 7.33. Thus if the gate length is divided by λ, the gate width, W, the gate oxide thickness, t_{ox}, the junction depth, r_j, and the width of the depletion layer, x_{dmax}, must all be divided by λ.

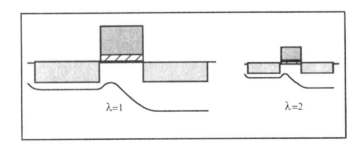

Figure 7.33: Scaling factor, λ.

In order to keep the electric field inside the device from increasing greatly, the supply voltage, V_{DD}, should be reduced by the same factor λ. This, unfortunately, poses a compatibility problem which would prevent one from using different generations of integrated circuits in the same system. For practical reasons, a supply voltage, V_{DD} of 5 volts was maintained for many years until the problems caused by high electric fields became unacceptable. Power supply voltage was then reduced to 3.3 V and lower. Ultimately the supply voltage of portable systems will be reduced to 0.9 volts or even 0.5 volts.

Table 7.1 shows the scaling factor by which different device parameters have to be multiplied when the MOSFET dimensions are divided by λ. The factor λ^{-1} used for physical dimensions speaks for itself. The same factor is used for the supply voltage, as a constant electric field must be maintained. The factor λ^{-1} for the current is obtained from the relationship $I_{Dsat} = \frac{1}{2}\frac{W}{L}\mu\, C_{ox}\,(V_G-V_{TH})^2$, where $V_G=V_{DD}$. The threshold voltage should be scaled the same way as the supply voltage. The λ factor for the doping concentration is not mathematically rigorous, but it shows that the depletion width must scale down with the device dimensions, while maintaining the threshold voltage constant. The capacitance of the gate electrode is equal to WLC_{ox}. The dissipated power is given by the $I_D V_{DD}$ product and the power density is obtained by dividing $I_D V_{DD}$ by the area of the device. The gate delay is obtained by dividing the capacitance of a gate by I_D and taking into account the reduction of signal dynamics (from 0 to V_{DD}). The delay is then proportional to $V_{DD}C_{ox}/I_D$.

Table 7.1: Constant-field scaling factor for different parameters.[28]

Parameter	Scaling factor
Dimension (t_{ox},r_j,W,L)	λ^{-1}
Voltage (V_{DD})	λ^{-1}
Threshold voltage (V_{TH})	λ^{-1}
Current (I_D)	λ^{-1}
Doping concentration (N_a)	λ
C_{ox}	λ
Capacitance of gate	λ^{-1}
Power density	1
Current density	λ
Power consumption	λ^{-2}
Gate delay	λ^{-1}
Power-delay product	λ^{-3}
Integration density (transistors/cm^2)	λ^2

7.12.2. Hot electrons

When scaling rules are not applied to the supply voltage intense electric fields can develop inside the MOS transistor, especially between the channel pinch-off point and the drain. In an n-channel MOSFET this electric field can accelerate electrons to high speeds such that the temperature which is equivalent to the electron energy, called electron temperature, can reach several thousand degrees centigrade, hence the name "hot electrons". Such electrons can be stopped by collision events, where the energy released can create electron-hole pairs. The created holes give rise to a substrate current. The electrons resulting from this generation mechanism can have enough energy to overcome the gate oxide potential barrier and thus be injected into the gate material, giving rise to a gate current. The evolution of substrate current and gate current with applied voltages is described next.

7.12.3. Substrate current

When the transistor is in saturation the large electric field near the drain substantially accelerates electrons. These electrons can undergo collision events in which energy is released and an electron-hole pairs are created. This generation mechanism is called "impact ionization". The created electrons are attracted by the positive bias at the drain. The generated holes diffuse towards the grounded substrate, giving rise to a substrate current. The magnitude of the substrate current is given by the relationship $I_{sub} = (M-1) I_{ch}$ where I_{ch} is the electron current in the channel, and M is called the "multiplication coefficient" $(M{\geq}1)$. The multiplication coefficient is strongly dependent on the electric field and is highest near the drain, where the electric field is highest. The amplitude of the lateral electric field in a saturated MOSFET from source to drain is shown in Figure 34.

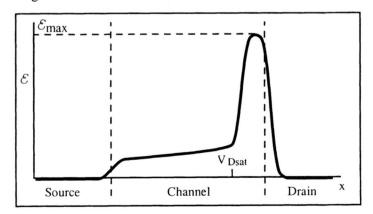

Figure 7.34: Lateral electric field, $\mathcal{E}$, at the surface of a MOSFET in saturation.[29]

The variation of the substrate current vs. gate voltage is shown in Figure 7.35 for a drain voltage is 5V. Below threshold ($V_G < V_{THo} = 0.7V$) the channel the substrate current increases with increasing channel current according to the relationship $I_{sub} = (M-1) I_{ch}$. If the gate voltage is increased beyond threshold voltage ($V_G = 1$ *to* 2 *volts*) an inversion channel is formed and the device operates in the saturation regime. The channel is thus pinched off and impact ionization produces a relatively large substrate current. The drain saturation voltage, V_{Dsat}, is equal to V_G-V_{TH}. Thus when the gate voltage is increased beyond 2 volts V_{Dsat} increases and the electric field near the drain, which is proportional to V_D-V_{Dsat}, decreases. As a result, the multiplication factor, M, is reduced and the substrate current decreases. Therefore, the substrate current reaches a maximum when gate voltage is slightly larger than the threshold voltage, *i.e.* when the current in the channel is sufficiently large to trigger impact ionization, and when the electric field near the drain is the largest.

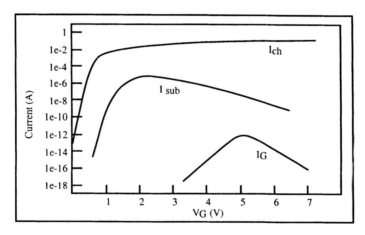

Figure 7.35: Logarithmic plot of channel, substrate and gate current as a function of gate voltage. $V_D = 5$ V.[30]

7.12.4. Gate current

Gate current is due to electrons which, as a result of acceleration by the electric field by collision or impact ionization generation, have acquired enough energy to overcome the potential barrier at the silicon-SiO$_2$ interface. In principle the energy required to overcome this barrier is 3.1 V. However, if the gate is positively biased it attracts electrons and there appears a barrier reduction effect similar to the Schottky effect. The value of the potential barrier is then equal to:

$$\Phi_b = 3.1\ V - C\mathcal{E}_{ox}^{1/2} - D\mathcal{E}_{ox}^{2/3} \quad (V_G > V_D) \qquad (7.12.1)$$

where $\mathcal{E}_{ox}$ is the electric field in the oxide, C is equal to 2.6×10^{-4} (cm V)$^{1/2}$, and D is equal to 10^{-5} (cm^2 V)$^{1/3}$. If the gate is negatively biased with respect to the channel, the potential barrier is increased and its magnitude is given by:

$$\Phi_b = 3.1\ V + \mathcal{E}_{ox}\ t_{ox} \quad (V_G < V_D) \tag{7.12.2}$$

Electrons gain energy from being accelerated by the electric field near the drain. Some electrons, often called "lucky electrons" can gather enough energy after a collision near the drain and be injected into the gate oxide, thereby giving rise to gate current.

The gate current, when plotted as a function of gate voltage, reaches a maximum around $V_G = V_D$ as shown in Figure 7.35. At the left of that maximum, for $V_G < V_D$, an electron near the drain locally sees a negatively biased gate, which increases the oxide potential barrier according to Equation 7.12.2 and thus lowers I_G. The current increases exponentially with V_G as Φ_b decreases with any increase of V_G. When $V_G > V_D$, the electric field in the oxide near the drain changes sign, which reduces Φ_b according to Relationship 7.12.1. However, the transistor is no longer saturated and the lateral electric field which accelerates the electrons decreases as V_G is increased. Thus any increase of V_G above V_D reduces impact ionization and the gate current decreases with increases of V_G.

7.12.5. Degradation mechanism

The flow of gate current through the gate oxide generates interface states at the silicon-SiO$_2$ interface. Electrons can be injected in oxide spacers at the gate sidewalls as well. These interface traps reduce the electron surface mobility and increase the local threshold voltage near the drain (Equation 7.3.6). Over a period of time, these two effects can become significant enough to cause a distinct reduction of the device current drive. With transistors unable to deliver the required current, the circuit may experience timing errors, and ultimately circuit failure may occur.

The degradation of the oxide is caused by the gate current, which is usually much smaller than a picoampere, and is therefore, difficult to measure. Since both gate current and substrate current are caused by similar mechanisms, *i.e.* high electric field near the drain and electron-hole pair generation by impact ionization, it is common practice to measure the substrate current and to assume that if the substrate current is low, the gate current must also be low. Therefore, transistor designs aimed at limiting the gate current to increase the lifetime of the device usually involve efforts to minimize the substrate current. One such design, called the "lightly doped drain" (LDD) structure, features lighter doping concentrations at the drain junction near the edges of the channel. This helps reduce the lateral drain electric field, and thus reduces impact ionization and increases device lifetime. The lightly-doped portions of the source and drain are commonly called "source and drain extensions" (SDE).

7.13. Terminal capacitances

In many circuit simulations it is important to know the capacitances between the different terminals of a MOSFET (source, gate, drain and substrate). The different capacitances include:

C_{Ssub}: source to substrate capacitance
C_{Dsub}: drain to substrate capacitance
C_{Gsub}: gate to substrate capacitance
C_{GS}: gate to source capacitance
C_{GD}: gate to drain capacitance

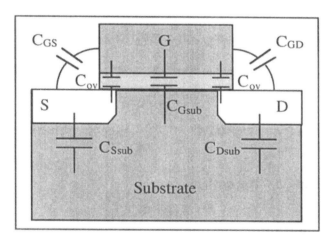

Figure 7.36: Capacitances in the MOSFET.

◊ C_{Ssub} and C_{Dsub} are simple PN-junction transition capacitances and their behavior has been described in the Chapter on the PN junction.

◊ If the gate voltage is positive but lower than the threshold voltage C_{Gsub} is equal to $W \times L \times C$ where C is the capacitance of a MOS capacitor in depletion and is given by Equation 7.2.33. If $V_G > V_{TH}$ the inversion layer acts as an electric shield between the gate and the substrate because the channel is connected to source and drain. As a result, $C_{Gsub} = 0$.

◊ If the gate voltage is positive but lower than the threshold voltage C_{GS} and C_{GD} are both equal to the overlap capacitance, C_{ov}. This capacitance arises from the fact that some of the source (or drain) junction extends somewhat under the gate due to device fabrication (Figure 36). If $V_G > V_{TH}$ and the device is not in saturation ($V_D < V_{Dsat}$) the inversion layer runs from source to drain, and the capacitance between the gate and the inversion channel can be equally divided into two parts: half of that capacitance connects the gate and the source, while the other half connects the gate to the drain. Keeping in mind that the overlap capacitances exist, one obtains: $C_{GS} = C_{GD} = C_{ov} + \frac{1}{2} W \times L \times C_{ox}$. When the device is in saturation the situation becomes more complicated. Since the channel is no longer connected to the drain, there is no influence of the gate to channel capacitance on the gate to drain capacitance, and $C_{GD} = C_{ov}$. Estimating the gate to source capacitance requires some calculation, as we will see next.

The electron charge in the channel is given by Equation 7.4.12. If $V_G >> V_{TH}$ the depletion charge can be neglected and we can write:

$$Q_{inv}(y) = -C_{ox} (V_G - V_T - V(y)) \quad \text{(unit: C cm}^{-2}) \quad (7.13.1)$$

The electron total charge in the channel is obtained by integrating Expression 7.13.1:

$$Q_{inv} = - W \int_0^L C_{ox} (V_G - V_T - V(y)) \, dy = - W \int_0^{V_D} C_{ox} (V_G - V_T - V(y)) \frac{dy}{dV(y)} \, dV(y)$$

$$\text{(unit: C)} \quad (7.13.2)$$

Using Equation 7.4.6 can write:

$$I_D = - W Q_{inv}(y) \, \mu_n \frac{dV(y)}{dy} \quad (7.13.3)$$

And thus, using Equation 7.13.1 we have:

$$I_D = - W Q_{inv}(y) \, \mu_n \frac{dV(y)}{dy} = C_{ox} W \, \mu_n \frac{dV(y)}{dy} (V_G - V_T - V(y)) \quad (7.13.4)$$

Therefore:

$$\frac{dV(y)}{dy} = \frac{I_D}{C_{ox} W \, \mu_n (V_G - V_T - V(y))} \quad (7.13.5)$$

Inserting Expression 7.13.5 into Equation 7.13.2 we obtain:

$$Q_{inv} = -\mu_n(WC_{ox})^2 \; \frac{\displaystyle\int_0^{V_D}(V_G - V_T - V(y))^2 \, dV(y)}{I_D} \qquad (7.13.6)$$

Using Equation 7.4.15 and neglecting the depletion charge once again, we have:

$$I_D = \mu_n C_{ox} \frac{W}{L} \int_{V_S}^{V_D}\left(V_G - V_T - V(y)\right) dV(y) \qquad (7.13.7)$$

Replacing I_D in Equation (7.13.6) yields:

$$Q_{inv} = -WLC_{ox} \; \frac{\displaystyle\int_0^{V_D}(V_G - V_T - V(y))^2 \, dV(y)}{\displaystyle\int_0^{V_D}\left(V_G - V_T - V(y)\right) dV(y)} \qquad (7.13.8)$$

Performing the integration we obtain:

$$Q_{inv} = -\frac{2}{3} WLC_{ox} \; \frac{(V_G - V_T - V_D)^3 - (V_G - V_T)^3}{(V_G - V_T - V_D)^2 - (V_G - V_T)^2} \qquad (7.13.9)$$

In saturation, $V_D = V_{Dsat} = V_G - V_T$, and thus Equation 7.13.9 simplifies to:

$$Q_{inv} = -\frac{2}{3} WLC_{ox} \, (V_G - V_T) \qquad (7.13.10)$$

Since, by definition, $C_{GS} = -\dfrac{dQ_{inv}}{dV_G}$ (Q_{inv} is the electron charge in the channel), we obtain:

$$C_{GS} = \frac{2}{3} W \times L \times C_{ox}$$

Adding C_{ov} to the latter equation we finally obtain:

$$C_{GS} = \frac{2}{3} W \times L \times C_{ox} + C_{ov} \qquad (7.13.11)$$

A summary of the values of the gate capacitance for the different modes of operation is presented in Table 7.2.

Table 7.2: Terminal capacitances in the different operation regimes.

Capacitance	$V_G < V_{TH}$	Non saturation	Saturation
C_{Gsub}	$W \times L \times C$	0	0
C_{GS}	C_{ov}	$C_{ov} + \frac{1}{2} W \times L \times C_{ox}$	$C_{ov} + \frac{2}{3} W \times L \times C_{ox}$
C_{GD}	C_{ov}	$C_{ov} + \frac{1}{2} W \times L \times C_{ox}$	C_{ov}
C_{Ssub}	PN junction capacitance	PN junction capacitance	PN junction capacitance
C_{Dsub}	PN junction capacitance	PN junction capacitance	PN junction capacitance

7.14. Particular MOSFET structures

There exist many variations of the standard MOSFET device. These include lateral and vertical MOS power devices and devices combining MOS and bipolar operation. Here we will focus on two MOSFET structures which are of practical interest for MOS integrated circuits: information-storing MOSFETs and silicon-on-insulator (SOI) MOSFETs.

7.14.1. Non-Volatile Memory MOSFETs

Information storage MOSFETs are primarily used in read-only memories. As suggested by the name "Read-Only Memory" (ROM), these devices were originally designed to contain information that could be read, but could neither be erased nor overwritten. Later on special MOSFETs were invented, which made it possible to fabricate Erasable Programmable Read-Only Memory chips (EPROM), Electrically Erasable Programmable Read-Only Memory circuits (EEPROM), and flash EEPROMs, also called flash memories.[31]

One of the most popular EPROM cells is based on the use of a Floating-gate Avalanche-injection MOS (FAMOS) device. This particular MOSFET comprises two gates stacked on one another, as shown in Figure 7.37. The top polysilicon gate electrode (poly 2) is a regular gate, called "control gate", which is connected to the outside world. The bottom electrode (poly 1), on the other hand, is completely surrounded by silicon dioxide and is electrically floating. It is called a "floating gate".

If there is no charge stored in the floating gate its potential is equal to:

$$V_{FG} = \frac{C_{FG} V_G + C_{FD} V_{DS}}{C_T} \qquad (7.14.1)$$

where C_{FG} is the capacitance between the floating gate and the control gate, C_{FS} is the capacitance between the floating gate and the source, C_{FD} is the capacitance between the floating gate and the drain, C_{FSub} is the capacitance between the floating gate and the substrate and $C_T = C_{FD} + C_{FG} + C_{FS} + C_{FSub}$. The latter equation can be rewritten:

$$V_{FG} = \frac{C_{FG}}{C_T} \left(V_{CG} + \frac{C_{FD}}{C_{FG}} V_{DS} \right)$$

or:

$$V_{FG} = \frac{C_{FG}}{C_T} \left(V_{CG} + f V_{DS} \right) \text{ with } f = \frac{C_{FD}}{C_{FG}} \qquad (7.14.2)$$

where V_{CG} is the control gate voltage.

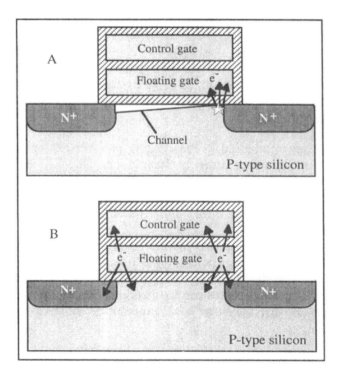

Figure 7.37: FAMOS transistor. A: Write operation by hot electron injection into the floating gate; B: Erase operation by exposure to ultraviolet light.

The equations for the floating-gate FAMOS transistor can thus be obtained from classic MOS theory provided that the gate voltage is replaced by V_{FG}, which in turn, can be expressed as a function of V_{CG} using Equation 7.14.2. In particular, when $V_{DS}=0$, the threshold voltage of the FAMOS device is given by:

$$V_{TH} \text{ (regular MOSFET)} = \frac{C_{FG}}{C_T} V_{TH} \text{ (FAMOS)}$$

or, in other words, the threshold voltage (the control gate being used as the device gate) is multiplied by a factor $\frac{C_T}{C_{FG}}$ when a floating gate is present between the control gate and the substrate.

Programming of a FAMOS transistor is achieved by applying a high voltage to both the drain and the control gate, such that the device is saturated and drives a high current. The high electric field near the drain provokes hot electron generation and impact ionization. In a mechanism similar to that describe in Section 7.12.4, some electrons acquire enough energy to overcome the gate oxide potential barrier and are injected through the gate oxide into the floating gate. This gives rise to a threshold voltage shift described by the relationship $\Delta V_{TH} = -Q_{FG}/C_{FSub}$, where Q_{FG} is the (negative) charge injected in the floating gate. The insulating quality of the gate oxide is so perfect that a charge stored in a floating gate can stay there for a period of 10 years without any detectable charge loss.

To "erase" the charge stored in the floating gate ultraviolet light is shone onto the device for approximately 30 minutes. The UV light gives the electrons stored in the floating gate enough energy to overcome the 3.1 V potential barrier between the polysilicon and the SiO_2, such that they can escape from the floating gate into either the silicon or the control gate.

The FAMOS transistor has two distinct modes of operation: one where the threshold voltage is low (no charge is stored on the floating gate), and one where the threshold voltage is high (electrons are stored on the floating gate). These two states can be distinguished by the sense amplifier of the chip, such that they can be interpreted as either a logic "0" or a logic "1".

EPROMs using FAMOS devices have an obvious disadvantage: during the erase operation, all memory cells are reset. Furthermore, this operation takes a long time and requires the memory circuit to be removed from the system in which it operates. Therefore, other information storage MOSFETs have been devised. One of them is the **FL**oating gate **T**unneling **OX**ide (FLOTOX) device in which each individual device can be electrically programmed or erased. Memory circuits using such devices are called Electrically Erasable Programmable Read-Only Memory

(EEPROM, or E^2PROM) circuits. The FLOTOX structure is shown in Figure 7.38. It contains a thin (5 nm) tunnel oxide above the drain. A polysilicon layer is used as floating gate. Programming is achieved by grounding the drain and applying a sufficiently large positive voltage to the control gate. This operation increases the potential of the floating gate such that electrons can tunnel from the drain into the floating gate. As in the case of the FAMOS transistor, the injected charge increases the threshold voltage of the device.

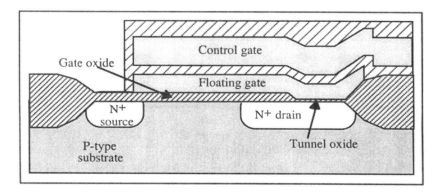

Figure 7.38: FLOTOX device.

To erase the FLOTOX cell, a sufficiently large positive bias is applied to the drain, while the control gate is grounded. This bias condition enables electrons to tunnel from the floating gate into the drain and to erase the information stored in the device. Equations 7.14.1 and 7.14.2 are applicable to the FLOTOX device. Because of the thin tunnel oxide between the floating gate and the drain, the value of C_{FG} is quite large, and the variation of floating gate bias with drain voltage is non negligible. As a result, the output resistance of FLOTOX devices is fairly low (they have a small Early voltage). Their saturation current is given by:

$$I_{Dsat} = \frac{1}{2} \mu \, C_{ox} \frac{W}{L} \, (\frac{C_{FG}}{C_T} \, (V_{CG} + f \, V_{DS}) - V_{TH})^2 \qquad (7.14.3)$$

An obvious dependence on drain voltage can be seen.

If the gate oxide of a FAMOS device is thin enough that tunneling of electrons can occur, programming and erase operations can be performed. This time tunneling takes place between the channel or source and the floating gate. Memory chips based on such devices are called flash memories.

7.14.2. SOI MOSFETs

In silicon-on-insulator (SOI) technology MOSFETs are realized in a thin layer of silicon sitting on top of an insulator, usually SiO_2, called "buried oxide". The thickness of the silicon film typically ranges between 50 and 200 nm, while the buried oxide thickness usually ranges between 80 and 400 nm. If the silicon film is thin enough the depletion zone below the gate extends all the way through the buried oxide, and the device is said to be "fully depleted" (Figure 7.39A). If this is not the case, the transistor is "partially depleted" (Figure 7.39B).

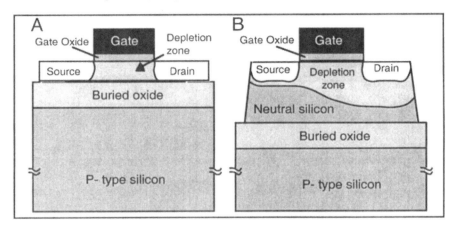

Figure 7.39: A: Fully depleted SOI MOSFET; B: Partially depleted SOI MOSFET.[32]

A partially depleted SOI MOSFET basically operates the same way as a regular "bulk" transistor does, especially if the neutral part of the silicon film is connected to ground. In a fully depleted device the vertical electric field extends through the entire silicon film. As a result, the surface potential at the top of the silicon film is coupled to the surface potential at the bottom of the device. If the doping concentration in the silicon film is uniform the potential is a parabolic function of depth, as shown in Figure 7.40. Because of the presence of both a gate oxide and a buried oxide, the SOI transistor has two gates, referred to as the front gate and the back gate.

The equations for a fully depleted SOI MOSFET are virtually identical to those for a bulk MOSFET. In particular, equations 7.4.31, 7.4.33 and 7.7.17 are applicable, such that the drain current and the subthreshold slope are given by:

$$I_D = \mu_n \, C_{ox} \frac{W}{L} \left\{ \left(V_G - V_{TH} \right) V_D - \frac{1}{2} n \, V_D^2 \right\}$$

$$I_{Dsat} = \mu_n \, C_{ox} \frac{W}{L} \frac{(V_G - V_{TH})^2}{2 \, n}$$

and

$$S = n \frac{kT}{q} \ln(10)$$

The remarkable feature of the fully depleted SOI MOSFET is that its body factor, n, is much smaller than that of a bulk MOSFET. Typical values for n are 1.5 and 1.05 in bulk and SOI devices, respectively. As a result, the current drive of SOI MOSFETs is higher than that of bulk devices, and their subthreshold slope is sharper (better) than that of bulk MOSFETs.

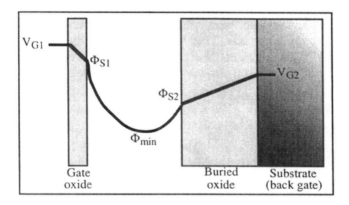

Figure 7.40: Potential in a fully depleted SOI MOSFET.

Using SOI technology, fully depleted double-gate MOSFETs can be made (Figure 7.41). In such a device the body factor is equal to 1. It has two channels (at the top and the bottom of the device) and is relatively free of short-channel effects.

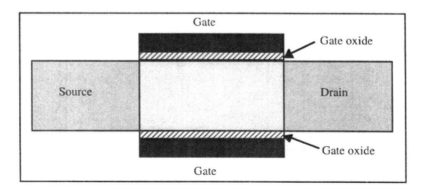

Figure 7.41: Double-gate SOI MOSFET. The two gates are connected together.

7.15. Advanced MOSFET concepts

As the features of MOS transistors are scaled to increasingly smaller dimensions some parasitic effects that were considered negligible in longer devices must be taken into account. This Section covers the most important of these effects.

7.15.1. Polysilicon depletion

Heavily doped polysilicon is the most widely used gate material in silicon MOSFETs. Typical doping concentrations are on the order of several times 10^{20} atoms cm^{-3}. Consider an N-type polysilicon gate used for an n-channel MOSFET. When a positive bias is applied to the gate the polysilicon in the gate "sees" the silicon underneath the gate oxide as a negatively biased electrode. This tends to deplete the bottom of the gate of electrons.[33] As a result, the capacitance between the quasi-neutral gate material and the silicon surface is no longer equal to C_{ox} but to $(1/C_{ox} + X_{dpoly}/\varepsilon_{si})^{-1}$ where X_{dpoly} is the thickness of the depleted polysilicon layer at the bottom of the gate. Consider an n-channel MOSFET with a 3 nm-thick gate oxide and an N^{+} polysilicon gate with a doping concentration $N_d = 10^{20}$ cm^{-3}. The maximum depletion depth in the polysilicon can be calculated using 7.2.37 and is equal to 3.8 nm at room temperature. Under these conditions, the measured gate oxide capacitance is 30% smaller than C_{ox}. The effect of polysilicon depletion on the C(V) curves of an MOS capacitor is illustrated in Figure 7.42. The reduction of gate oxide capacitance reduces the current drive of the MOSFET, according to Equations 7.4.38 and 7.4.40.

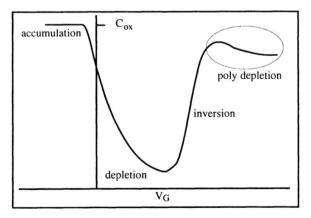

Figure 7.42: Quasistatic current-voltage characteristic of a MOS capacitor showing the polysilicon depletion effect.

7.15.2. High-k dielectrics

To achieve a large current drive in a MOSFET a large C_{ox} (small t_{ox}) is desirable. As the thickness of the gate oxide is reduced below a few nanometers, however, tunnel current can flow between the gate and the substrate. One method used to increase C_{ox} without generating excess gate current is to use materials other than SiO_2 as gate the dielectric. These materials have a high dielectric constant value, κ, compared to SiO_2 (κ_{SiO2} = 3.9) and are called "high-k dielectrics". Table 7.3 lists some materials being studied for use as a MOS gate dielectric.

Table 7.3: Dielectric constant of different metallic oxides.[34,35]

Material	Al_2O_3	ZrO_2	HfO_2	Ta_2O_5	CeO_2	TiO_2	$SrTiO_3$
κ	10	20	22	25	52	60	100

7.15.3. Drain-induced barrier lowering (DIBL)

The source and drain of a MOSFET form PN junctions within the substrate. The width of the depletion regions associated with the junctions increase with applied reverse bias. Consider an n-channel MOSFET with grounded source and substrate. If the channel is long enough (Figure 7.43A) the application of a drain bias does not modify the potential barrier of the source junction. In a short-channel device (Figure 7.43B), on the other hand the potential barrier at the source can be reduced by a value $\Delta\Phi$ depending on the drain bias. This reduction of the potential barrier reduces the threshold voltage. The magnitude of the drain-induced barrier lowering effect is usually defined by the following relationships:

$$DIBL = V_{TH}\Big|_{V_{DS}=V_1} - V_{TH}\Big|_{V_{DS}=V_2} \qquad \text{(unit: V)} \qquad (7.15.1)$$

or

$$DIBL = \frac{V_{TH}\Big|_{V_{DS}=V_1} - V_{TH}\Big|_{V_{DS}=V_2}}{(V_2 - V_1)} \qquad \text{(dimensionless)} \qquad (7.15.2)$$

where V_1 is usually equal to 50 or 100 mV and V_2 to 1 or 1.5 V.

In extreme cases the potential barrier at the source can become so small that the current between source and drain is no longer controlled by the gate. This phenomenon is called "punch-through".

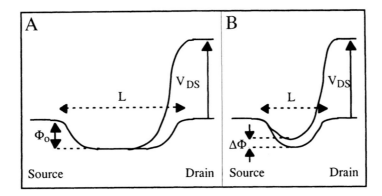

Figure 7.43: Potential from source to drain in A) a long-channel MOSFET and B) a short-channel MOSFET.

7.15.4. Gate-induced drain leakage (GIDL)

When a negative gate bias is applied to an n-channel MOSFET a depletion region can be created in the drain region overlapped by the gate (Figure 7.44A). This effect is also seen when the drain voltage is positive while the gate is grounded. Since the doping concentration in the drain is typically high the depletion region is very thin and therefore, an intense vertical electric field occurs at the drain. Under these conditions electron-hole pairs are generated through band-to-band tunneling of electrons from the valence band into the conduction band, as shown in Figure 7.44B. The generated holes create a substrate current and the electrons a drain current that increases with increased negative gate voltage (Figure 7.45).

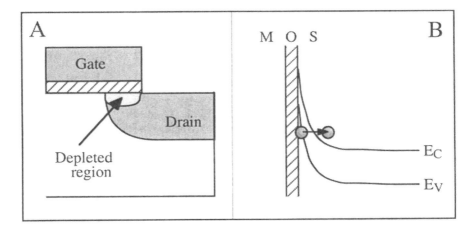

Figure 7.44: A: Formation of a depletion region in the drain and B: Energy band curvature in the depleted drain region and electron tunneling from the valence band into the conduction band. [36]

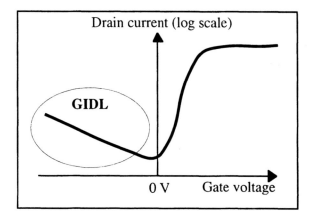

Figure 7.45: Drain current vs. gate voltage in an n-channel MOSFET illustrating the gate-induced drain leakage (GIDL) effect.

7.15.5. Reverse short-channel effect

To reduce the DIBL effect in a short-channel MOSFET the substrate doping concentration can be increased at the edges of the source and drain junctions. These regions with increased doping concentration are commonly called "halos" (Figure 7.46). When the channel length is reduced in halo devices the average channel doping concentration (per gate unit length) increases. This causes the threshold voltage to increase when gate length is reduced. This phenomenon is called the "reverse short-channel effect".[37] At shorter gate lengths, however, the regular short-channel effect described in Section 7.11 becomes dominant and the threshold voltage drops.

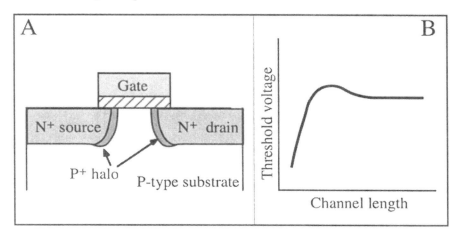

Figure 7.46 N-channel MOSFET with halo structure (A) and threshold voltage variation with gate length (B).

7.15.6. Quantization effects in the inversion channel

When derived using Poisson's equation, the electron profile in the channel is an exponential function of depth with a maximum at x=0. When the derivation of the electron profile is carried out taking quantum mechanical effects into consideration, *i.e.* using both the Poisson and Schrödinger equations, it is observed that the electron wave function is close to zero at the oxide/silicon interface and that the electron concentration peaks at a depth approximately equal to one nanometer from the Si/SiO$_2$ interface (Figure 7.47). As a result the distance between the inversion charge centroid and the gate electrode is larger than the physical gate oxide thickness t_{ox}. The equivalent, "effective" gate oxide thickness is given by:

$$t_{ox,effective} = t_{ox,physical} + \frac{\varepsilon_{ox}}{\varepsilon_{si}} \Delta x \qquad (7.15.3)$$

where Δx is the depth of the peak electron concentration. The increase of effective oxide thickness reduces C_{ox}, and therefore, reduces the current drive of the MOSFET, according to Equations 7.4.38 and 7.4.40.

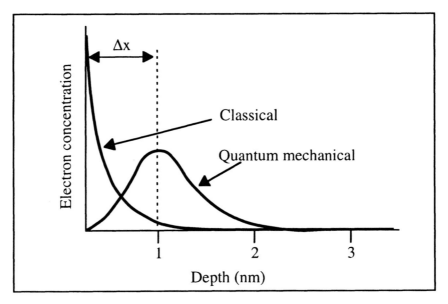

Figure 7.47: Electron concentration vs. depth in the inversion layer.[38]

Important Equations

MOS Capacitor (Depletion)

$$C = \frac{C_{ox}}{\sqrt{1 + \dfrac{2C_{ox}^2 V_G}{qN_a \varepsilon_{si}}}} \tag{7.2.33}$$

Maximum Depletion Depth

$$x_{dmax} = \sqrt{\frac{4\varepsilon_{si}\Phi_F}{qN_a}} \tag{7.2.37}$$

Threshold Voltage

$$V_{THo} = \Phi_{ms} - \frac{Q_{ox}}{C_{ox}} + \frac{2qN_{it}\Phi_F}{C_{ox}} + 2\Phi_F + \frac{\sqrt{4q\varepsilon_{si}N_a\Phi_F}}{C_{ox}} \tag{7.3.8}$$

Simplified Current-Voltage Relationship; Saturation Drain Voltage and Current

$$I_D = \mu_n C_{ox} \frac{W}{L} \left\{ \left(V_G - V_{THS}\right)V_D - \frac{1}{2}n\, V_D^2 \right\} \tag{7.4.31}$$

$$V_{Dsat} = \frac{V_G - V_{THS}}{n} \tag{7.4.32}$$

$$I_{Dsat} = \mu_n C_{ox} \frac{W}{L} \frac{(V_G - V_{THS})^2}{2n} \tag{7.4.33}$$

These expressions can be further simplified by excluding the body effect ($n = 1$).

Subthreshold Current / Subthreshold Slope

$$I_D = \mu_n \frac{W}{L} q \left(\frac{kT}{q}\right)^2 \frac{n_i^2}{N_a} \left[1 - exp\,(-qV_D/kT)\right] \frac{exp\,(q\Phi_S/kT)}{-\dfrac{d\Phi_S}{dx}} \tag{7.7.4}$$

$$S = \frac{kT}{q}\, ln(10) \left(1 + \frac{C_D + C_{it}}{C_{ox}}\right) \quad (mV/decade) \tag{7.7.21}$$

EKV Model

$$I_D = I_F - I_R = 2n\,\mu_n\,C_{ox}\,\frac{W}{L}\left(\frac{kT}{q}\right)^2$$

$$\times\left[\left\{ln\left[1 + exp\left(\frac{V_P-V_S}{2kT/q}\right)\right]\right\}^2 - \left\{ln\left[1 + exp\left(\frac{V_P-V_D}{2kT/q}\right)\right]\right\}^2\right]\quad(7.8.3)$$

$$\text{with } V_P = \frac{(V_G - V_{TH})}{n}$$

Terminal Capacitances

Capacitance	$V_G < V_{TH}$	Non saturation	Saturation
C_{Gsub}	$W \times L \times C$	0	0
C_{GS}	C_{ov}	$C_{ov} + \frac{1}{2}\,W \times L \times C_{ox}$	$C_{ov} + \frac{2}{3}\,W \times L \times C_{ox}$
C_{GD}	C_{ov}	$C_{ov} + \frac{1}{2}\,W \times L \times C_{ox}$	C_{ov}
C_{Ssub}	PN junction capacitance	PN junction capacitance	PN junction capacitance
C_{Dsub}	PN junction capacitance	PN junction capacitance	PN junction capacitance

Problems

 Problem 7.1:

If you have not already done so, now is a good time to solve Problem 2.4. In this problem the potential, electric field and charge distribution in a MOS capacitor are analyzed.

 Problem 7.2:

1) Plot X_{dmax} (maximum depletion depth) in silicon, at room temperature, as a function of the substrate doping concentration (from 10^{15} to 10^{18} cm^{-3}). Plot the x-axis on a log scale and the y-axis on a linear scale.

2) We have an MOS capacitor. The silicon substrate is P-type. The area of the MOS capacitor is 1 mm^2. We measure the low-frequency (quasistatic) C-V curve shown in Problem Figure 7.2 with $C_{max} = 6.9\times10^{-10}$ F and $C_{min} = 3\times10^{-10}$ F. What is the gate oxide thickness, and what is the P-type doping concentration (assume the doping concentration is uniform in the silicon)?

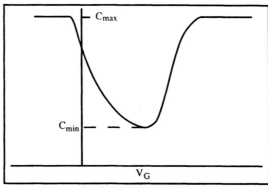

Problem Figure 7.2

Problem 7.3:
◊ Consider a MOSFET fabricated 15 years ago. The parameters are:

Gate oxide thickness: t_{ox} = 200 nm

Substrate doping concentration (P-type): $N_a = 1.7 \times 10^{15}$ cm^{-3}

Gate material: N^{++} (poly)silicon

Charge in the oxide: $Q_{ox} = q \times 10^{10}$ C cm^{-2}

Interface trap density: $N_{it} = 0$ cm^{-2} V^{-1}

$n_i = 1.45 \times 10^{10}$ cm^{-3}

1) Calculate the threshold voltage.
2) What happens to the threshold voltage if the device gets contaminated during fabrication, and the oxide charge is increased by a factor 5 such that $Q_{ox} = q \times 5 \times 10^{10}$ C cm^{-2}

◊ Consider a MOSFET fabricated today. The parameters are:

Gate oxide thickness: t_{ox} = 20 nm

Substrate doping concentration (P-type): $N_a = 8.7 \times 10^{16}$ cm^{-3}

Gate material: N^{++} (poly)silicon

Charge in the oxide: $Q_{ox} = q \times 10^{10}$ C cm^{-2}

Interface trap density: $N_{it} = 0$ cm^{-2} V^{-1}

$n_i = 1.45 \times 10^{10}$ cm^{-3}

3- Calculate the threshold voltage.
4- What happens to the threshold voltage if the device gets contaminated during fabrication, and the oxide charge is increased by a factor 5 such that $Q_{ox} = q \times 5 \times 10^{10}$ C cm^{-2}
5- Compare results from 2 and 4 (the variation of threshold voltage due to contamination), and explain the differences between the "old" device and the "new" one.

 Problem 7.4:
Consider a MOSFET having the following parameters:

Gate oxide thickness: t_{ox} = 20 nm

Substrate doping concentration (P-type): $N_a = 1 \times 10^{17}$ cm^{-3}
$\mu_n = 600$ cm^{2}V^{-1}s^{-1} ; W = 1 µm; L = 1 µm
$n_i = 1.45 \times 10^{10}$ cm^{-3}
$V_{FB} = -0.97$ V and $V_{TH} = 0.77$ V

Using the following equations:

$$V_G = V_{FB} + \Phi_S - \frac{Q_d}{C_{ox}} = V_{FB} + \Phi_S + \frac{\sqrt{2q\varepsilon_{si}N_a\Phi_S}}{C_{ox}} = V_{FB} + \Phi_S + \gamma\sqrt{\Phi_S}$$

$$I_D = \mu_n \frac{W}{L} q \left(\frac{kT}{q}\right)^2 \frac{n_i^2}{N_a} [1 - \exp(-qV_D/kT)] \frac{\exp(q\Phi_S/kT)}{-\frac{d\Phi_S}{dx}} \qquad (7.7.4)$$

$$-\frac{d\Phi_S(x)}{dx}\bigg|_{x=0} = \mathcal{E}_s = \sqrt{\frac{2qN_a\Phi_S}{\varepsilon_{si}}} \qquad (7.7.5)$$

$$I_D = \mu_n C_{ox} \frac{W}{L} \left\{ (V_G - V_{THo})V_D - \frac{1}{2} V_D^2 \right\} \qquad (7.4.38)$$

$$I_{Dsat} = \mu_n C_{ox} \frac{W}{L} \frac{(V_G - V_{THo})^2}{2} \qquad (7.4.40)$$

Plot I_D as a function of V_G for $V_{sub} = 0$ V, $V_S = 0$ V and $V_D = 0.1$ V. V_G ranges from 0 to 3 volts per steps of 10 mV. Use <u>both</u> a linear and a logarithmic scale for I_D.

 Problem 7.5:
Consider the following circuit (One resistor plus one n-channel MOSFET). Consider that the current in the transistor is equal to zero if $V_G < V_{TH}$.

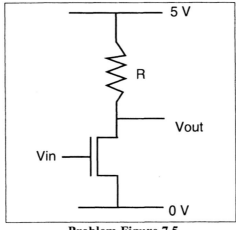

Problem Figure 7.5

The transistor parameters are:

$$\mu_n = 600 \text{ cm}^2\text{V}^{-1}\text{s}^{-1}; \quad W = 10 \ \mu m; \quad L = 1 \ \mu m;$$
$$t_{ox} = 200\text{e-8 cm} \ ;$$
$$\varepsilon_0 = 8.854\text{e-14 F/cm}; \quad \varepsilon_{si} = 11.7 \ast \varepsilon_0; \quad \varepsilon_{ox} = 3.9 \ast \varepsilon_0;$$
$$V_{TH} = 0.7 \text{ V};$$

Plot V_{out} as a function of V_{in} for V_{in} ranging from 0 to 5 V per steps of 0.01 V, and for R = 100Ω, 1kΩ, 10kΩ, 100kΩ and 1MΩ. Plot the 5 curves on a single graph. Use the simplified current model with n=1 (Equations 7.4.38-40).

 Problem 7.6:

Consider a MOSFET having the following parameters:
Gate oxide thickness: t_{ox} = 20 nm
Substrate doping concentration (P-type): $N_a = 10^{17}$ cm^{-3}
Gate material: N^{++} (poly)silicon
Charge in the oxide: Q_{ox} = 0 cm^{-2}
Interface trap density: N_{it} = 0 cm^{-2} V^{-1}
n_i = 1.45x10^{10} cm^{-3}
W=2 μm and L=1 μm
μ_n = 600 cm^{2}V^{-1}s^{-1}
$V_S = V_{sub} = 0V$

On a single graph plot I_D as a function of V_D where V_D ranges from 0 to 5V. Let V_G = 0, 1, 2, 3, 4 and 5 V for each I_D vs. V_D plot. Use the following equations:
- the complete model; Equations 7.4.17 - 7.4.18
- the simplified model; Equations 7.4.31 - 7.4.33
- the simplified model with n=1 (Equations 7.4.38 - 7.4.40)

 Problem 7.7:

Using Matlab, plot the threshold voltage as a function of gate length (short channel effect, Equation 7.11.7) in an n-channel MOSFET with the following parameters: t_{ox} = 25 nm, N_a = 6x10^{16} cm^{-3}, V_{FB} = -1 V and r_j = 300 nm. Let the gate length range from 0.1μm to 5μm.

 Problem 7.8:

Using Matlab, calculate the evolution of threshold voltage in an n-channel MOSFET with temperature, from 0 to 300°C. The p-type substrate doping concentration is equal to 5x10^{16} cm^{-3}. The gate oxide thickness is 25 nm. The gate material is degenerately doped N-type polycrystalline silicon. Under that doping condition, E_F = E_C in the polycrystalline material. The flatband voltage is given by the difference in Fermi levels between the gate material and the silicon substrate, Φ_{MS}:

$$\Phi_{MS} = V_{FB} = \frac{(E_{Fsilicon} - E_{Fgate})}{q} = -\Phi_F - \frac{E_C - E_i}{q} = -\Phi_F - \frac{E_g}{2q}$$

Assume there are no charges in the oxide and no interface states ($Q_{ox} = N_{it} = 0$).

 Problem 7.9:

The Figure below shows a CMOS inverter. It contains an n-channel and a p-channel transistor. Using Matlab, plot the transfer characteristics of this inverter (*i.e.*: plot V_{out} as a function of V_{in}). Use the simplified current model with n=1 (Equations 7.4.38-40). The transistors have the following parameters:

The supply voltage, V_{DD}, is equal to 5 V.
The threshold voltage of the N and P-channel devices are $V_{THN} = 0.7$ V and $V_{THP} = -0.7$ V, respectively.
$\mu_n = 540$ cm^{2}V^{-1}s^{-1}, $\mu_p = 180$ cm^{2}V^{-1}s^{-1}, (W/L)$_n$=2, (W/L)$_p$=6 and C_{ox}=2.3x10^{-7} F cm^{-2}.

Note that $\mu_n C_{ox} (W/L)_N = \mu_p C_{ox} (W/L)_P$.

Comment on the differences between the inverter in problem 7.5 and the inverter of this problem.

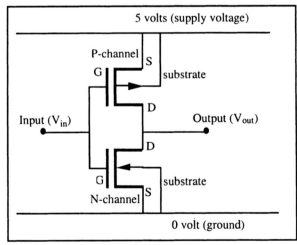

Problem Figure 7.9

 Problem 7.10:

Using Matlab and the EKV model (see Section 7.8), plot the following curves for an n-channel MOSFET:
$I_D(V_G)$ for $0 < V_G < 3$V and $V_D = 100$ mV (on both linear and log scale for I_D)
$I_D(V_D)$ for $0 < V_D < 5$V and $V_G = 0, 1, 2, 3, 4$ and 5 V
$g_m(V_D)$ for $0 < V_G < 3$V and $V_D = 100$ mV

Use the following parameters:
W=10 µm, L=1 µm, t_{ox} =10 nm, V_{TH}=0.7 V, and n=1.4. The leakage current of the drain junction is 0.1 pA (add 0.1 pA to the drain current obtained from Equation 7.8.3). Include mobility degradation effects (μ_o=600 cm^2/Vs and θ=0.2 V^{-1})

Problem 7.11:
The threshold voltage of an n-channel MOSFET having a degenerately doped N$^+$ polysilicon gate is 0.7 V. What would the threshold voltage be if the gate material was degenerately doped P$^+$ polysilicon, all other fabrication parameters being unchanged?

Problem 7.12:
An n-channel MOSFET has the following parameters:
$$t_{ox}=100 \text{ nm}, N_a= 5x10^{15} \text{ cm}^{-3}$$
There are no charges in the oxide and no interface traps and the gate is made out of degenerately doped N$^+$ polysilicon.
Calculate the threshold voltage when the source voltage is 0 V and 5 V. The substrate is grounded and taken as voltage reference.

Problem 7.13:
These $I_D(V_D)$ characteristics were measured on an n-channel MOSFET (Problem Figure 7.13). Source and substrate are grounded. The gate oxide thickness is 25 nm. W=10 µm and L=2 µm. Calculate the mobility of the electrons in the channel, μ_n.

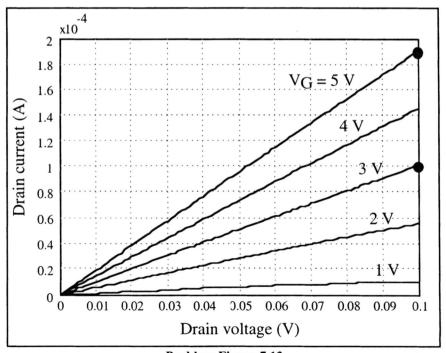

Problem Figure 7.13

Problem 7.14:
A silicon n-channel MOSFET has the following parameters:
$$W = L = 1 \ \mu m$$
Gate material is N^+ polysilicon
$$\mu_n = 600 \ cm^2 V^{-1} s^{-1}$$
$$V_G = V_D = 3.3 \ V; \ V_S = V_{sub} = 0V$$
$$t_{ox} = 10 \ nm$$
$$N_a = 2.8 \times 10^{17} \ cm^{-3}$$
$$Q_{ox} = q \times 10^{10} \ C/cm^2$$

1: Calculate the current in the transistor.

2: Because of unavoidable fabrication parameter fluctuations, t_{ox}, N_a and Q_{ox} can vary by ±5%. Each of these parameters is independent of the others. The increase of some of these parameters will increase the current and the increase of other parameters will decrease the current. Calculate the maximum "worst case" increase and decrease of current that can result from the ±5% variation of t_{ox}, N_a and Q_{ox}. Use the simplified model (Equations 7.4.31-33)

 Problem 7.15:
An n-channel MOSFET is used in an integrated circuit operating in a satellite. This MOSFET is continuously exposed to ionizing radiations, at a dose rate of 0.1 rad(SiO_2) per second. The rad(SiO_2) is the unit of energy deposition in SiO_2 and is equivalent to the deposition of 1 erg of energy in that material. The dose of radiation absorbed in SiO_2, D, is equal to the dose rate times the duration of the exposure to radiation. Upon irradiation positive charges are created in the oxide. Passed a given absorbed dose, however, the creation of charge saturates according to the following expression:
$$Q_{ox} = q \times N_{ox} = q \times 10^{12} \times \frac{D/D_c}{D/D_c + 1} \ (C/cm^2)$$
where D is the absorbed dose (in rad(SiO_2)) and D_c is a critical dose equal to 2×10^4 rads(SiO_2). In parallel to that process interface traps are created. The density of these traps, N_{it}, increases linearly with the radiation dose according to the following law:
$$N_{it} = Dose \ (in \ rads(SiO_2)) \times 30,000 \ (cm^{-2} \ V^{-1})$$
The threshold voltage of the device is given by an equation similar to 7.3.8:
$$V_{THo} = \Phi_{ms} - \frac{Q_{ox}}{C_{ox}} + \frac{2qN_{it}\Phi_F}{C_{ox}} + 2\Phi_F + \frac{\sqrt{4q\varepsilon_{si}N_a\Phi_F}}{C_{ox}}$$
where $\Phi_{ms} = -0.9$ V, $t_{ox} = 0.025 \times 10^{-4}$ cm, $N_a = 8 \times 10^{16}$ cm^{-3}. Both N_{it} and Q_{ox} are equal to zero before irradiation. We assume T = 300K.

1) Plot the threshold voltage as a function of the irradiation time for times ranging from 1 second to 10^9 seconds using a log scale for time.

2) Plot the saturation current, for a supply voltage of 5 V, as a function of time ($V_G=V_D=5V$, $V_S=V_{sub}=0V$). The mobility of the electrons in the channel is 650 $cm^2V^{-1}s^{-1}$, W= 10 μm and L = 1 μm. Use the simplified current model with n=1 (Equations 7.4.38-40).

3) The saturation current of the transistor before irradiation is 7.4 mA. The circuit will fail operating properly if the drain saturation current falls below 6 mA. How long will the circuit be able to operate properly (in years)?

 Problem 7.16:

Using a numerical solution for Poisson's equation (see Problem 2.4):

1) Plot the charge in the silicon of an MOS capacitor as a function of surface potential and gate voltage in order to obtain curves similar to that of Figure 7.10. The gate insulator material is SiO_2 (thickness = 15 nm), and the flat-band voltage is 0V.

Plot the curves for $\dfrac{-5kT}{q} < \Phi_S < 2\Phi_F + \dfrac{4kT}{q}$. The silicon is P-type and the doping concentration, N_a, is equal to 5×10^{16} cm^{-3}. T=300K.

The gate voltage, V_G, is equal to: $V_G = \Phi_S - Q_{silicon}/C_{ox}$ where the total charge in silicon, $Q_{silicon}$, is equal to the accumulation charge + the depletion charge + the inversion charge.

2) Plot the MOS capacitance $C = -\dfrac{dQ_{silicon}}{dV_G}$ as a function of V_G to obtain a quasi-static capacitance-voltage characteristics similar to that of Figure 7.8.

Use the following data:

q=1.6e-19;	% electron charge (C)
epsil=8.854e-14;	% Permittivity of vacuum (F/cm)
esi=epsil*11.7;	% Permittivity of silicon (F/cm)
k=1.3805e-23;	% Boltzmann constant (J K-1)
ni=1.45e10;	% Intrinsic carrier concentration in
	% silicon at room temperature (cm^{-3})
Na=5e16;	% substrate doping (cm^{-3})
T=300;	% temperature (K)
eox=epsil*3.9;	% Permittivity of SiO2 (F/cm)
tox=150e-8;	% Gate oxide thickness (cm)

<u>Tip</u>: In this problem we recommend using a sample thickness different than $2x_{dmax}$ since there is no depletion zone when the device is in accumulation or in flat-band situation. The recommended sample thickness is $5L_D + 2x_d$ where L_D is the Debye length (Expression 7.2.10) and where $x_d = \sqrt{\dfrac{2\varepsilon_{si}\left|\Phi_s\right|}{qN_a}}$ (adapted from Equation 7.2.26). To avoid convergence problems when $\Phi_S \cong 0$, we suggest linearizing the right-hand term of the Poisson equation. This can be done the following way: the result of the n-th iteration is used as an initial solution for the n+1 iteration, such that $\Phi_{n+1} = \Phi_n + \Delta\Phi_{n+1}$, where $\Delta\Phi_{n+1}$ is small. Since $\Delta\Phi_{n+1}$ is small the exponential terms in Poisson's equation

$$\frac{d^2\Phi(x)}{dx^2} = -\frac{q}{\varepsilon_s}\left\{N_a\exp\left[\frac{-q\Phi(x)}{kT}\right] - \frac{n_i^2}{N_a}\exp\left[\frac{q\Phi(x)}{kT}\right] - N_a\right\}$$

can be developed in a series. In a discrete form this linearization step yields:

$$N_a\exp\left(\frac{-q(\Phi(i)+\Delta\Phi(i))}{kT}\right) = N_a\exp\left(\frac{-q(\Phi(i))}{kT}\right)\left(1 - \frac{q\Delta\Phi(i)}{kT}\right)$$

and

$$\frac{n_i^2}{N_a}\exp\left(\frac{q(\Phi(i)+\Delta\Phi(i))}{kT}\right) = \frac{n_i^2}{N_a}\exp\left(\frac{q(\Phi(i))}{kT}\right)\left(1 + \frac{q\Delta\Phi(i)}{kT}\right)$$

and the discrete Poisson equation, which was originally

$$\left[-\Phi(i-1)+2*\Phi(i)-\Phi(i+1)\right]_{n+1} = \left[\frac{q(\Delta x)^2}{\varepsilon_s}\left\{N_a\exp\left[\frac{-q\Phi(i)}{kT}\right] - \frac{n_i^2}{N_a}\exp\left[\frac{q\Phi(i)}{kT}\right] - N_a\right\}\right]_n$$

now becomes:

$$\left[-\Delta\Phi(i-1) + \Delta\Phi(i)*\left(2+\frac{(q\Delta x)^2}{\varepsilon_s kT}\left[N_a\exp\left[\frac{-q\Phi(i)}{kT}\right]+\frac{n_i^2}{N_a}\exp\left[\frac{q\Phi(i)}{kT}\right]\right]\right) - \Delta\Phi(i+1)\right]_{n+1}$$

$$= \left[\frac{q(\Delta x)^2}{\varepsilon_s}\left[N_a - N_a\exp\left[\frac{-q\Phi(i)}{kT}\right]+\frac{n_i^2}{N_a}\exp\left[\frac{q\Phi(i)}{kT}\right]\right]+\Phi(i-1)-2\Phi(i)+\Phi(i+1)\right]_n$$

Once a $\Delta\Phi_{n+1}$ is found the corresponding Φ_{n+1} is obtained by adding $\Delta\Phi_{n+1}$ to Φ_n.

Problem 7.17:
Using a numerical solution for Poisson's equation (see Problem 2.4), plot the subthreshold current of an n-channel MOS transistor. Plot the curve for $\frac{\Phi_F}{2} < \Phi_S < 2\Phi_F + \frac{4kT}{q}$. Calculate the subthreshold slope (in millivolts per decade) using V_D = 50 mV. Note that Equation 7.7.1 can be rewritten:

$$I_D = q\,A\,D_n\frac{n(0) - n(L)}{L} = q\frac{W}{L}D_n\left(\int n(x,y=0)dx - \int n(x,y=L)dx\right)$$

For any value of the surface potential $\Phi(x=0)$ the electron concentration at $x=L$ corresponds to a surface potential $\Phi(x=L) = \Phi(x=0) - V_D$. The gate insulator material is SiO_2 (thickness = 20 nm), and the flat-band voltage is -0.8V. The silicon is P-type and the doping concentration, N_a, is equal to 5×10^{16} cm^{-3}. Assume T=300K and W=L= 1 μm. The gate voltage, V_G, is equal to: $V_G = \Phi_S$ - $Q_{silicon}/C_{ox}$ with the total charge in the silicon, $Q_{silicon}$, equal to the accumulation charge + the depletion charge + the inversion charge. Use the following data:

```
q=1.6e-19;           % electron charge (C)
epsil=8.854e-14;     % Permittivity of vacuum (F/cm)
esi=epsil*11.7;      % Permittivity of silicon (F/cm)
```

```
k=1.3805e-23;          % Boltzmann constant (J/K)
ni=1.45e10;            % Intrinsic carrier concentration (cm-3)
                       % in silicon at room temperature (cm-3)
Na=5e16;               % substrate doping (cm-3)
T=300;                 % temperature (K)
eox=epsil*3.9;         % Permittivity of SiO2 (F/cm)
tox=200e-8;            % Gate oxide thickness (cm)
mu=600;                % Electron mobility (cm2/Vs)
W=1e-4;L=1e-4;         % Gate width and length (cm)
VFB=-0.8;              % Flat-band voltage (V)
```

 Problem 7.18:

Consider the MOS capacitor shown in Problem Figure 7.18a. The doping concentration in the P-type silicon is 5×10^{15} cm^{-3}. The width of the silicon sample is 1 μm and its thickness is 0.9 μm. The oxide thickness is 0.1 μm. The gate is 0.3 μm wide and surrounded by a grounded electrode called a "guard ring". The back of the sample is grounded.

To solve the two-dimensional Poisson equation the structure is represented by t × t mesh points (t=11 is the maximum mesh points allowed by the Student Edition of MATLAB, but a larger number of mesh points can be used with the Professional Version of MATLAB). The distance between mesh points is $\Delta x = \Delta y = 0.1$ μm (Problem Figure 7.18b).

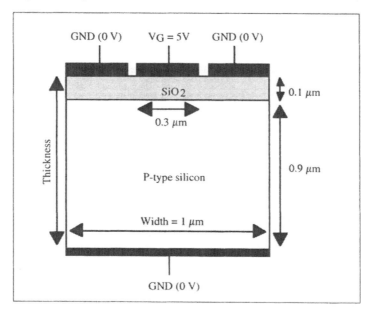

Problem Figure 7.18a: Two-dimensional (2D) MOS capacitor.

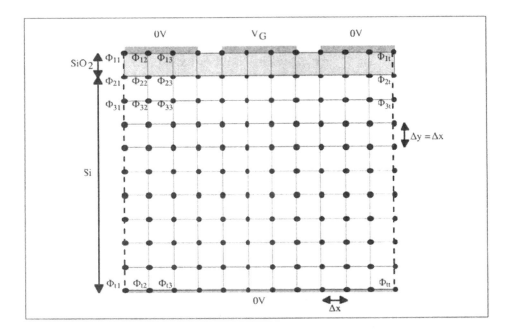

Problem Figure 7.18b: Two-dimensional mesh definition.

The two-dimensional Poisson equation is:

$$\frac{d^2\Phi(x,y)}{dx^2} + \frac{d^2\Phi(x,y)}{dy^2} = -\frac{q}{\varepsilon(x,y)}\left(p(x,y) - n(x,y) + N_d^+(x,y) - N_a^-(x,y)\right) \tag{1}$$

In a discrete form, second derivatives at node (i,j) are given by:

$$\frac{d^2\Phi}{dx^2} \cong \frac{\Phi_{(i-1,j)} + \Phi_{(i+1,j)} - 2\Phi_{(i,j)}}{(\Delta x)^2} \tag{2}$$

and

$$\frac{d^2\Phi}{dy^2} \cong \frac{\Phi_{(i,j-1)} + \Phi_{(i,j+1)} - 2\Phi_{(i,j)}}{(\Delta y)^2} \tag{3}$$

Since $\Delta x = \Delta y$, we have:

$$\frac{d^2\Phi}{dx^2} + \frac{d^2\Phi}{dy^2} \cong \frac{\Phi_{(i-1,j)} + \Phi_{(i+1,j)} + \Phi_{(i,j-1)} + \Phi_{(i,j+1)} - 4\Phi_{(i,j)}}{(\Delta x)^2} \tag{4}$$

The latter expression must be solved at every mesh point, except at nodes (1,j) and (t,j) where the potential is known (boundary conditions) and at nodes (i,1) and (i,t) where one has to solve the 1D Poisson equation given by Equation (3) (see Problem 2.4).

The discrete Poisson equation has the following matrix form:

$$A*\Phi = R$$

where A is a $t^2 \times t^2$ matrix representing the Laplace operator, Φ is a vector containing the potential at each mesh point, and R is the right term of the equation.

R is a vector containing both the boundary conditions and the values $-\dfrac{\rho(\Phi_{ij})}{\varepsilon_{ij}}(\Delta x)^2$

If we were using a 4×4 mesh instead of a 11×11 mesh the discrete Poisson equation would be:

$$
\begin{bmatrix}
1 & 0 & 0 & 0 & 0 & 0 & 0 & 0 & 0 & 0 & 0 & 0 & 0 & 0 & 0 & 0 \\
1 & -2 & 1 & 0 & 0 & 0 & 0 & 0 & 0 & 0 & 0 & 0 & 0 & 0 & 0 & 0 \\
0 & 1 & -2 & 1 & 0 & 0 & 0 & 0 & 0 & 0 & 0 & 0 & 0 & 0 & 0 & 0 \\
0 & 0 & 0 & 1 & 0 & 0 & 0 & 0 & 0 & 0 & 0 & 0 & 0 & 0 & 0 & 0 \\
0 & 0 & 0 & 0 & 1 & 0 & 0 & 0 & 0 & 0 & 0 & 0 & 0 & 0 & 0 & 0 \\
0 & 1 & 0 & 0 & 1 & -4 & 1 & 0 & 0 & 1 & 0 & 0 & 0 & 0 & 0 & 0 \\
0 & 0 & 1 & 0 & 0 & 1 & -4 & 1 & 0 & 0 & 1 & 0 & 0 & 0 & 0 & 0 \\
0 & 0 & 0 & 0 & 0 & 0 & 0 & 1 & 0 & 0 & 0 & 0 & 0 & 0 & 0 & 0 \\
0 & 0 & 0 & 0 & 0 & 0 & 0 & 0 & 1 & 0 & 0 & 0 & 0 & 0 & 0 & 0 \\
0 & 0 & 0 & 0 & 0 & 1 & 0 & 0 & 1 & -4 & 1 & 0 & 0 & 1 & 0 & 0 \\
0 & 0 & 0 & 0 & 0 & 0 & 1 & 0 & 0 & 1 & -4 & 1 & 0 & 0 & 1 & 0 \\
0 & 0 & 0 & 0 & 0 & 0 & 0 & 0 & 0 & 0 & 0 & 1 & 0 & 0 & 0 & 0 \\
0 & 0 & 0 & 0 & 0 & 0 & 0 & 0 & 0 & 0 & 0 & 0 & 1 & 0 & 0 & 0 \\
0 & 0 & 0 & 0 & 0 & 0 & 0 & 0 & 0 & 0 & 0 & 0 & 1 & -2 & 1 & 0 \\
0 & 0 & 0 & 0 & 0 & 0 & 0 & 0 & 0 & 0 & 0 & 0 & 0 & 1 & -2 & 1 \\
0 & 0 & 0 & 0 & 0 & 0 & 0 & 0 & 0 & 0 & 0 & 0 & 0 & 0 & 0 & 1
\end{bmatrix}
\begin{bmatrix}
\Phi_{11} \\ \Phi_{21} \\ \Phi_{31} \\ \Phi_{41} \\ \Phi_{21} \\ \Phi_{22} \\ \Phi_{23} \\ \Phi_{24} \\ \Phi_{31} \\ \Phi_{32} \\ \Phi_{33} \\ \Phi_{34} \\ \Phi_{41} \\ \Phi_{42} \\ \Phi_{43} \\ \Phi_{44}
\end{bmatrix}
=
\begin{bmatrix}
\Phi_{11} \\ R_{21} \\ R_{31} \\ \Phi_{41} \\ \Phi_{21} \\ R_{22} \\ R_{23} \\ \Phi_{24} \\ \Phi_{31} \\ R_{32} \\ R_{33} \\ \Phi_{34} \\ \Phi_{41} \\ R_{42} \\ R_{43} \\ \Phi_{44}
\end{bmatrix}
$$

where Φ_{i1} and Φ_{i4} are the boundary conditions and $R_{ij} = -\dfrac{\rho(\Phi_{ij})}{\varepsilon_{ij}}(\Delta x)^2$

Note that matrix A is composed of 4 types of $t \times t$ blocs:

$$
\begin{bmatrix}
1 & 0 & 0 & 0 \\
1 & -2 & 1 & 0 \\
0 & 1 & -2 & 1 \\
0 & 0 & 0 & 1
\end{bmatrix} ,
\begin{bmatrix}
1 & 0 & 0 & 0 \\
1 & -4 & 1 & 0 \\
0 & 1 & -4 & 1 \\
0 & 0 & 0 & 1
\end{bmatrix} ,
\begin{bmatrix}
0 & 0 & 0 & 0 \\
0 & 1 & 0 & 0 \\
0 & 0 & 1 & 0 \\
0 & 0 & 0 & 0
\end{bmatrix}
\text{ and }
\begin{bmatrix}
0 & 0 & 0 & 0 \\
0 & 0 & 0 & 0 \\
0 & 0 & 0 & 0 \\
0 & 0 & 0 & 0
\end{bmatrix}
$$

Use the MATLAB function "repmat" to assemble these different blocks and build matrix A.

Question: Using the following data:

$$t = 11, \ V_G = 5 \text{ V}, \ q = 1.6 \times 10^{-19} \text{ C} \ \text{ and } T = 300 \text{ K}$$
$$N_a = 5 \times 10^{15} \text{ cm}^{-3} \ \text{ and } \ n_i = 1.45 \times 10^{10} \text{ cm}^{-3}$$

$$\varepsilon_{si} = \text{permittivity of silicon} = 11.7 \times 8.854 \times 10^{-14} \text{ F cm}^{-1}$$

$$\varepsilon_{ox} = \text{permittivity of silicon dioxide} = 3.9 \times 8.854 \times 10^{-14} \text{ F cm}^{-1}$$

produce the following 3D plots:

◊ Potential in the silicon and silicon dioxide *vs.* x and y
◊ Log of hole concentration in the silicon *vs.* x and y
◊ Log of electron concentration in the silicon *vs.* x and y
◊ Arrow plot of the electric field in the silicon *vs.* x and y (using the "quiver" plot function).

Problem 7.19:
Using Relationship 7.2.18 plot the surface potential Φ_S and the potential drop in the gate oxide V_{ox} in an MOS capacitor in accumulation using the following parameters:

$$N_a = 10^{16} \text{ cm-3}$$
$$t_{ox} = 10 \text{ nm}$$
$$T = 300 \text{ K}$$
$$V_G \text{ ranges from 0 to -5 volts}$$
$$\Phi_{MS} = 0 \text{ V}, Q_{ox} = 0 \text{ C cm}^{-2}, N_{it} = 0 \text{ C cm}^{-2} \text{ V}^{-1}$$

 Problem 7.20:
Consider the CMOS inverter shown in Problem Figure 7.20. Using the EKV model plot the output characteristics $V_{out}(V_{in})$ for $0 \text{ V} \le V_{in} \le 5 \text{ V}$. On a separate graph plot the current going through the transistors as a function of input voltage. Consider the output terminal an open connection. Therefore, the current in the n-channel transistor is always equal to the current in the p-channel transistor.

The n-channel transistor parameters are:
$V_{TH} = 0.7 \text{ V}$; $n = 1.4$; $t_{ox} = 10 \text{ nm}$;$W_n = 1 \text{ μm}$; $L_n = 1 \text{ μm}$; $\mu_n = 600 \text{ cm}^2/\text{Vs}$
The p-channel transistor parameters are:
$V_{TH} = - 0.7 \text{ V}$; $n = 1.4$; $t_{ox} = 10 \text{ nm}$;$W_p = 3 \text{ μm}$; $L_p = 1 \text{ μm}$; $\mu_p = 200 \text{ cm}^2/\text{Vs}$

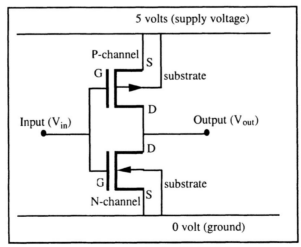

Problem Figure 7.20: CMOS inverter.

Problem 7.21:
Derive 7.4.17 from 7.4.15

Problem 7.22:
Derive 7.4.31 from 7.4.30

References

1   G. Moore, "Cramming more components onto integrated circuits", *Electronics*, Vol. 38, No. 8, p. 114, 1965

2   T. Sakata, M. Horiguchi, T. Sekiguchi, S. Ueda, H. Tanaka, E. Yamasaki, Y. Nakagome, M. Aoki, T. Kaga, M. Ohkura, R. Nagai, F. Murai, T. Tanaka, S. Iijima, N. Yokoyama, Y. Gotoh, I. Shoji, T. Kisu, H. Yamashita, T. Nishida, and E. Takeda, "An experimental 220-MHz 1-Gb DRAM with a distributed-column-control architecture", *IEEE Journal of Solid-State Circuits*, Vol. 30, No. 11, p.1165, 1995

3   J.E. Lilienfield, U.S. Patent 1,745,175, 1930

4   O. Heil, British Patent 439,457, 1935

5   D. Kahng and .M. Atalla, "Silicon-silicon dioxide field induced surface devices", *IRE Solid-State Device Research Conference*, Carnegie Institute of Technology, Pittsburg, 1960

6   S.M. Sze, *Physics of semiconductor devices*, J. Wiley & Sons, p. 365, 1981

7   J.R. Hauser and M.A. Littlejohn, "Approximation for accumulation and inversion space-charge layers in semiconductors", *Solid-State Electronics*, Vol. 11, p. 667, 1968

8   J.P Raskin, A. Viviani, D. Flandre, and J.P Colinge, "Substrate crosstalk reduction using SOI technology", *IEEE Transactions on Electron Devices*, Vol. 44, No. 12, p. 2252, 1987

9   R.S. Muller and T.I. Kamins, *Device electronics for integrated circuits*, J. Wiley and Sons, p. 390, 1986

10   S.M. Sze, *Physics of semiconductor devices*, J. Wiley & Sons, p. 369, 1981

11   A.S. Grove, *Physics and technology of semiconductor devices*, J. Wiley & Sons, p. 279, 1967

12   A.S. Grove, *Physics and technology of semiconductor devices*, J. Wiley & Sons, pp. 279-282, 1967

13   R.S. Muller and T.I. Kamins, *Device electronics for integrated circuits*, J. Wiley and Sons, p. 429, 1986

14   R.S. Muller and T.I. Kamins, *Device electronics for integrated circuits*, J. Wiley and Sons, p. 437, 1986

15   A.G. Sabnis and J.T. Clemens, "Characterization of the electron mobility in the inverted <100> Si surface", *Technical Digest of the International Electron Devices Meeting*, p. 18, 1979

16   F.M. Klaasen, "A MOS model for computer-aided design", *Philips Research Reports*, Vol. 31, p. 71, 1976

17   S.M. Sze, *Physics of semiconductor devices*, J. Wiley & Sons, p. 449, 1981

18   C.C. Enz, "The EKV model: a MOST model dedicated to low-current and low-voltage analogue circuit design and simulation", in *Low-power HF microelectronics: a unified approach*, edited by G.A.S. Machado, IEE Circuits and Systems Series 8, the Institution of Electrical Engineers, p. 247, 1996

19   R.S. Muller and T.I. Kamins, *Device electronics for integrated circuits*, J. Wiley and Sons, p. 480, 1986

20   D.J. Wouters, J.P. Colinge, and H.E. Maes, "Subthreshold current in thin-film SOI MOSFET transistors", *IEEE Transactions on Electron Devices*, Vol. 37, p. 2022, 1990
     and

R.J. Van Overstraeten, G. Declerck, and G.L. Broux, "Inadequacy of the classical theory of the MOS transistor operating in weak inversion", *IEEE Trans. on Electron Devices*, Vol. 20, p. 1150, 1973

21 S.M. Sze, *Physics of semiconductor devices*, J. Wiley & Sons, p. 470, 1981

22 C.C. Enz, "The EKV model: a MOST model dedicated to low-current and low-voltage analogue circuit design and simulation", in *Low-power HF microelectronics: a unified approach*, edited by G.A.S. Machado, IEE Circuits and Systems Series 8, the Institution of Electrical Engineers, p. 247, 1996

23 C. Enz, F. Krummenacher, and E.A. Vittoz, "An analytical MOS transistor model valid in all regions of operation and dedicated to low-voltage and low-current applications", *Analog Integrated Circuit and Signal Processing*, Vol. 8, No. 1, p. 83, 1995

24 R.S. Muller and T.I. Kamins, *Device electronics for integrated circuits*, J. Wiley and Sons, p. 441, 1986

25 Simulations generated by "SUPREM-IV", AVANT! Corporation

26 Simulations generated by "MEDICI", AVANT! Corporation

27 R.S. Muller and T.I. Kamins, *Device electronics for integrated circuits*, J. Wiley and Sons, pp. 486-488, 1986

28 P.K.K. Ko, "Approaches to scaling", *Advanced MOS device physics*, VLSI electronics microstructure science, Vol. 18, Academic Press, pp. 1-37, 1989

29 J.M. Pimbley, M. Ghezzo, H.G. Parks and D.M. Brown, *Advanced CMOS process technology*, VLSI electronics microstructure science, Vol. 19, Academic Press, pp.181-198, 1989

30 C. Hu, "Hot-carrier effects", *Advanced MOS device physics*, VLSI electronics microstructure science, Vol. 18, Academic Press, pp. 119-160, 1989

31 D.W. Greve, *Field-Effect Devices and Applications*, Prentice Hall Series in Electronics and VLSI, pp. 239-249, 1998

32 J.P. Colinge, *Silicon-on-Insulator Technology: Materials to VLSI*, 2nd Edition, Kluwer Academic Publishers, 1997

33 J.-M. Sallese, M. Bucher and C. Lallement, "Improved analytical modeling of polysilicon depletion in MOSFETs for circuit simulation", *Solid-State Electronics*, Vol. 44, p. 905, 2000

34 S.I. Lee, "Recent progress in high-k dielectric films for ULSIs", *Extended Abstracts of the International Conference on Solid-State Devices and Materials*, p.8, 2001

35 Y. Nishikawa, N. Fukushima and N. Yasuda, "Direct growth of single crystalline CeO2 high-k gate dielectric", *Extended Abstracts of the International Conference on Solid-State Devices and Materials*, p.174, 2001

36 T. Hori, *Gate dielectrics and MOS ULSIs: principles, technologies and applications*, Springer, pp. 126-130, 1997

37 C. Masuré and M. Orlowski, "Guidelined for reverse short-channel behavior", *IEEE Electron Device Letters*, Vol. 10, N0. 12, p. 556, 1989

38 S. A. Hareland, S. Krishnamurthy, S. Jallepalli, C.F. Yeap, K. Hasnat, A.F. Tasch, Jr. and C.M. Maziar, "A computationally efficient model for inversion layer quantization effects in deep submicron N-channel MOSFET's", *IEEE Transactions on Electron Devices*, Vol. 43, No. 1, p. 90, 1996

Chapter 8

THE BIPOLAR TRANSISTOR

8.1. Introduction and basic principles

The first bipolar transistor was realized in 1947 by Brattain, Bardeen and Shockley.[1] The three of them received the Nobel prize in 1956 for their invention. In a bipolar transistor current is due to transport of both electrons and holes, unlike unipolar devices such as the JFET and the MOSFET where current is due to transport of one type of carrier only. The bipolar transistor is composed of two PN junctions and hence is also called the "Bipolar Junction Transistor" (BJT).

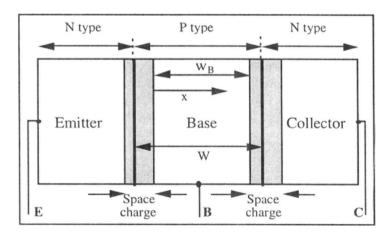

Figure 8.1: NPN bipolar junction transistor. [2]

There are two types of bipolar transistors: the NPN transistor, in which a P-type region is sandwiched between two N-type regions, and the PNP transistor, where N-type silicon is confined between two P-type regions. Here, we will consider only the case of an NPN device shown in Figure 8.1. The equations for a PNP transistors can easily be obtained from the expressions derived for the NPN transistors, provided that the appropriate sign changes are made. In an NPN device the two N-type regions are called "emitter" and "collector", and the P region is called "base". The distance between the two metallurgical junctions is noted W, and the length of the neutral base, defined as the distance between the two space-charge regions generated by the junctions, is noted w_B (Figure 8.1).

8.1.1. Long-base device

If no bias is applied to the device terminals ($V_E=V_B=V_C=0$) both junctions are at thermal equilibrium and there is no current flow (Figure 8.2).

If the emitter-base junction is forward biased ($V_{BE} = V_B-V_E \cong 0.7 \, V$ for a silicon device) current flows through the emitter-base junction. Holes are injected from the base into the emitter where they recombine with majority carriers (electrons). Similarly, electrons are injected from the emitter into the base where they recombine with the local majority carriers (holes). If the collector-base junction is reverse-biased, only a small reverse current (the collector-base junction saturation current) flows between base and collector.

If the width of the neutral base, w_B, is large enough, all the electrons injected by the emitter into the base recombine in the P-type material, because the base width is larger than the electron diffusion length in the base ($w_B >> L_{nB}$). There is no interaction between both junctions and therefore no current flowing between emitter and collector. Neglecting the small reverse current in the collector-base junction, the only current flowing through the device is between the base and the emitter: $I_E = - I_B$.

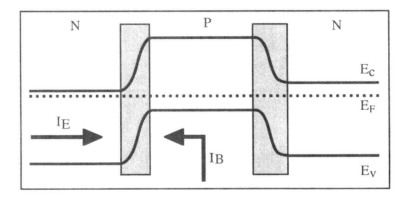

Figure 8.2: Energy bands at thermodynamic equilibrium ($V_E=V_B=V_C$).

8.1.2. Short-base device

Consider now a device with a short base. The term "short base" implies that the neutral base width is smaller than the electron diffusion length: $w_B < L_{nB}$. Let the emitter-base junction be forward biased ($V_{BE} = V_B\text{-}V_E > 0$) and the collector-base junction be reverse biased ($V_{BC} = V_B\text{-}V_C < 0$). Because the length of the neutral base is smaller than the diffusion length for electrons in the base, a number of electrons injected from the emitter into the base can diffuse to the collector-base junction depletion region, at $x=w_B$. Once there, they are accelerated by the electric field of the depletion region and transported into the collector (Figure 8.3).

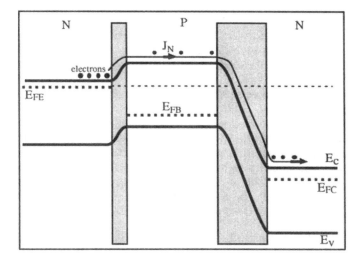

Figure 8.3: Energy bands for a device biased in active mode.[3]

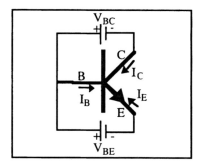

Figure 8.4: Symbolic representation, applied bias, and currents in an NPN bipolar transistor.

In modern bipolar transistors a large portion (99% or more) of the electrons injected by the emitter into the base reach the collector. It is worth noting that the magnitude of current flowing in the collector does not depend on magnitude of the collector voltage; the collector-base junction simply needs to be reverse biased. Rather, the collector current is fixed by the bias applied to the emitter-base diode. This effect, in which the current in a junction is controlled by the bias applied to another junction, is called "transistor effect". An NPN bipolar transistor with a forward-biased emitter-base junction and a reverse-biased collector-base junction is said to operate in the *forward active mode*. The symbolic representation of an NPN bipolar transistor in Figure 8.4 shows the conventions for current direction and applied voltages in the device.

It is possible to bias the device differently than in the forward active mode:

   If both junctions are forward biased the transistor is said to be in saturation. In that case electrons are injected from the emitter through the base into the collector *and* from the collector through the base into the emitter.

   If both junctions are reverse biased there is no current flow at all and the device is in the cut-off mode.

   If the emitter junction is reverse biased and the collector junction is forward biased the transistor operates in the reverse active mode. Although this mode of operation appears to be very similar to the forward active mode, poor performances are obtained from transistors operating in the reverse biased mode. As we will see later, this is due to the use of different doping concentrations in the emitter and the collector.

Let us consider a bipolar transistor biased in the forward active mode. The current flowing through the emitter junction is given by the sum of the hole current injected from the base into the emitter and the electron current injected from the emitter into the base (Figure 8.5). The ratio between these two current components can be obtained using Equation (4.4.23) at $x = l_n$ and Equation (4.4.24) at $x = -l_p$:

$$\frac{I_{nE}}{I_{pE}} = \frac{D_n n_{po} L_p}{D_p p_{no} L_n} = \sqrt{\frac{\mu_n \tau_p}{\mu_p \tau_n}} \frac{N_{dE}}{N_{aB}} \qquad (8.1.1)$$

where N_{aB} and N_{dE} are the doping concentrations in the base and the emitter, respectively.

The collector current, I_{nC}, is due to the diffusion through the base of electrons injected by the emitter into the base. A very small portion of the electrons injected in the base are lost due to inevitable recombination in the base. I_{nC} is equal to $I_{nE} - I_{rB}$, where I_{rB} is the current due to the recombination of electrons in the base (Figure 8.5). The collector current is directly proportional to the electron current injected by the emitter in the base, and the base current is proportional to the hole current injected by the base into the emitter. As we will see later the gain of a bipolar transistor is defined as the collector current divided by the base current. Since high gain values are desirable, a higher doping concentration is used in the emitter than in the base, which yields a high electron to hole current ratio in the emitter-base junction, according to Equation 8.1.1.

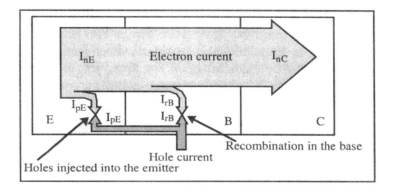

Figure 8.5: Carrier flows in a bipolar transistor biased in the forward active mode.[4]

When the transistor is operating in the forward active mode the collector junction is reverse biased. Any current flowing through the collector can, therefore, not originate from that junction. Figure 8.5 shows the electron and hole currents in the device. A hole current, I_{pE} is injected by the base into the emitter. Once inside the emitter these holes recombine with majority carriers (electrons). A larger electron current, I_{nE}, is injected from the emitter into the base. Some of these electrons recombine with holes in the base, giving rise to another hole current, I_{rB}. The majority of the electrons, however, go through the base without recombining and give rise to a collector current, I_{nC}. The base current is equal to $I_B = I_{pE} + I_{rB}$. Using the convention for current direction of Figure 8.4 and Kirchoff's current law we can write:

$$I_C + I_B + I_E = 0 \qquad (8.1.2)$$

Since the transistor is designed in such a way that $I_B << I_E$, the emitter and the collector current are almost equal in magnitude. One can define a parameter called the "common-base gain", noted α_F:

$$I_C = -\alpha_F I_E \qquad (8.1.3)$$

or:

$$I_C = \frac{\alpha_F}{1 - \alpha_F} I_B \qquad (8.1.4)$$

and thus:

Common-Emitter Current Gain

$$I_C = \beta_F I_B \quad \text{with} \quad \beta_F \equiv \frac{\alpha_F}{1 - \alpha_F} \qquad (8.1.5)$$

The common-base current gain, α_F, describes the relationship between emitter and collector currents when the base is grounded. It represents the ratio of the number of electrons reaching the collector to the number of electrons leaving the emitter. Parameter β_F is the "common-emitter gain" which describes the relationship between collector and base currents when the emitter is grounded. Most of the time, β_F is simply called "current gain". Common-base and common-emitter configurations are shown in Figure 8.6.

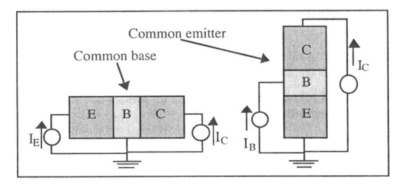

Figure 8.6: Common-base and common-emitter configurations.

The value of α_F in typical bipolar transistors is approximately 0.99. As a result, the value of the current gain, β_F, usually ranges between 50 and 300. There are, however, transistors called "super-β transistors" which have current gains higher than 1,000 or even 10,000.

The common-emitter configuration illustrates the amplification effect created by the bipolar transistor: any current I_B supplied to the base corresponds to a collector current I_C which is β_F times larger than I_B. From the PN junction theory we know that the potential drop in a forward-biased junction can be considered as a constant, which is approximately equal to 0.7 V in silicon (see Section 4.6.1). Therefore, the base-emitter voltage, V_{BE}, in a silicon bipolar transistor biased in the forward active mode is assumed equal to 0.7 V (0.35 V in germanium).

8.1.3. Fabrication process

Before investigating the physics of the bipolar transistor it is interesting to understand how it is fabricated. To fabricate an NPN device the starting material is a P-type silicon substrate. A heavily doped N-type region is locally formed at the surface of the silicon. This region is called a "buried collector" and its function is to create a low-resistance path between the lightly doped collector underneath the active region and the collector contact at the surface of the device (Figure 8.7). Making use of a P-type substrate insures that transistors on a same chip are electrically insulated from one another by reverse-biased PN junctions (the buried collector-substrate junctions). A layer of single-crystal, N-type silicon is then grown in an operation called "epitaxy". Silicon dioxide (SiO_2) is then used to isolate the BJTs from one another laterally. An N-type region is diffused to connect the buried collector to the surface. The active region (where the transistor effect takes place) of the device is formed next using the diffusion of P-type impurities to form the base and N-type doping atoms to form the emitter.

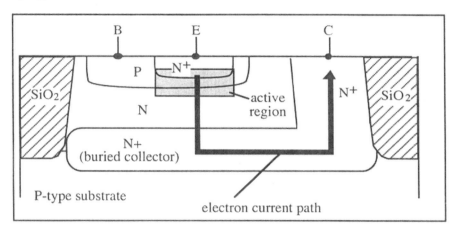

Figure 8.7: Cross section of an NPN bipolar transistor. The active region is the region where the actual transistor effect takes place.

Figure 8.8 shows the doping impurity profile in the bipolar transistor as a function of depth in the silicon along a cut through the center of the active region. In this example impurities in the emitter and the base are diffused from the silicon surface. The impurity concentration in the epitaxial collector remains constant. The base is located where the P-type impurity concentration is larger than the N-type impurity concentration. The emitter-base metallurgical junction is located at the depth where the emitter arsenic profile and the base boron profile intersect. The collector junction is located at the point where the base P-type concentration is equal to the doping concentration in the N-type collector. It is worth noting that $N_{dE} \gg N_{aB}$, which insures that the electron current injected by the emitter into the base is much larger than the hole current injected by the base into the emitter (Equation 8.1.1).

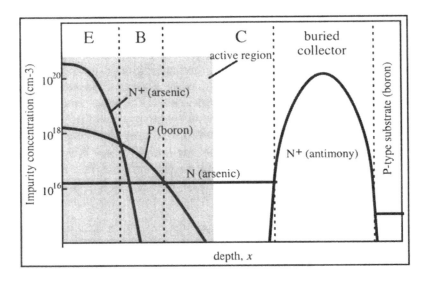

Figure 8.8: Doping impurity profile in an NPN transistor. [5]

8.2. Amplification using a bipolar transistor

Consider the simple amplifier composed of an NPN bipolar transistor and two resistors shown in Figure 8.9. The power supply is held at a constant positive voltage, V_{cc}. The input signal is delivered by the voltage source V_{in}. The output signal, V_{out}, is measured between collector and emitter. The transistor is biased in the forward active mode due to its configuration with the supply voltage.

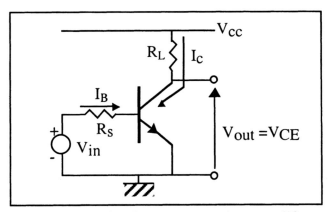

Figure 8.9: Simple common-emitter amplifier

The relationship between the output voltage and the input voltage can be obtained using basic circuit theory. Using Ohm's and Kirchoff's laws one finds:

$$V_{in} = R_S I_B + V_{BE} \quad \text{and} \quad V_{CC} = R_L I_C + V_{CE}$$

Since the transistor operates in the forward active mode we have:

$$V_{BE} = 0.7 \, V \quad \text{and} \quad I_C = \beta_F I_B$$

Combining these relationships we obtain:

$$V_{out} = V_{CE} = V_{CC} - \beta_F \frac{R_L}{R_S} (V_{in} - 0.7V)$$

Thus any variation of the input voltage ΔV_{in} corresponds to a variation of the output voltage. That variation is proportional to ΔV_{out} since:

$$\Delta V_{out} = -\beta_F \frac{R_L}{R_S} \Delta V_{in}$$

Therefore the output signal is equal to the input signal multiplied by a voltage amplification factor $\beta_F \frac{R_L}{R_S}$. Note that there is a 180° phase difference between the output and input signals indicated by the minus sign between ΔV_{in} and ΔV_{out}. If we multiply the equation $V_{CC} = R_L I_C + V_{CE}$ by I_C, we obtain:

$$I_C V_{CC} = R_L I_C^2 + I_C V_{CE}$$

In this expression $I_C V_{CC}$ is the power supplied by the power supply, $R_L I_C^2$ is the power dissipated in the load resistor and $I_C V_{CE}$ is the power dissipated in the transistor. The later term is the price one has to pay to obtain amplification by the transistor.

Example

Calculate the small-signal voltage gain and dc power dissipation of the circuit in Figure 8.9 for $V_{CC} = 10V$, $V_{in} = 2 \, V + \Delta V_{in}$, where ΔV_{in} is considered small compared to V_{in}, $R_S = 1000\Omega$, $R_L = 50\Omega$, and $\beta_F = 100$. Verify that the collector-base junction is reverse biased.

$$\text{gain} = \frac{\Delta V_{out}}{\Delta V_{in}} = -\beta_F \frac{R_L}{R_S} = 100 \frac{50\Omega}{1000\Omega} = -5$$

dc power dissipation $= I_C \times V_{CC} = \beta_F I_B \times V_{CC} = 100(2V-0.7V)/R_S \times 10V = 1.3 \, W$
$V_{CE} = 10V - I_C R_L = 3.5 \, V \Rightarrow V_{BC} = -(V_{CE} - V_{BE}) = -2.8V$: C-B junction is reverse biased.

8.3. Ebers-Moll model

In 1954 J.J. Ebers and J.L. Moll developed a model for the bipolar transistor which is still used in modern circuit simulators.[6]

Consider the NPN bipolar transistor in Figure 8.10. The width of the quasi-neutral regions in the emitter, base, and collector are noted w_E, w_B, and w_C, respectively. The boundaries of the space-charge (depletion) regions are noted l_{E1} and l_{B1} for the emitter-base junction and l_{B2} and l_{C2} for the collector-base junction. To simplify the study of the device we will assume that the impurity concentrations in the emitter, base and collector are constant.

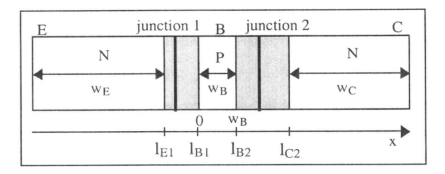

Figure 8.10: Transistor geometry.

To calculate current in the transistor one must use the continuity equation in the base:

$$\frac{\partial n}{\partial t} = \frac{1}{q}\frac{\partial J_n}{\partial x} - U = \frac{1}{q}\frac{\partial J_n}{\partial x} - \frac{n(x)-n_{oB}}{\tau_{nB}} \qquad (8.3.1)$$

where U is the SRH (Shockley-Read-Hall) generation/recombination term for minority carriers (Equation 3.5.20). The equilibrium concentration of electrons in the P-type base is given by:

$$n_{oB} = \frac{n_i^2}{N_{aB}} \qquad (8.3.2)$$

In the absence of an electric field, current in the base is strictly due to diffusion. Electrons injected by the emitter at $x=0$ diffuse until they reach $x=w_B$. The electron current density in the base is equal to:

$$J_n = qD_{nB}\frac{\partial n}{\partial x} \qquad (8.3.3)$$

If we assume steady-state ($\frac{\partial n}{\partial t} = 0$) Equations 8.3.1 and 8.3.3 can be combined and yield the following relationship:

$$\frac{d^2n}{dx^2} = \frac{n(x)-n_{oB}}{D_{nB}\tau_{nB}} = \frac{n(x)-n_{oB}}{L_{nB}^2} \qquad (8.3.4)$$

where $L_{nB} = \sqrt{D_{nB}\tau_n}$ is the diffusion length of the electrons in the base, which represents the average distance along which electrons can diffuse in the base before recombining.

The solution to Equation 8.3.4 has the following form:

$$n(x) = n_{oB} + A\ exp\left(\frac{x}{L_{nB}}\right) + B\ exp\left(\frac{-x}{L_{nB}}\right) \qquad (8.3.5)$$

where A and B are integration constants which will be calculated using the Boltzmann relationships 2.7.1. and 2.7.2 as boundary conditions. The Boltzmann relationships give us the electron concentration in the base at the edge of the space-charge regions of the emitter and collector junctions, *i.e.* at $x=0$ and $x=w_B$:

$$n(0) = n_{oB}\ exp\left(\frac{qV_{BE}}{kT}\right)\ \text{and}\ n(w_B) = n_{oB}\ exp\left(\frac{qV_{BC}}{kT}\right) \qquad (8.3.6)$$

from which A and B can be extracted:

Using the boundary condition $n(0) = n_{oB} + A + B \Rightarrow B = n(0) - n_{oB} - A$

and

$$n(w_B) = n_{oB} + A\ exp\left(\frac{w_B}{L_{nB}}\right) + B\ exp\left(\frac{-w_B}{L_{nB}}\right)$$

$$= n_{oB} + A\ exp\left(\frac{w_B}{L_{nB}}\right) + [n(0) - n_{oB} - A]\ exp\left(\frac{-w_B}{L_{nB}}\right)$$

we find:

$$A = \frac{n(w_B) - n_{oB} - [n(0) - n_{oB}]exp\left(\frac{-w_B}{L_{nB}}\right)}{2\ sinh\left(\frac{w_B}{L_{nB}}\right)}$$

and since

$$B = n(0) - n_{oB} - A$$

$$B = \frac{2[n(0) - n_{oB}]sinh\left(\frac{w_B}{L_{nB}}\right) - [n(w_B) - n_{oB}] + [n(0) - n_{oB}]exp\left(\frac{-w_B}{L_{nB}}\right)}{2\ sinh\left(\frac{w_B}{L_{nB}}\right)}$$

we find:

$$B = \frac{[n(0) - n_{oB}]exp\left(\frac{w_B}{L_{nB}}\right) - [n(w_B) - n_{oB}]}{2\ sinh\left(\frac{w_B}{L_{nB}}\right)}$$

Knowing integration constants A and B we can now write the electron concentration as a function of x in the base:

$$n(x) = n_{oB} + \frac{[n(w_B) - n_{oB}] - [n(0) - n_{oB}] \exp\left(\dfrac{-w_B}{L_{nB}}\right)}{2 \sinh\left(\dfrac{w_B}{L_{nB}}\right)} \, \exp\left(\frac{x}{L_{nB}}\right)$$

$$+ \frac{[n(0) - n_{oB}] \exp\left(\dfrac{w_B}{L_{nB}}\right) - [n(w_B) - n_{oB}]}{2 \sinh\left(\dfrac{w_B}{L_{nB}}\right)} \, \exp\left(\frac{-x}{L_{nB}}\right)$$

which can be rewritten:

$$n(x) = n_{oB} + \frac{[n(w_B) - n_{oB}] \sinh\left(\dfrac{x}{L_{nB}}\right) + [n(0) - n_{oB}] \sinh\left(\dfrac{w_B - x}{L_{nB}}\right)}{\sinh\left(\dfrac{w_B}{L_{nB}}\right)}$$

Expressing $n(w_B)$ and $n(0)$ as a function of the applied voltages using 8.3.6 we finally obtain:

$$n(x) = n_{oB} \left\{ 1 + \left[\exp\left(\frac{qV_{BE}}{kT}\right) - 1 \right] \frac{\sinh\left(\dfrac{w_B - x}{L_{nB}}\right)}{\sinh\left(\dfrac{w_B}{L_{nB}}\right)} + \left[\exp\left(\frac{qV_{BC}}{kT}\right) - 1 \right] \frac{\sinh\left(\dfrac{x}{L_{nB}}\right)}{\sinh\left(\dfrac{w_B}{L_{nB}}\right)} \right\}$$

$$(8.3.7)$$

The electron concentration profile, $n(x)$, is shown in Figure 8.11.

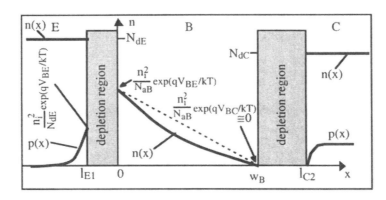

Figure 8.11: Electron concentration in the emitter, base and collector, and hole concentration in the emitter in an NPN BJT biased in the forward active mode ($V_{BE} > 0$ and $V_{CB} > 0$).

The diffusion current at the emitter-side edge of the neutral base ($x=0$) is equal to:

$$J_{nE} = qD_{nB} \left. \frac{dn}{dx} \right|_{x=0}$$

$$= -\frac{qD_{nB}n_{oB}}{L_{nB}tanh\left(\frac{w_B}{L_{nB}}\right)} \left[exp\left(\frac{qV_{BE}}{kT}\right) - 1 \right] + \frac{qD_{nB}n_{oB}}{L_{nB}sinh\left(\frac{w_B}{L_{nB}}\right)} \left[exp\left(\frac{qV_{BC}}{kT}\right) - 1 \right] \quad (8.3.8)$$

At the collector-side edge of the neutral base ($x = w_B$), the diffusion current is equal to:

$$J_{nC} = qD_{nB} \left. \frac{dn}{dx} \right|_{x=w_B}$$

$$= -\frac{qD_{nB}n_{oB}}{L_{nB}sinh\left(\frac{w_B}{L_{nB}}\right)} \left[exp\left(\frac{qV_{BE}}{kT}\right) - 1 \right] + \frac{qD_{nB}n_{oB}}{L_{nB}tanh\left(\frac{w_B}{L_{nB}}\right)} \left[exp\left(\frac{qV_{BC}}{kT}\right) - 1 \right] \quad (8.3.9)$$

The hole concentration profile in the emitter and the collector can be found using the PN junction theory, assuming that the width of the quasi-neutral N-type regions are much larger than the hole diffusion length. The hole current injected by the base into the emitter can be found using Relationship 4.4.23 for $x = l_{E1}$:

$$J_{pE} = -qD_{pE} \left. \frac{dp}{dx} \right|_{x=l_{E1}} = -\frac{q\,D_{pE}\,p_{oE}}{L_{pE}} \left[exp\left(\frac{qV_{BE}}{kT}\right) - 1 \right] \quad (8.3.10)$$

where p_{oE} is the equilibrium hole concentration in the emitter. Similarly, the hole current flowing from the base into the collector at $x = l_{C2}$ is equal to:

$$J_{pC} = -qD_{pC} \left. \frac{dp}{dx} \right|_{x=l_{C2}} = \frac{q\,D_{pC}\,p_{oC}}{L_{pC}} \left[exp\left(\frac{qV_{BC}}{kT}\right) - 1 \right] \quad (8.3.11)$$

where p_{oC} is the equilibrium hole concentration in the collector. The emitter current encompasses both the current of electrons injected by the emitter into the base and the current of holes injected by the base into the emitter.

If the area of the cross section of transistor is noted A, we can write:

$$I_E = A(J_{pE} + J_{nE}) \quad (8.3.12)$$

which, using 8.3.8 and 8.3.10, yields the emitter current for the Ebers-Moll model:

$$I_E = -\left[\frac{AqD_{pE}p_{oE}}{L_{pE}} + \frac{AqD_{nB}n_{oB}}{L_{nB}tanh\left(\frac{w_B}{L_{nB}}\right)}\right]\left(exp\left(\frac{qV_{BE}}{kT}\right)-1\right)$$

$$+ \frac{AqD_{nB}n_{oB}}{L_{nB}sinh\left(\frac{w_B}{L_{nB}}\right)}\left(exp\left(\frac{qV_{BC}}{kT}\right)-1\right) \qquad (8.3.13a)$$

Similarly the collector current is given by:

$$I_C = -A(J_{pC} + J_{nC})$$

which, using 8.3.9 and 8.3.11, yields:

$$I_C = \frac{AqD_{nB}n_{oB}}{L_{nB}sinh\left(\frac{w_B}{L_{nB}}\right)}\left(exp\left(\frac{qV_{BE}}{kT}\right)-1\right)$$

$$- \left[\frac{AqD_{nB}n_{oB}}{L_{nB}tanh\left(\frac{w_B}{L_{nB}}\right)} + \frac{AqD_{pC}p_{oC}}{L_{pC}}\right]\left(exp\left(\frac{qV_{BC}}{kT}\right)-1\right) \qquad (8.3.13b)$$

These expressions can be simplified by defining the emitter junction reverse saturation current, I_{ES}, as the current that flows in the emitter when the emitter-base junction is reverse biased ($V_{BE}<0$) and the collector is short-circuited to the base ($V_{BC}=0$):

$$I_{ES} = \frac{AqD_{pE}p_{oE}}{L_{pE}} + \frac{AqD_{nB}n_{oB}}{L_{nB}tanh\left(\frac{w_B}{L_{nB}}\right)} \qquad (8.3.14)$$

In a similar way one can define the collector junction reverse saturation current, I_{CS}, as the current that flows in the collector when the collector-base junction is reverse biased ($V_{BC}<0$) and the emitter is short-circuited to the base ($V_{BE}=0$):

$$I_{CS} = \frac{AqD_{nB}n_{oB}}{L_{nB}tanh\left(\frac{w_B}{L_{nB}}\right)} + \frac{AqD_{pC}p_{oC}}{L_{pC}} \qquad (8.3.15)$$

The forward common-base gain, α_F, is defined as the ratio of collector to emitter current when the collector is shorted to the base ($V_{BC} = 0$):

$$\alpha_F = -\left.\frac{I_C}{I_E}\right|_{V_{BC}=0} = \frac{\dfrac{AqD_{nB}n_{oB}}{L_{nB}sinh\left(\dfrac{w_B}{L_{nB}}\right)}}{\dfrac{AqD_{pE}p_{oE}}{L_{pE}} + \dfrac{AqD_{nB}n_{oB}}{L_{nB}tanh\left(\dfrac{w_B}{L_{nB}}\right)}}$$

which can be rewritten:

$$\alpha_F = \frac{1}{cosh\left(\dfrac{w_B}{L_{nB}}\right)\left[1 + \dfrac{p_{oE}}{n_{oB}}\dfrac{D_{pE}}{D_{nB}}\dfrac{L_{nB}}{L_{pE}}tanh\left(\dfrac{w_B}{L_{nB}}\right)\right]} \qquad (8.3.16)$$

In a similar way the reverse common-base gain, α_R, is defined as the ratio of emitter to collector current when the emitter is shorted to the base ($V_{BE} = 0$):

$$\alpha_R = -\left.\frac{I_E}{I_C}\right|_{V_{BE}=0} = \frac{\dfrac{AqD_{nB}n_{oB}}{L_{nB}sinh\left(\dfrac{w_B}{L_{nB}}\right)}}{\dfrac{AqD_{nB}n_{oB}}{L_{nB}tanh\left(\dfrac{w_B}{L_{nB}}\right)} + \dfrac{AqD_{pC}p_{oC}}{L_{pC}}}$$

which can be rewritten:

$$\alpha_R = \frac{1}{cosh\left(\dfrac{w_B}{L_{nB}}\right)\left[1 + \dfrac{p_{oC}}{n_{oB}}\dfrac{D_{pC}}{D_{nB}}\dfrac{L_{nB}}{L_{pC}}tanh\left(\dfrac{w_B}{L_{nB}}\right)\right]} \qquad (8.3.17)$$

Finally, the Ebers-Moll Equations 8.3.13a and 8.3.13b can be written in a compact form using the parameters defined in Expressions 8.3.14 to 8.3.17, as a function of applied biases V_{BE} and V_{BC}:

$$I_E = -I_{ES}\left[exp\left(\frac{qV_{BE}}{kT}\right) - 1\right] + \alpha_R I_{CS}\left[exp\left(\frac{qV_{BC}}{kT}\right) - 1\right]$$

and

$$I_C = \alpha_F I_{ES}\left[exp\left(\frac{qV_{BE}}{kT}\right) - 1\right] - I_{CS}\left[exp\left(\frac{qV_{BC}}{kT}\right) - 1\right] \qquad (8.3.18)$$

or, in a matrix form:

$$\begin{bmatrix} I_E \\ I_C \end{bmatrix} = \begin{bmatrix} -I_{ES} & \alpha_R I_{CS} \\ \alpha_F I_{ES} & -I_{CS} \end{bmatrix} \begin{bmatrix} exp(qV_{BE}/kT)-1 \\ exp(qV_{BC}/kT)-1 \end{bmatrix} \qquad (8.3.19)$$

In the case of a PNP transistor the Ebers-Moll equations are:

$$\begin{bmatrix} I_E \\ I_C \end{bmatrix} = \begin{bmatrix} I_{ES} & -\alpha_R I_{CS} \\ -\alpha_F I_{ES} & I_{CS} \end{bmatrix} \begin{bmatrix} exp(qV_{EB}/kT)-1 \\ exp(qV_{CB}/kT)-1 \end{bmatrix}$$

These equations accurately describe the current of a bipolar transistor for any mode of operation, *i.e.* they predict the current for all permutations of biasing of V_{BE} and V_{BC}. By adding Kirchoff's current law, $I_E+I_B+I_C=0$ the base current can be derived as well. The four parameters used in the Ebers-Moll equations (α_F, α_R, I_{ES} and I_{CS}) are not independent from one another, and any of these parameters can be calculated if three are known using the so-called *reciprocity relationship*:

$$\alpha_F I_{ES} = \alpha_R I_{CS} = \frac{AqD_{nB}n_{oB}}{L_{nB}sinh\left(\dfrac{w_B}{L_{nB}}\right)} \qquad (8.3.20)$$

In the forward active region the transistor encompasses two diodes, the first of which is the forward-biased emitter-base junction in which flows a current given by:

$$I'_F = I_{ES}\left[exp\left(\frac{qV_{BE}}{kT}\right)-1\right] \qquad (8.3.21)$$

The second diode (the collector-base junction) is reverse biased and the current flowing through it is:

$$I'_R = I_{CS}\left[exp\left(\frac{qV_{BC}}{kT}\right)-1\right] \qquad (8.3.22)$$

Combining the two latter Relationships with Expression 8.3.19 we can write:

$$I_E = -I'_F + \alpha_R I'_R \quad \text{and} \quad I_C = -I'_R + \alpha_F I'_F \qquad (8.3.23)$$

and an equivalent circuit of the transistor can be drawn (Figure 8.12).

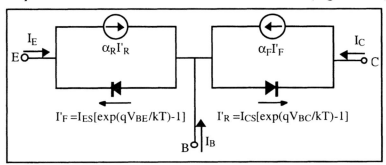

Figure 8.12: Equivalent circuit of the Ebers-Moll model for the NPN transistor.

Note: α_R is called the "reverse active, common-base gain". It is defined in a similar way to α_F. α_R represents the common-base gain of a device biased in the reverse active mode, where $V_{BE}<0$ and $V_{BC}>0$. The reverse active gain, α_R, is much smaller than the forward active gain, α_F, because the collector doping concentration is smaller then the doping concentration in the emitter $(N_{dC}<<N_{dE} \Rightarrow p_{oE}<<p_{oC})$ (see Section 8.2). The bipolar transistor is not a symmetrical device, unlike the MOS transistor where the source and drain are interchangeable without modifying device operation. It can be noted, however, that if the doping impurity concentration in the collector, N_{dC}, is equal to that in the emitter, N_{dE}, and if the base concentration, N_{aB}, is constant as a function of x, then the device becomes symmetrical $(\alpha_F=\alpha_R$ and $I_{CS}=I_{ES})$.

The model presented in Figure 8.12 is not often used in practice because it calls for two parameters, I'_F and I'_R, that cannot be easily measured. To circumvent that problem the Ebers-Moll equations can be written in a different form.

Let us note I_{C0} the saturation current flowing in the collector when the collector junction is reverse biased and the emitter is left open $(I_E=0)$. In that case the Ebers-Moll equations become:

$$\begin{bmatrix} 0 \\ I_{C0} \end{bmatrix} = \begin{bmatrix} -I_{ES} & \alpha_R I_{CS} \\ \alpha_F I_{ES} & -I_{CS} \end{bmatrix} \begin{bmatrix} exp(qV_{BE}/kT)-1 \\ -1 \end{bmatrix}$$

from which we conclude:

$$I_{ES} [exp(qV_{BE}/kT)-1] = - \alpha_R I_{CS} \text{ and } I_{C0} = \alpha_F I_{ES} [exp(qV_{BE}/kT)-1] + I_{CS}$$
$$\Rightarrow I_{C0} = -\alpha_F \alpha_R I_{CS} + I_{CS} \Rightarrow I_{C0} = (1 - \alpha_F \alpha_R) I_{CS} \qquad (8.3.24)$$

Similarly we can define I_{E0} as the saturation current flowing in the emitter when the emitter junction is reverse biased and the collector is left open $(I_C=0)$, in which case we have:

$$\begin{bmatrix} I_{E0} \\ 0 \end{bmatrix} = \begin{bmatrix} -I_{ES} & \alpha_R I_{CS} \\ \alpha_F I_{ES} & -I_{CS} \end{bmatrix} \begin{bmatrix} -1 \\ exp(qV_{BC}/kT)-1 \end{bmatrix}$$

from which we conclude:

$$I_{E0} = I_{ES} + \alpha_R I_{CS} [exp(qV_{BC}/kT)-1] \text{ and } \alpha_F I_{ES} = - I_{CS} [exp(qV_{BC}/kT)-1]$$
$$\Rightarrow I_{E0} = -\alpha_F \alpha_R I_{ES} + I_{ES} \Rightarrow I_{E0} = (1 - \alpha_F \alpha_R) I_{ES} \qquad (8.3.25)$$

It is worth noting that there exists a reciprocity relationship between I_{E0} and I_{C0} that is similar to that defined in Expression 8.3.20 since we have:

$$\frac{I_{E0}}{I_{C0}} = \frac{I_{ES}}{I_{CS}} \Rightarrow \alpha_F I_{E0} = \alpha_R I_{C0} \qquad (8.3.26)$$

The Ebers-Moll equations can, therefore, be re-written in the following form::

$$\begin{bmatrix} I_E \\ I_C \end{bmatrix} = \frac{1}{1-\alpha_F\alpha_R}\begin{bmatrix} -I_{E0} & \alpha_R I_{C0} \\ \alpha_F I_{E0} & -I_{C0} \end{bmatrix}\begin{bmatrix} exp(qV_{BE}/kT)-1 \\ exp(qV_{BC}/kT)-1 \end{bmatrix}$$

Eliminating $\dfrac{I_{C0}[exp(qV_{BC}/kT)-1]}{1-\alpha_F\alpha_R}$ between the equations for I_E and I_C

one obtains:

$$I_E = -\alpha_R I_C - I_{E0}\ [exp(qV_{BE}/kT)-1] \qquad (8.3.27)$$

and the elimination of $\dfrac{I_{E0}[exp(qV_{BE}/kT)-1]}{1-\alpha_F\alpha_R}$ between the expression of

I_E and I_C yields:

$$I_C = -\alpha_F I_E - I_{C0}\ [exp(qV_{BC}/kT)-1] \qquad (8.3.28)$$

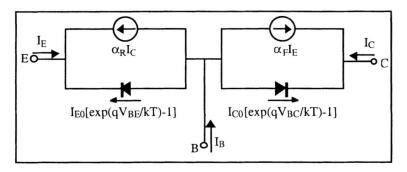

Figure 8.13: Equivalent circuit of the Ebers-Moll model for the NPN transistor. [7,8]

Equations 8.3.27 and 8.3.28 show that emitter and collector currents are each made up of two components: a diode-like junction current (a reverse current for the collector junction and a forward current for the emitter junction when the device is biased in the forward active mode) and a current imposed by a current source ($-\alpha_R I_C$ or $-\alpha_F I_E$). It is important to note that each of these currents can be obtained by a direct measurement of the device. This new formulation of the Ebers-Moll equations can be represented by the equivalent circuit of Figure 8.13.

8.3.1. Emitter efficiency

In an "ideal" bipolar transistor the base current should be much smaller than the emitter and collector currents. Similarly, the hole current injected by the base into the emitter should be much smaller than the electron current injected from the emitter into the base, and from there, into the collector. One defines the emitter efficiency, γ_F, as the ratio of the electron current injected from the emitter into the base to the total current in the emitter-base junction. The latter current is the sum of the electron current injected from the emitter into the base and the hole current injected by the base into the emitter with the collector shorted to the base:

$$\gamma_F = \left.\frac{I_{nE}}{I_{nE} + I_{pE}}\right|_{V_{BC}=0} = \left.\frac{1}{1 + \dfrac{I_{pE}}{I_{nE}}}\right|_{V_{BC}=0} \tag{8.3.29}$$

Using 8.3.8 and 8.3.10 one can write:

$$\gamma_F = \frac{1}{1 + \dfrac{D_{pE}\,p_{oE}\,L_{nB}}{D_{nB}\,n_{oB}\,L_{pE}}\,tanh\!\left(\dfrac{w_B}{L_{nB}}\right)} \tag{8.3.30}$$

In modern bipolar transistors the width of the neutral base is much smaller than the diffusion length of the electrons in the base, such that $w_B \!<\!< L_{nB}$. In that case the term $tanh(w_B/L_{nB})$ can be approximated by w_B/L_{nB} and one obtains:

Emitter Efficiency

$$\gamma_F \cong \frac{1}{1 + \dfrac{D_{pE}\,p_{oE}\,L_{nB}}{D_{nB}\,n_{oB}\,L_{pE}}\dfrac{w_B}{L_{nB}}} = \frac{1}{1 + \dfrac{\mu_{pE}\,p_{oE}\,w_B}{\mu_{nB}\,n_{oB}\,L_{pE}}} \tag{8.3.31}$$

The latter relationship explains why a higher doping concentration is used in the emitter than in the base: the emitter efficiency is large (close to 1) if the following inequalities are met:

$$n_{oE} \cong N_{dE} >> N_{aB} \cong p_{oB} \quad \Rightarrow \quad \frac{n_i^2}{n_{oE}} = p_{oE} << n_{oB} = \frac{n_i^2}{p_{oB}}$$

A similar conclusion has already been drawn from analyzing the different parameters in Relationship 8.1.1.

8.3.2. Transport factor in the base

The success rate at which the electrons injected into the base reach the collector is measured by a parameter called "transport factor in the base" and noted α_T. It represents the percentage of electrons which have "escaped" recombination with holes (majority carriers) during their journey through the base and is defined as the current of electrons reaching the collector after crossing the base divided by the current of electrons injected by the emitter into the base:

$$\alpha_T = - \left.\frac{I_{nC}}{I_{nE}}\right|_{V_{BC}=0} = \left.\frac{J_{nC}}{J_{nE}}\right|_{V_{BC}=0} \tag{8.3.32}$$

Using 8.3.8 and 8.3.9 one can write:

$$\alpha_T = \frac{1}{cosh\left(\dfrac{w_B}{L_{nB}}\right)} \tag{8.3.33}$$

From this Relationship it is clear that α_T is large if w_B/L_{nB} is small, *i.e.* if the base width is small or if the diffusion length of the electrons in the base is large. In modern bipolar transistors the following relationship is verified: $w_B<<L_{nB}$. One can, therefore approximate $cosh(w_B/L_{nB})$ by $1 + \dfrac{(w_B/L_{nB})^2}{2}$ and one obtains:

Transport Factor in the Base

$$\alpha_T = \frac{1}{cosh\left(\dfrac{w_B}{L_{nB}}\right)} \cong \frac{1}{1 + \dfrac{(w_B/L_{nB})^2}{2}} \cong 1 - \frac{(w_B/L_{nB})^2}{2} \tag{8.3.34}$$

It is easy to verify that when the collector is shorted to the base the common-base gain, α_F, is given by the product of the emitter efficiency by the transport factor in the base:

Common-Base Current Gain

$$\alpha_F = - \left.\frac{I_C}{I_E}\right|_{V_{BC}=0} = - \left.\frac{I_{nC}}{I_{nE} + I_{pE}}\right|_{V_{BC}=0}$$

$$= \left.\frac{I_{nE}}{I_{nE} + I_{pE}}\right|_{V_{BC}=0} \times - \left.\frac{I_{nC}}{I_{nE}}\right|_{V_{BC}=0} = \gamma_F \, \alpha_T \tag{8.3.35}$$

The common-emitter gain, β_F, is given by 8.1.5: $\beta_F = \dfrac{\alpha_F}{1 - \alpha_F}$. Because most analog amplifiers use the common-emitter configuration we might ask what can be done to achieve a high common-emitter gain. Large gain transistors can be achieved by varying some processing parameters during device fabrication, such as:

◊   A reduction of base width which yields devices with higher transport factor in the base, and hence higher gain. The base width of bipolar transistors has been reduced from tens of micrometers in 1954 to 0.1 μm or less today.

◊   A higher doping concentration in the emitter than in the base ($N_{dE} >> N_{aB}$) to increase the emitter efficiency.

◊ Polysilicon can be used as the emitter material. In that case the interface between the silicon base and the polysilicon decreases the hole current injected by the base into the emitter, which increases the emitter efficiency.[9]

◊ Alternative base materials, such as silicon-germanium alloys, can be used to decrease the hole current injected by the base into the emitter, which increases the emitter efficiency (see Section 9.2).

If the base width is small ($w_B <<L_{nB}$) the hyperbolic functions in Relationship 8.3.7 can be linearized and the electron concentration in the base becomes a linear (straight line) function of the position, x. Equation 8.3.7 therefore becomes:

$$n(x) = \frac{n_i^2}{N_{aB}} \left\{ \left[exp\left(\frac{qV_{BE}}{kT}\right) - exp\left(\frac{qV_{BC}}{kT}\right) \right] [1 - \frac{x}{w_B}] + exp\left(\frac{qV_{BC}}{kT}\right) \right\} \qquad (8.3.36)$$

In such a case the electron current density is constant and independent of the position in the base (*i.e.*, the slope of the concentration gradient dn/dx is constant):

$$J_n = qD_{nB}\frac{dn}{dx} = \frac{-q\, D_{nB}\, n_i^2}{N_{aB}\, w_B} \left\{ exp\left(\frac{qV_{BE}}{kT}\right) - exp\left(\frac{qV_{BC}}{kT}\right) \right\} \qquad (8.3.37)$$

Figure 8.14 show the distribution of carriers in a thin-base transistor biased in the forward active mode.

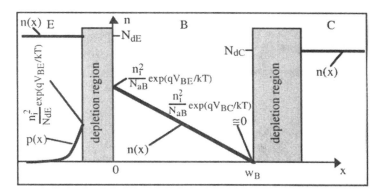

Figure 8.14: Distribution of minority carriers in the base (forward active mode). [10]

It is worth noting that assuming that the minority carrier distribution in the base is linear is equivalent to neglecting recombination in the base. This can easily be verified by using the drift-diffusion and steady-state continuity equation for electrons in the base:

$$J_n = qD_{nB}\frac{dn}{dx} = constant \Rightarrow \frac{dJ_n}{dx} = 0 \text{ and thus } \frac{1}{q}\frac{dJ_n}{dx} - U = 0 \Rightarrow U = 0$$

It is also worthwhile noting that the transport factor in the base, α_T, is equal to 1 when there is no recombination of electrons in the base.

8.4. Regimes of operation

The Ebers-Moll Model can be used to describe the different possible regimes of operation of the bipolar transistor, which depend on the bias applied to the different device terminals. Figure 8.15 shows these different regimes of operation as a function of the two junction biases.

◊ If both emitter-base and collector-base junctions are reverse biased, the transistor is in cut-off and no carriers are injected into the base.

◊ If both emitter-base and collector-base junctions are forward biased, the transistor is said to be in saturation and minority carriers are injected into the base by both the emitter and the collector.

◊ If the emitter-base junction is forward biased and the collector-base junction is reverse biased the device is operating in forward active mode. Electrons are injected by the emitter into the base and most of them are collected by the collector. If the transistor is a silicon device the potential drop across the emitter junction is equal to 0.7 V.

◊ If the emitter-base junction is reverse biased and the collector-base junction is forward biased the device is operating in reverse active mode. Electrons are injected by the collector into the base and are collected by the emitter. Since the doping concentration in the collector is lower than that in the base the gain of the transistor is very low (it is usually less than unity).

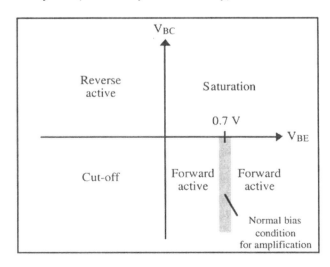

Figure 8.15: Regimes of operation for an NPN silicon bipolar transistor.

Figure 8.16 shows the electron concentration profile in the base for each operation regime, neglecting recombination in the base. Note that the profile in saturation corresponds to the superposition of the forward active and reverse active profiles where both emitter-base and collector-base junctions are forward biased.

The distribution of minority carriers in the base can be calculated for all regions of operation by using Relationship 8.3.36:

$$n(x) = \frac{n_i^2}{N_{aB}} \left\{ \left[exp\left(\frac{qV_{BE}}{kT}\right) - exp\left(\frac{qV_{BC}}{kT}\right) \right] [1 - \frac{x}{w_B}] + exp\left(\frac{qV_{BC}}{kT}\right) \right\}$$

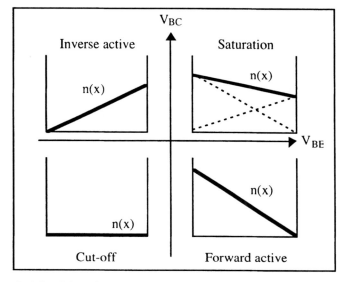

Figure 8.16: Distribution of minority carriers in the base, neglecting recombination in the base, for the different regimes of operation.

8.5. Transport model

The Ebers-Moll equations for an NPN transistor are given by Relationship 8.3.19:

$$\begin{bmatrix} I_E \\ I_C \end{bmatrix} = \begin{bmatrix} -I_{ES} & \alpha_R I_{CS} \\ \alpha_F I_{ES} & -I_{CS} \end{bmatrix} \begin{bmatrix} exp(qV_{BE}/kT)-1 \\ exp(qV_{BC}/kT)-1 \end{bmatrix}$$

Using the reciprocity relationship 8.3.20 a saturation current, I_S, can be defined:

$$\alpha_F I_{ES} = \alpha_R I_{CS} = I_S \qquad (8.5.1)$$

Using this saturation current the Ebers-Moll equations can be rewritten in the following form:

$$\begin{bmatrix} I_E \\ I_C \end{bmatrix} = I_S \begin{bmatrix} -\dfrac{1}{\alpha_F} & 1 \\ 1 & -\dfrac{1}{\alpha_R} \end{bmatrix} \begin{bmatrix} exp(qV_{BE}/kT)-1 \\ exp(qV_{BC}/kT)-1 \end{bmatrix}$$

The model can be optimized for use in the common-emitter configuration by expressing the common-base gains, α_F and α_R, as functions of the common-emitter gains, β_F and β_R, which gives:

$$\beta_F = \frac{\alpha_F}{1-\alpha_F} \Rightarrow \alpha_F = \frac{\beta_F}{1+\beta_F} \Rightarrow \frac{1}{\alpha_F} = 1+\frac{1}{\beta_F} \tag{8.5.2}$$

and
$$\beta_R = \frac{\alpha_R}{1-\alpha_R} \Rightarrow \alpha_R = \frac{\beta_R}{1+\beta_R} \Rightarrow \frac{1}{\alpha_R} = 1+\frac{1}{\beta_R} \tag{8.5.3}$$

Using the above equations one can write:

Ebers-Moll Equation (Transport Model)

$$\begin{bmatrix} I_E \\ I_C \end{bmatrix} = I_S \begin{bmatrix} -\dfrac{1}{\beta_F} - 1 & 1 \\ 1 & -1 - \dfrac{1}{\beta_R} \end{bmatrix} \begin{bmatrix} exp(qV_{BE}/kT)-1 \\ exp(qV_{BC}/kT)-1 \end{bmatrix} \tag{8.5.4}$$

If we now define:

$$I_F \equiv I_S \, [exp(qV_{BE}/kT)-1] \ \text{ and } \ I_R \equiv I_S \, [exp(qV_{BC}/kT)-1] \tag{8.5.5}$$

we obtain the following relationships:

$$I_E = I_R - I_F - \frac{I_F}{\beta_F} \ \text{ and } \ I_C = I_F - I_R - \frac{I_R}{\beta_R} \tag{8.5.6}$$

Equation 8.5.6 highlights the fact that the emitter and the base share a common current component, $I_F - I_R$, corresponding to the electrons injected by the emitter and collected by the collector. The equivalent circuit for the transport model is shown in Figure 8.17. This circuit represents the "transport model" of the transistor since it illustrates the transport of electrons from the emitter to the collector, apparently without passing through the base. This is, of course, incorrect from a device physics point of view, but perfectly valid from an equivalent circuit point of view.

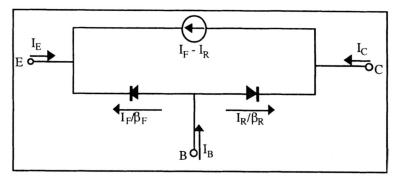

Figure 8.17: Equivalent circuit for the transport model.

When the device is operated in the forward active mode the equivalent circuit corresponding to the transport model can be simplified, as shown in Figure 8.18. In that case the coupling between the input of the device (the base) and its output (the emitter) disappears.

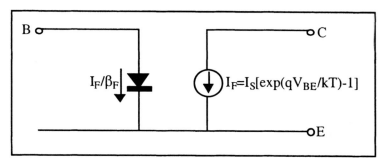

Figure 8.18: Equivalent circuit for the transport model in forward active mode. [11]

8.6. Gummel-Poon model

The doping concentration in the base and the emitter of a real bipolar transistor is not constant, as shown in Figure 8.8. The so-called Gummel-Poon model accounts for inhomogenous distributions of doping concentrations in the device.[12] We will use the same notations for the device as previously, as shown in Figure 8.19. The different electron and hole fluxes as well as the current components in the transistor are shown in Figure 8.20.

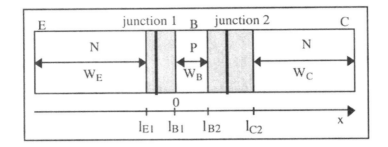

Figure 8.19: Transistor geometry.

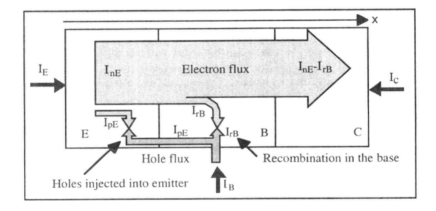

Figure 8.20: Current components in the transistor in the forward active mode.

In Figure 8.20 the different current components have the following signs:

$I_{nE} < 0$ (electron flux in the $+x$ direction)
$I_{pE} < 0$ (electron flux in the $+x$ direction and/or hole flux in the $-x$ direction)
$I_{rB} < 0$ (electron flux in the $+x$ direction and/or hole flux in the $-x$ direction)

Giving positive values to I_E, I_B and I_C in Figure 8.20 one obtains the following general relationships:

◊   The current flowing into the emitter terminal is equal to the sum of the magnitude of the electron current injected by the emitter into the base and the hole current injected by the base into the emitter:

$$I_E = AJ_{nE} + AJ_{pE} = I_{nE} + I_{pE} \qquad (8.6.1.a)$$

(Note that J_{nE} and J_{pE} have negative values referred to the direction of I_E (See Equations 8.3.8 and 8.3.10)). Therefore in the forward active mode the current flows out of the emitter.

◊ The base current is equal to the hole current injected by the base into the emitter plus the current due to recombination into the base:

$$I_B = - I_{rB} - I_{pE} \qquad (8.6.1.b)$$

(In the forward active mode the base current flows in the base).

◊ The collector current is given by:

$$I_C = - A J_{nE} + I_{rB} = - I_{nE} + I_{rB} \qquad (8.6.1.c)$$

(In the forward active mode the base current flows into the collector).

One can readily verify that $I_E + I_B + I_C = 0$.

Calculations for I_{nE}, I_{rB} and I_{pE} (or J_{nE}, J_{rB} and J_{pE}) will enable us to determine the terminal currents, I_E, I_B and I_C using Relationships 8.6.1.a, b and c. J_{nE} will be calculated in the following section using Expression 8.6.12.a. J_{rB} will be calculated in Section 8.6.1.1 and J_{pE} in Section 8.6.1.2.

Calculation of J_{nE}

In a non uniformly doped semiconductor, such as the emitter and the base of a bipolar transistor, the presence of an impurity concentration variation gives rise to an electric field in the semiconductor. When no external bias is applied ($V_{BE} = V_{BC} = 0$) the equilibrium electric field, $\mathcal{E}_{0x}$, can be calculated using the drift-diffusion equation for the majority carriers (holes):

$$J_{pB} = 0 \implies q\mu_{pp}\mathcal{E}_{0x} - qD_{pB}\frac{dp}{dx} = 0 \qquad (8.6.2.a)$$

$$\implies \mathcal{E}_{0x} = \frac{D_{pB}}{\mu_p}\frac{1}{p}\frac{dp}{dx} \cong \frac{D_{pB}}{\mu_p}\frac{1}{N_{aB}}\frac{dN_{aB}}{dx} = \frac{kT}{q}\frac{1}{N_{aB}}\frac{dN_{aB}}{dx} \qquad (8.6.2.b)$$

where $\mathcal{E}_{0x}$ is the electric field in the base in the absence of external bias. Note that N_{aB}, p and $\mathcal{E}_{0x}$ are functions of x, where $x=0$ at the boundary between the emitter-base transition region and the quasi-neutral base, and continues in the positive direction toward the collector (Figure 8.19)

Let us now analyze what happens when an external bias is applied to the junctions. Assuming that the majority carrier concentration is not perturbed by the injection of minority carriers in the base (low injection condition) one can write in the neutral base:

$$p(x) - n(x) = N_{aB}(x) \implies \frac{dp}{dx} - \frac{dn}{dx} = \frac{dN_{aB}}{dx}$$

Noting $\mathcal{E}_x$ the electric field resulting from the applied bias the majority carrier (hole) current density injected by the base into the emitter is given by:

$$J_{pB} = q\,\mu_p\,p\,\mathcal{E}_x - qD_{pB}\frac{dp}{dx} = q\left(\mu_p\,p\,\mathcal{E}_x - D_{pB}\frac{dn}{dx} - D_{pB}\frac{dN_{aB}}{dx}\right) \quad (8.6.3)$$

The minority carrier (electron) current density in the base is equal to the electron current density injected by the emitter into the base, and is given by:

$$J_{nB} = q\left(\mu_n\,n\,\mathcal{E}_x + D_{nB}\frac{dn}{dx}\right) = J_{nE} \quad (8.6.4)$$

Eliminating $\mathcal{E}_x$ between Equations 8.6.3 and 8.6.4 (see Problem 8.8) one obtains:

$$\frac{J_{nB}}{qD_{nB}} - \frac{n}{p}\frac{J_{pB}}{qD_{pB}} = \frac{dn}{dx}\left(1 + \frac{n}{p}\right) + \frac{n}{p}\frac{dN_{aB}}{dx} \quad (8.6.5)$$

Since $n(x) << p(x)$ in the base, and since $p(x) \cong N_{aB}(x)$, Relationship 8.6.5 can be simplified into:

$$\frac{J_{nB}}{qD_{nB}} = \frac{dn}{dx} + \frac{n}{N_{aB}}\frac{dN_{aB}}{dx} \quad (8.6.6)$$

Eliminating J_{nB} from 8.6.4 and 8.6.6, we obtain:

$$\mathcal{E}_x = \frac{kT}{q}\frac{1}{N_{aB}}\frac{dN_{aB}}{dx}$$

Comparing the latter equation with 8.6.2.b we conclude that

$$\mathcal{E}_x = \mathcal{E}_{0x} \quad (8.6.7)$$

Relationship 8.6.7 shows that the electric field in the base, $\mathcal{E}_x$, is not modified by the flow of electrons through the base. Similarly, substituting $\mathcal{E}_x$ for $\mathcal{E}_{0x}$ in Equation 8.6.2a one finds that $J_{pB} = 0$. It is, however, worthwhile noting that such a conclusion can be drawn only in the low-level injection regime.

Owing to Relationship 8.6.2b, the electric field in the base, $\mathcal{E}_x = \mathcal{E}_{0x}$, can be replaced by $\dfrac{kT}{q}\dfrac{1}{N_{aB}}\dfrac{dN_{aB}}{dx} = \dfrac{kT}{q}\dfrac{1}{p}\dfrac{dp}{dx}$ in Expression 8.6.4, which yields:

$$J_{nB} = \frac{qD_{nB}}{p}\left(n\frac{dp}{dx} + p\frac{dn}{dx}\right) = \frac{qD_{nB}}{p}\frac{d(pn)}{dx} \quad (8.6.8)$$

If recombination in the base can be neglected, which is the case if the base is thin, $J_{nB} = J_{nE}$ is constant and the latter equation can be integrated between arbitrary positions in the neutral base, x and x':

$$J_{nB} \int_x^{x'} \frac{p}{q} \frac{dx}{D_{nB}} = \int_x^{x'} \frac{d(pn)}{dx} dx = p(x')n(x') - p(x)n(x) \qquad (8.6.9)$$

The *pn* product at the edges of the neutral base region is given by the pn junction theory (Equation 4.4.29):

$$p(0)n(0) = n_i^2 \, exp\left(\frac{qV_{BE}}{kT}\right) \quad and \quad p(w_B)n(w_B) = n_i^2 \, exp\left(\frac{qV_{BC}}{kT}\right)$$

Integrating Relationship 8.6.9 over the neutral base ($x=0$ and $x'=w_B$) one obtains:

$$J_{nB} = \frac{qn_i^2 D_{nB}\left[exp\left(\frac{qV_{BC}}{kT}\right) - exp\left(\frac{qV_{BE}}{kT}\right)\right]}{w_B \int_0^{w_B} p \, dx} \qquad (8.6.10)$$

We will consider D_{nB} is constant and independent of the position, x. If we define the total charge of majority carriers per unit area in the base, Q_{pB}, as:

$$Q_{pB} = q \int_0^{w_B} p(x) \, dx = q \int_0^{w_B} N_{aB}(x) \, dx \qquad (8.6.11)$$

we finally obtain:

$$J_{nE} = J_{nB} = J_S\left[exp\left(\frac{qV_{BC}}{kT}\right) - exp\left(\frac{qV_{BE}}{kT}\right)\right] \qquad (8.6.12a)$$

with:

$$J_S = \frac{q^2 n_i^2 D_{nB}}{Q_{pB}} \qquad (8.6.12b)$$

Note that $J_{nE} \cong 0$ when V_{BC} and V_{BE} are negative, *i.e.* when both junctions are reverse biased, and that J_{nE} is independent of the position, x, in the base. This is due to the fact that electron recombination in the base is neglected. As a result, $I_{nE}=AJ_{nE}$ and $I_{nC} = -A J_{nE}$. The injection of electrons by the emitter into the base gives rise to an electron concentration at $x=l_{B1}$ that increases exponentially with V_{BE}, and the electron concentration at $x=l_{B2}$ is virtually equal to zero since the collector junction is reverse biased.

The total charge of majority carriers in the base is given by the following

expression: $Q_{pB} = q \int\limits_{0}^{w_B} N_{aB}(x)dx \equiv qG_B$, where G_B is called the "Gummel

number" in the base.[13] Using the Gummel number, Relationship 8.6.12b can be rewritten in the following form:

$$J_S = \frac{q \, n_i^2 \, D_{nB}}{G_B} \qquad (8.6.13)$$

Note: The current obtained using Expressions 8.6.12a and 8.6.13 is equal to the current given by Expression 8.3.37 if the doping concentration in the base is uniform. In that case the base Gummel number is simply equal to: $G_B = N_{aB} \, w_B$. The Ebers-Moll model can, therefore, be considered as a subset of the Gummel-Poon model.

8.6.1. Current gain

To calculate the current gain of a transistor one needs to know the value of the current in the base. The base current can be divided into three current components: the hole current injected by the base into the emitter, the base current due to the recombination with electrons in the base, and a base current component due to recombination of holes in the emitter junction transition region. In a device biased in the forward active mode, the latter component is much smaller than the two others and is typically neglected in a first-order analysis. It will be dealt with, however, when we study the variation of gain with current, in Section 8.8.1.

8.6.1.1. Recombination in the base

The recombination of electrons in the neutral base was neglected in the calculation of J_{nE} in the previous Section. Recombination can be accounted for using the SRH recombination theory developed in Section 3.5. We will maintain the assumption of low-level injection ($n-n_o << p$ in the base). The recombination rate is therefore simplified by Expression (3.5.20):

$$U_n = \frac{n'}{\tau_n} \qquad (8.6.14)$$

where $n'=n-n_o$ is the electron *excess* concentration in the base and τ_n is their lifetime. Using the continuity equation 2.6.6a in absence of external generation we have:

$$\frac{\partial n}{\partial t} = \frac{1}{q} \frac{\partial J_n}{\partial x} - U_n$$

The equilibrium electron concentration is $n_{po} = \dfrac{n_i^2}{N_{aB}}$. In steady-state $(\dfrac{\partial n}{\partial t} = 0)$ one can thus write:

$$\frac{dJ_n}{dx} = q\,\frac{n'}{\tau_n} = q\,\frac{n - \dfrac{n_i^2}{N_{aB}}}{\tau_n} \qquad (8.6.15)$$

Noting that both n and N_{aB} are functions of x, we can calculate the current due to recombination by integrating the electron current density variation over the base:

$$I_{rB} = A\,(J_{nE} - J_{nC}) = -q\,A \int\limits_0^{w_B} \frac{[n - \dfrac{n_i^2}{N_{aB}}]\,dx}{\tau_n} \qquad (8.6.16)$$

Since the concentration of the electrons injected by the emitter into the base is much larger than the equilibrium electron concentration in the base $(n \gg \dfrac{n_i^2}{N_{aB}})$, the latter equation can be simplified:

$$I_{rB} = -\frac{qA}{\tau_n} \int\limits_0^{w_B} n(x)\,dx \qquad (8.6.17)$$

If the minority carrier concentration in the base is linearized, which is a valid practice if the base is thin, one obtains, in the forward active mode $(V_{BC} \ll 0)$:

$$n(x) \cong n(x)' = n_{po}(0)\left[1 - \frac{x}{w_B}\right]\left(\exp\left(\frac{qV_{BE}}{kT}\right) - 1\right) \qquad (8.6.18)$$

Once the value of $n(x)$ is known, Equation 8.6.17, can be used to calculate the recombination current in the base:

$$I_{rB} = -\frac{qA}{\tau_n}\,\frac{n_i^2}{N_{aB}(0)}\,\frac{w_B}{2}\left(\exp\left(\frac{qV_{BE}}{kT}\right) - 1\right) \qquad (8.6.19)$$

As previously calculated the loss of minority carriers in the base can be represented by the transport factor in the base, α_T, which is defined as the electron current reaching the collector divided by the electron current injected into the base by the emitter, I_{nE}:

$$\alpha_T = \left.\frac{I_{nE} - I_{rB}}{I_{nE}}\right|_{V_{BC}=0} = \left.1 - \frac{I_{rB}}{I_{nE}}\right|_{V_{BC}=0} \qquad (8.6.20)$$

Using Relationship 8.6.12a we obtain:

$$J_{nE} = \frac{q^2 n_i^2 D_{nB}}{Q_{pB}} \left[exp\left(\frac{qV_{BC}}{kT}\right) - exp\left(\frac{qV_{BE}}{kT}\right) \right]$$

and, therefore,

$$I_{nE}\Big|_{V_{BC}=0} = A \, J_{nE}\Big|_{V_{BC}=0} = \frac{-A \, q^2 n_i^2 D_{nB}}{Q_{pB}} \left[exp\left(\frac{qV_{BE}}{kT}\right) - 1 \right]$$

Using the latter Relationship in conjunction with Equations 8.6.19 and 8.6.20, one finally obtains:

$$\alpha_T = 1 - \frac{w_B \, Q_{pB}}{2q D_{nB} \tau_n N_{aB}(0)} = 1 - \frac{w_B \, G_B}{2 \, N_{aB}(0) \, L_{nB}^2} \qquad (8.6.21)$$

where $L_{nB} = \sqrt{D_{nB}\tau_n}$ is the diffusion length of the electrons in the base.

Note: If the base doping concentration is homogeneous the base transport factor derived in Equation 8.6.21 is identical to that of Expression 8.3.34 since, in that case, $G_B = N_{aB} \, w_B$, and, therefore, the transport factor equals:

$$\alpha_T = 1 - \frac{w_B^2}{2L_{nB}^2}$$

The Ebers-Moll model can, therefore, be considered as a subset of the Gummel-Poon model.

8.6.1.2. Emitter efficiency and current gain

It is possible to calculate the hole current in the emitter using the same technique as that used to derive the electron current in the base (Equations 8.6.2a to 8.6.12b). If the doping impurity concentration in the emitter is not homogeneous there exists an electric field, $\mathcal{E}_{0x}$, within the quasi-neutral emitter at equilibrium. Using the drift-diffusion equation for the electrons in the emitter, and noting that the electron concentration $n = N_{dE}$ is a function of x, one obtains:

$$J_{nE} = 0 = q\mu_n N_{dE} \mathcal{E}_{0x} + q D_{nE} \frac{dN_{dE}}{dx} \Rightarrow \mathcal{E}_{0x} = -\frac{D_{nE}}{\mu_n} \frac{1}{N_{dE}} \frac{dN_{dE}}{dx} \quad (8.6.22)$$

which yields, using Einstein's relationship $D = \frac{kT}{q}\mu$:

$$\mathcal{E}_{0x} = -\frac{kT}{q} \frac{1}{N_{dE}} \frac{dN_{dE}}{dx} \qquad (8.6.23)$$

In the quasi-neutral emitter region we can write:

$$n(x) - p(x) = N_{dE}(x) \Rightarrow \frac{dn}{dx} - \frac{dp}{dx} = \frac{dN_{dE}}{dx} \qquad (8.6.24)$$

If the non-equilibrium electric field is noted $\mathcal{E}_x$ the majority carrier current density in the emitter, when an external bias is applied, is given by:

$$J_{nE} = q\,\mu_n\,n\,\mathcal{E}_x + qD_{nE}\frac{dn}{dx} = q\left(\mu_n\,n\,\mathcal{E}_x + D_{nE}\frac{dp}{dx} + D_{nE}\frac{dN_{dE}}{dx}\right) \quad (8.6.25)$$

and the minority carrier current density is:

$$J_{pE} = q\left(\mu_p\,p\,\mathcal{E}_x - D_{pE}\frac{dp}{dx}\right) \quad (8.6.26)$$

Eliminating $\mathcal{E}_x$ between Equations 8.6.25 and 8.6.26 one obtains:

$$\frac{J_{pE}}{qD_{pE}} - \frac{p}{n}\frac{J_{nE}}{qD_{nE}} = -\frac{dp}{dx}\left(1 + \frac{p}{n}\right) - \frac{p}{n}\frac{dN_{dE}}{dx} \quad (8.6.27)$$

Using the low-level injection condition in the emitter $p(x) << n(x)$ and writing $n(x) \cong N_{dE}(x)$, Equation 8.6.27 can be written as:

$$\frac{J_{pE}}{qD_{pE}} = -\frac{dp}{dx} - \frac{p}{N_{dE}}\frac{dN_{dE}}{dx} \quad (8.6.28)$$

Eliminating J_{pE} from 8.6.26 and 8.6.28 we obtain:

$$\mathcal{E}_x = -\frac{kT}{q}\frac{1}{N_{dE}}\frac{dN_{dE}}{dx} \quad (8.6.29)$$

Comparing the latter equation with 8.6.23 we conclude that

$$\mathcal{E}_x = \mathcal{E}_{0x} \quad (8.6.29)$$

According to Equation 8.6.29 the electric field in the emitter $\mathcal{E}_x$ remains equal to its equilibrium value, $\mathcal{E}_{0x}$, even when an external bias is applied. It is, however, worthwhile noting that such a conclusion can be drawn only in the low-level injection regime.

Using Relationship 8.6.23, $\mathcal{E}_x\,(= \mathcal{E}_{0x})$ can be replaced by $-\dfrac{kT}{q}\dfrac{1}{N_{dE}}\dfrac{dN_{dE}}{dx}$ in Equation 8.6.29, which yields:

$$J_{pE} = -q\mu_p p\frac{kT}{q}\frac{1}{N_{dE}}\frac{dN_{dE}}{dx} - qD_{pE}\frac{dp}{dx}$$

or:

$$J_{pE} = -q\,D_{pE}\,p\frac{1}{N_{dE}}\frac{dN_{dE}}{dx} - qD_{pE}\frac{dp}{dx} = -\frac{qD_{pE}}{N_{dE}}\frac{d(p\,N_{dE})}{dx} \quad (8.6.30)$$

This equation is similar to that obtained previously for the electron current in the base (Expression 8.6.8). To render the integration easier we will consider the case where the length of the quasi-neutral emitter is small ($w_E << L_{pE}$). In that case the recombination of holes in the emitter can be neglected. Let us also assume that there is a metallic, ohmic contact at the surface of the emitter which induces an infinite surface recombination. This condition imposes that the hole concentration at $x = X_E$ is equal to its equilibrium value (Figure 8.21). In other words, $p(x_E) = n_i^2/N_{dE}(x_E)$. Under such conditions the hole distribution in the emitter is a linear function of depth, and the hole current is constant throughout the emitter.

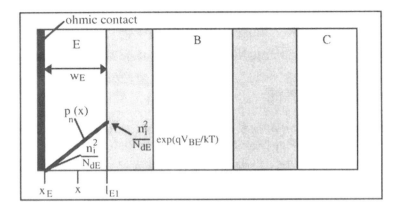

Figure 8.21: Hole distribution in the emitter ($w_E << L_{pE}$)

The integration of $J_{pE} = -\dfrac{qD_{pE}}{N_{dE}(x)} \dfrac{d\left(p(x)N_{dE}(x)\right)}{dx}$ between the edge of

the emitter depletion region, l_{E1}, where $p(l_{E1})N_{dE}(l_{E1}) = n_i^2 \, exp\left(\dfrac{qV_{BE}}{kT}\right)$,

and at the ohmic contact, x_E, where $p(x_E) = \dfrac{n_i^2}{N_{dE}(x_E)}$ yields the following

relationships:

$$J_{pE} = -\frac{qD_{pE}}{N_{dE}(x)} \frac{d\left(p(x)N_{dE}(x)\right)}{dx} \Rightarrow J_{pE} \frac{N_{dE}(x)}{qD_{pE}} = -\frac{d\left(p(x)N_{dE}(x)\right)}{dx}$$

$$\Downarrow$$

$$J_{pE} \int_{l_{E1}}^{x_E} \frac{N_{dE}(x)\, dx}{D_{pE}} = \int_{l_{E1}}^{x_E} \frac{d\left(p(x)N_{dE}(x)\right)}{dx} = q \, n_i^2 \left[exp\left(\frac{qV_{BE}}{kT}\right) - 1\right] \quad (8.6.31)$$

from which we obtain:

$$I_{pE} = AJ_{pE} = \cfrac{q\,D_{pE}\,n_i^2\,A}{x_E \atop \displaystyle\int_{l_{El}} N_{dE}(x)dx}\left[exp\left(\frac{qV_{BE}}{kT}\right) - 1\right]$$

$$= -\cfrac{q\,D_{pE}\,n_i^2\,A}{l_{El} \atop \displaystyle\int_{x_E} N_{dE}(x)dx}\left[exp\left(\frac{qV_{BE}}{kT}\right) - 1\right] \qquad (8.6.32)$$

We can now define the "Gummel number" in the emitter as the total concentration of doping impurities in the emitter: $G_E = \displaystyle\int_{x_E}^{l_{El}} N_{dE}(x)dx$, where G_E is expressed in cm^{-2}.

The emitter efficiency was defined in Expression 8.3.29:

$$\gamma_F = \left.\frac{I_{nE}}{I_{nE} + I_{pE}}\right|_{V_{BC}=0} = \left.\frac{1}{1 + \dfrac{I_{pE}}{I_{nE}}}\right|_{V_{BC}=0} \qquad (8.6.33)$$

Relationship 8.6.12a gives us the electron current injected into the base:

$$J_{nE} = \frac{q^2\,n_i^2\,D_{nB}}{Q_{pB}}\left[exp\left(\frac{qV_{BC}}{kT}\right) - exp\left(\frac{qV_{BE}}{kT}\right)\right]$$

When the collector is shorted to the base ($V_{BC}=0$) we have:

$$\left.I_{nE}\right|_{V_{BC}=0} = \left.A\,J_{nE}\right|_{V_{BC}=0} = \frac{-A\,q^2\,n_i^2\,D_{nB}}{Q_{pB}}\left[exp\left(\frac{qV_{BE}}{kT}\right) - 1\right]$$

Using these relationships γ_F can be calculated for non uniformly doped devices utilizing the Gummel numbers:

$$\gamma_F = \frac{1}{1 + \dfrac{D_{pE}\,G_B}{D_{nB}\,G_E}} \qquad (8.6.34)$$

It is worth noting that, as in the uniform doping case, the common-base current gain, α_F, is equal to the $\gamma_F\,\alpha_T$ product, which can easily be shown using Equations (8.6.1.a) and (8.6.1.c):

$$\alpha_F = -\left.\frac{I_C}{I_E}\right|_{V_{BC}=0} = \left.\frac{I_{nE} - I_{rB}}{I_{nE} + I_{pE}}\right|_{V_{BC}=0} = \left.\left(\frac{I_{nE}}{I_{nE} + I_{pE}} \frac{I_{nE} - I_{rB}}{I_{nE}}\right)\right|_{V_{BC}=0} = \gamma_F \alpha_T$$

$$(8.6.35)$$

Note:

1. The higher the doping concentration in the emitter, the higher the Gummel number in the emitter, and, therefore, the higher the emitter efficiency.

2. The emitter efficiency described by Relationship 8.6.34 is equivalent to that obtained in Equation 8.3.31. Replacing L_{pE} by w_E and substituting $\frac{n_i^2}{N_{dE}}$ for p_{oE} and $\frac{n_i^2}{N_{aB}}$ for n_{oB} in Equation 8.3.31 one obtains:

$$\gamma_F \cong \frac{1}{1 + \dfrac{D_{pE}\, p_{oE}\, w_B}{D_{nB}\, n_{oB}\, w_E}} = \frac{1}{1 + \dfrac{D_{pE}\, N_{aB}\, w_B}{D_{nB}\, N_{dE}\, w_E}}$$

3. In addition if the doping concentrations are homogeneous, $G_E = N_{dE}\, w_E$, $G_B = N_{aB}\, w_B$, and $\gamma_F = \dfrac{1}{1 + \dfrac{D_{pE}\, G_B}{D_{nB}\, G_E}}$, which is equivalent to Relationship 8.6.34. The Ebers-Moll model can, therefore, be considered as a subset of the Gummel-Poon model.

8.7. Early effect

We have so far considered that the collector current is independent of the collector-base voltage when the device operates in the forward active mode. This is not completely true, and the collector current actually increases with the collector-base bias. This effect was explained by J. Early in 1952 and is due to the modulation of the neutral base width, w_B, by the applied collector-base reverse bias.[14]

Let us consider a bipolar transistor operating in the common-emitter configuration. According to the models developed hitherto the relationship between the collector current and the base current is: $I_C = \beta_F I_B$, which shows no dependence of the collector current on the collector-base voltage, as long as $V_{BC} \leq 0$. Under such conditions the bipolar transistor behaves as a perfect current source with infinite output impedance, as can be seen on the output characteristics sketched in Figure 8.22. In an actual device the output impedance is finite because of the Early effect caused by the modulation of the neutral base width. The mechanism giving rise to the Early effect is the following: any variation of V_{CB} induces a variation of the width of the depletion region in the base, at the collector-base junction (Equation 4.3.2). That variation induces a variation of the neutral base width, and therefore, a variation of the current gain. Since an increase of V_{CB} increases the width of the depletion width, and therefore, a decrease of the neutral base width, the collector current increases with V_{CB} accordingly.

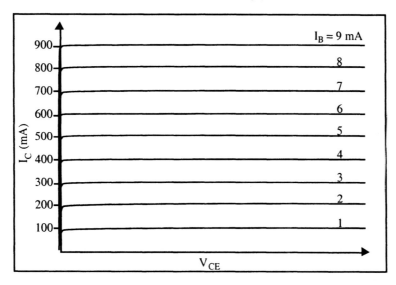

Figure 8.22: Output characteristics of a silicon bipolar transistor in forward active mode, $I_C(V_{CE})$, in the absence of Early effect. $V_{BE} = 0.7$ V in the forward active mode and $V_{CE} = V_{CB} + V_{BE}$.

The Early effect can be derived from the expression of the current derived previously (Equation 8.6.10):

$$I_C = -\frac{AqD_{nB}n_i^2\left[exp\left(\frac{qV_{BC}}{kT}\right) - exp\left(\frac{qV_{BE}}{kT}\right)\right]}{\int_0^{w_B}pdx}$$

$$\cong \frac{AqD_{nB}n_i^2\ exp\left(\frac{qV_{BE}}{kT}\right)}{\int_0^{w_B}pdx} \tag{8.7.1}$$

in the forward active regime. The variation of collector current resulting from the base width modulation induced by V_{CB} can be expressed as follows:

$$\frac{dI_C}{dV_{CB}} = - \frac{AqD_{nB}n_i^2 \exp\left(\frac{qV_{BE}}{kT}\right)}{\left[\int_0^{w_B} pdx\right]^2} p(w_B) \frac{dw_B}{dV_{CB}} = - \frac{I_C \, p(w_B)}{w_B} \frac{dw_B}{dV_{CB}} \qquad (8.7.2)$$

or, if we define the Early voltage, V_A ($V_A < 0$):

$$V_A = \frac{\int_0^{w_B} pdx}{p(w_B) \dfrac{dw_B}{dV_{CB}}} \qquad (8.7.3)$$

we obtain the output conductance: $\dfrac{dI_C}{dV_{CB}} = - \dfrac{I_C}{V_A}$ (8.7.4)

Note that $\displaystyle\int_0^{w_B} pdx = G_B$ (8.7.5)

and $p(w_B) \dfrac{dw_B}{dV_{CB}} = \dfrac{dG_B}{dV_{CB}}$ (8.7.6)

In practice the base Gummel number, G_B, shows little variation with V_{CB}. As a result, the Early voltage can be considered constant in a given device. In practice the output conductance, $\dfrac{dI_C}{dV_{CB}}$, is measured when $V_{CB}=0$, *i.e.* for $V_{CE}=0.7$ V. Note that $V_{CB} = V_{CE}-V_{BE}$ and that $V_{BE} = 0.7$ V in the forward active mode in a silicon device. We also have $\dfrac{dI_C}{dV_{CB}} = \dfrac{dI_C}{dV_{CE}}$. The equation for the output characteristics, $I_C(V_{CE})$, are, therefore, given by :

$$I_C(V_{CE}) = I_C(V_{CE}=0.7V) + \frac{dI_C}{dV_{CB}} (V_{CE} - 0.7V)$$

$$= I_C(V_{CE}=0.7V) - \frac{I_C(V_{CE} = 0.7V)}{V_A} (V_{CE} - 0.7V)$$

All these characteristics intersect the *x*-axis ($I_C=0$) at the same voltage, V_A. It is, therefore, very easy to extract the Early voltage of a bipolar transistor from its output characteristics, $I_C(V_{CE})$, as shown in Figure 8.23.

It is easy to understand that the reduction of base width caused by an increase of V_{CB} gives rise to an increase of collector current. We know that the electron current, AJ_n, flowing from emitter to collector is proportional to the gradient, or the slope, of the minority carrier concentration in the base. Since the electron concentration at the emitter-side of the base is fixed by V_{BE}, and since $n(w_B)=0$, the electron concentration gradient, *dn/dx*, must increase when the width of the neutral base is reduced from w_{B1} to w_{B2} (Figure 8.24).

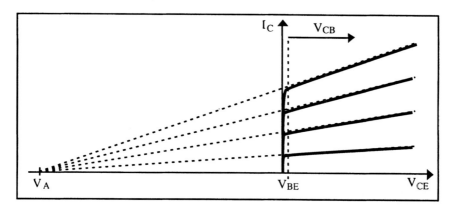

Figure 8.23: Output characteristics, $I_C(V_{CE})$, in the presence of Early effect.

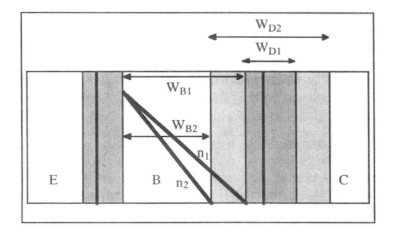

Figure 8.24: Early effect: minority carrier concentration in the quasi-neutral base. w_{D1} and w_{D2}, w_{B1} and w_{B2}, and n_1 and n_2 are the width of the collector-base transition region (the width of the neutral base) and the electron concentration, for two different collector-emitter voltages, V_{CE1} and V_{CE2}, where $V_{CE1} < V_{CE2}$.

If the base width modulation is pushed to the limit, such that $w_B = 0$ the transistor is in punchthrough and the emitter and collector junction space-charge regions touch one another. In such a case a large current can flow from emitter to collector. This current is, however, no longer controlled by the base current.

8.8. Dependence of current gain on collector current

We have so far considered that the current gain in the transistor was constant. In reality, it depends on the current level, although it remains constant over a wide range of current values. The common emitter current gain defined by the relationship $\beta_F = \dfrac{I_C}{I_B}$ is actually quite small when the collector current is small. It then increases up to its nominal value where it remains until the current in the device becomes quite large. At that point, the current gain decreases again. The reduction of gain at low current levels is due to recombination in the emitter-base transition region. The reduction of gain at high current levels is due to high-level injection and to the Kirk effect. For high-level injection all previous assumptions may be invalid.

8.8.1. Recombination at the emitter-base junction

So far we have considered that there was no recombination in the emitter-base junction space-charge region. Since the lifetime of the carriers is not infinite the number of carriers exiting the space-charge region is lower than the number that were injected into it. One can associate a current to this loss of carriers. This recombination current, I_{rBE}, is negligible under usual operation conditions, but it cannot be neglected if the current level in the transistor is low, as shown in the PN junction chapter.

The total electron current injected at the emitter-base junction is equal to the sum of the electrons injected into the base, I_{nE}, and the recombination current in the junction depletion zone, I_{rBE} (Figure 8.25). When V_{BE} is sufficiently high the diffusion current, which varies as $exp(qV_{BE}/kT)$ is sufficiently large to completely overshadow the recombination current, which varies as $exp(qV_{BE}/2kT)$. However, at low current levels, *i.e.* when V_{BE} is small, the diffusion current becomes smaller than the recombination current. The base current is given by $I_B = -I_{pE}-I_{rBE}-I_{rB}$, the emitter current is equal to $I_E=I_{pE}+I_{nE}+I_{rBE}$, and the collector current is given by $I_C = -I_{nE} + I_{rB}$. If one defines $\alpha_F\big|_{recomb}$ and $\beta_F\big|_{recomb}$ as the common-base and common-emitter gain taking into account the recombination current one obtains:

$$\alpha_F\big|_{recomb} = -\frac{I_C}{I_E} = \frac{I_{nE}-I_{rB}}{I_{pE}+I_{nE}+I_{rBE}} = \frac{\alpha_F}{1+\dfrac{I_{rBE}}{I_{nE}+I_{pE}}} \qquad (8.8.1)$$

and

$$\beta_F\Big|_{recomb} = \frac{I_C}{I_B} = \frac{I_{nE} - I_{rB}}{I_{pE} + I_{rBE} + I_{rB}} = \frac{\beta_F}{1 + \dfrac{I_{rBE}}{I_{pE} + I_{rB}}} \qquad (8.8.2)$$

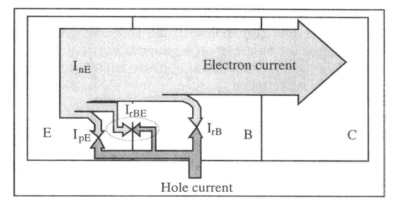

Figure 8.25: Currents in the transistor, including the current due to recombination in the space-charge region of the emitter-base junction.

As a result the current gain, β_F, rolls off when the recombination current in the emitter-base junction space-charge region becomes comparable to the diffusion current.

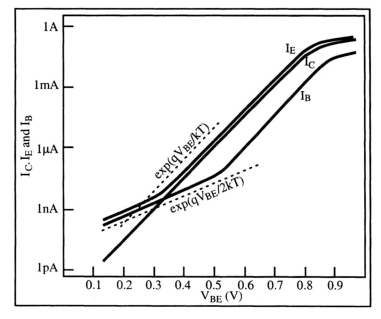

Figure 8.26: Gummel plot.[15]

Figures 8.26 shows the base, emitter, and collector currents on a logarithmic scale, as a function of the emitter-base voltage. Such a plot is called a "Gummel plot". The current gain, $\beta_F = I_C/I_B$, is constant over the part of the plot where the base current is proportional to $exp(qV_{BE}/kT)$. At low current levels, the collector current varies as $exp(qV_{BE}/kT)$, but the base current is proportional to $exp(qV_{BE}/2kT)$, which reduces the gain, β_F. The effects observed at high current levels will be described in the next Section.

8.8.2. Kirk effect

The Kirk effect is a result of the widening of the base under high-level injection conditions. A reduction of the current gain occurs from the base widening. [16]

Consider a transistor operating in the forward active mode and under high-level injection. The base-collector junction is reverse biased. The density of charges in the depletion region of the collector-base junction is normally equal to qN_{dC} on the collector side, and to $-qN_{aB}$ on the base side. If a high electron current density flows through the junction the charges in those depletion regions will be modified. If we note $N(x) = N_d(x) - N_a(x)$, and recall that $I_C \cong I_{nC} = -AJ_{nC}$ (>0), the charge density in the depletion regions become:

$$\rho(x) = q\,N(x) - \frac{I_C}{A\,v(x)} \qquad (8.8.3)$$

where $v(x)$ is the velocity of the electrons passing through the depletion regions and A is the area of the junctions. Since the junction is reverse biased one can assume that the electric field is large enough for the electron velocity to be equal to the electron saturation velocity, v_{max}, which is equal to 10^7 cm/s in silicon.

Integrating Poisson's equation in the base-collector space-charge region yields the electric field:

$$\frac{d\mathcal{E}}{dx} = \frac{1}{\varepsilon_{si}}\left[qN(x) - \frac{I_C}{A\,v_l}\right] \qquad (8.8.4)$$

A second integration yields the voltage drop across the base-collector space-charge region. Noting Φ_o the junction potential and l_{B2} and l_{C2} the position of the left and right edges of the collector-base transition region (Figure 8.27) one obtains:

$$V_{CB} + \Phi_o = -\int_{l_{B2}}^{l_{C2}}\mathcal{E}dx \qquad (8.8.5)$$

One can define a critical current, $I_1 = q\, N_{dC}\, A\, v_{vmax}$, above which the charge in the space charge region on the collector side changes sign.

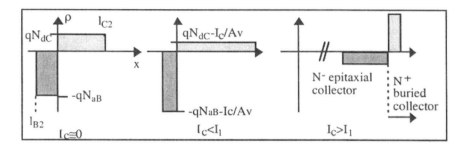

Figure 8.27: Distribution of charges in the collector-base junction for different levels of collector current.

We will not make a complete analysis of this phenomenon, but rather qualitatively describe what happens when the collector current increases (Figure 8.27).

◊ When the level of I_C is low, the injection of electrons does not affect the space-charge region. The transistor operates in a standard manner and the width of the neutral base has its "normal" value.

◊ When I_C is increased, while remaining smaller than I_1 the charge in the space-charge region on the base side increases from $\rho(x)=-qN_{aB}$ to $\rho(x) = -q\, N_{aB} - \dfrac{I_C}{Av(x)}$. At the same time the space-charge region on the collector side sees its charge decrease from $\rho(x)=qN_{dC}$ to $\rho(x)=qN_{dC} - \dfrac{I_C}{Av(x)}$. As a result l_{B2} shifts to the right, which increases the width of the quasi-neutral base. Therefore, the transport factor in the base, α_T, and thus the current gain, decrease.

◊ When the collector current becomes larger than the critical current, I_1, the space charge on the collector side becomes negative. Poisson's equation imposes that the whole space charge region shifts to the right until it reaches the heavily doped buried collector, where the doping concentration is very high (the result of the double integration of the Poisson equation 8.8.4 and 8.8.5 must be equal to $V_{CB} + \Phi_0$. In the buried collector a positive space charge is formed ($qN_{dBuried\_Collector} - \dfrac{I_C}{Av(x)} > 0$) while the lightly doped collector region carries the opposite negative charge. As a result l_{B2} is shifted far to the right, which leads to an increase of the quasi-neutral base width, and therefore, a decrease of the transport factor in the base, α_T, and a decrease of the current gain. The point I_{C2} is now positioned in the buried collector (Figure 8.28).

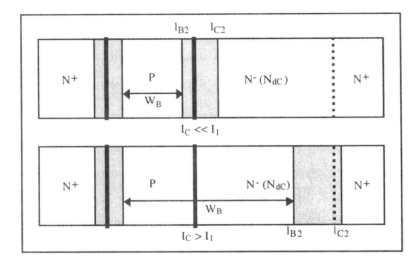

Figure 8.28: Extension of the space-charge regions for collector currents below and above I_1.

As a consequence of both the recombination in the emitter-base space-charge region and the Kirk effect a decrease of the current gain of the transistor is observed at low and high current levels, as shown in Figure 8.29. The gain, however, is constant over a wide range of current values where transistors typically operate.

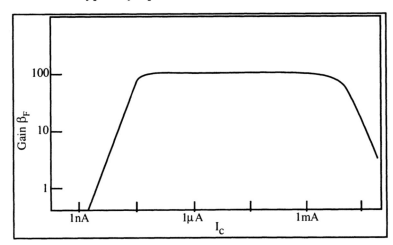

Figure 8.29: Common-emitter gain *vs.* collector current showing gain reduction due to recombination in the base-emitter transition region for small I_C and to the Kirk effect for large I_C.[17]

8.9. Base resistance

One might think that since the base current in a bipolar transistor is much smaller than the emitter and collector currents, the presence of a finite base resistance has little impact on the device characteristics. This is not true, since any potential drop in the base has an exponential influence on the collector current. The emitter is usually heavily doped, such that the potential drop across the quasi-neutral emitter is negligible. The base, on the other hand, is more lightly doped, and therefore has a non-negligible resistance which gives rise to a potential drop between the base contact and the active region of the base. Taking base resistance into account one can write:

$$V_{BE}\Big|_{across\ the\ junction} = V_{BE}\Big|_{at\ the\ contacts} - I_B R_B \qquad (8.9.1)$$

and, therefore:

$$I_C = I_S \exp\left[\frac{q(\ V_{BE}\big|_{at\ the\ contacts} - I_B R_B)}{kT}\right] \qquad (8.9.2)$$

The potential drop in the base causes the curves of the Gummel plot in Figure 8.26 to deviate from the ideal exponential dependence of currents on the base-emitter voltage for high current levels.

8.10. Numerical simulation of the bipolar transistor

It is possible to simulate the characteristics of a bipolar transistor on a computer. These simulations are based on the solution of the transport equations (Poisson, drift-diffusion and continuity) at the nodes of a mesh representing the device. These simulations are based on the discretization of the device into a series of nodes connected together by mesh elements.

Figure 8.30 shows the cross section of a bipolar transistor, and Figure 8.31 represents the mesh generated by a computer code which will be used for simulating the device. Figure 8.31 was generated by a process simulator software code which emulated the device fabrication steps and produced an output file containing the topology of the device, the different materials used for fabrication, and the doping type and concentration at every simulation node. In this example the collector contact is placed at the bottom to simplify the transistor structure.

Figures 8.32 and 8.33 show 1) the hole current, J_p, flowing from the base contact into the base-emitter junction and 2) the electron current density, J_n. Each arrow represents the magnitude and direction of the current at each node of the mesh.

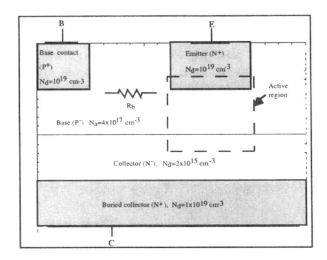

Figure 8.30: Cross section of the transistor.

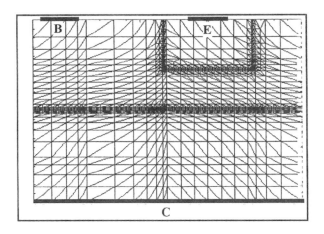

Figure 8.31: Mesh generated by the simulator. [18]

The simulation results allow one to visualize the currents at the transistor terminals. Figure 8.34 shows the base current and the collector current as a function of V_{BE} (Gummel plot). The distance between the two curves represents the common-emitter gain, β_F. The gain increases with collector current up to $I_C \cong 1$ nA, is constant for 1nA $< I_C <$ 100μA. At $I_C >$ 100μA decreases. The common-emitter current gain is shown in Figure 8.35.

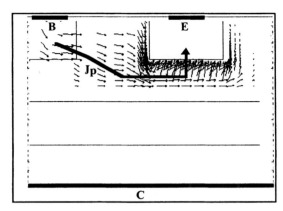

Figure 8.32: Hole current density, J_p. V_{BE}=0.7V, V_{CE}=3V.

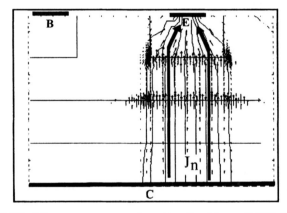

Figure 8.33: Electron current density, J_n. V_{BE}=0.7V, V_{CE}=3V.

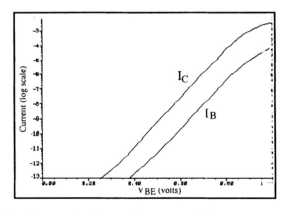

Figure 8.34: Gummel plot: collector and base currents vs. V_{BE}. V_{CE}=3V.

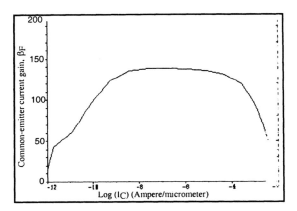

Figure 8.35: Common-emitter current gain, β_F , as a function of log(I_C). V_{CE}=3V.

8.11. Collector junction breakdown

8.11.1. Common-base configuration

When V_{CB} is large the collector junction can undergo avalanche breakdown similar to what was observed in a simple PN junction. As in Relationship 8.3.24 the current flowing in the reverse collector junction, in the common-base configuration, will be noted I_{C0} when the emitter terminal is open (I_E=0). Since the emitter is left floating the collector current is equal to I_{C0} in the absence of avalanche multiplication. When multiplication takes place the collector current is equal to $I_C = M\,I_{C0}$ where M is the multiplication factor which can be related to the applied voltage using Equation 4.4.38:

$$M = \frac{1}{1 - \left(\dfrac{V_{CB}}{BV}\right)^n} \tag{8.11.1}$$

where BV is the junction breakdown voltage, where $M \to \infty$ when $V_{CB} \to BV$, and where n ranges between 4 and 6, depending on the impurity concentration profile.

In the common-base configuration we have:

$$M = \frac{1}{1 - \left(\dfrac{V_{CB}}{BV_{CB0}}\right)^n} \tag{8.11.2}$$

where BV_{CB0} is the common-base collector breakdown voltage when the emitter terminal is open.

8.11.2. Common-emitter configuration

The avalanche phenomenon is due to the creation of electron-hole pairs due to impact ionization caused by a large electric field, such as in a reverse-biased junction. The pairs are separated by the electric field in the junction; the electrons are swept into the collector, and the holes into the base. In the common-base configuration the holes injected into the base exit the device through the grounded base contact. In the common-emitter configuration with open base the holes injected into the base by impact ionization constitute a base current which gives rise to injection of electrons from the emitter through the base, and into the collector. This increases the flow of carriers through the collector depletion region, and therefore, the rate of avalanche. Avalanche and amplification by transistor effect produce a positive feedback loop. Because of the amplification effect due to the transistor, the collector breakdown voltage will be reduced compared to the common-base case.

The base voltage is different from zero when the base is left floating. Its value can be obtained from the Ebers-Moll equation where $I_E = -I_C$. Using:

$$\begin{bmatrix} -I_C \\ I_C \end{bmatrix} = \begin{bmatrix} -I_{ES} & \alpha_R I_{CS} \\ \alpha_F I_{ES} & -I_{CS} \end{bmatrix} \begin{bmatrix} exp(qV_{BE}/kT)-1 \\ -1 \end{bmatrix}$$

one readily finds

$$(I_{ES}-\alpha_F I_{ES})[(exp(qV_{BE}/kT)-1] - I_{CS} + \alpha_R I_{CS} = 0$$

Solving the latter equation for V_{BE} yields the base voltage.

When avalanche multiplication is activated the emitter current is equal to the sum of the hole current originating from the reverse-biased collector-base junction, which flows through the base and reaches the emitter junction, and the electron current injected from the emitter through the base into the collector, both currents being multiplied by M. In addition, $I_E=-I_C$ since the base contact is open. We can thus write:

$$I_E = -I_C = M (\alpha_F I_E - I_{C0}) \Rightarrow I_E = -\frac{M I_{C0}}{1-M\alpha_F} \ or \ I_C = \frac{M I_{C0}}{1-M\alpha_F} \quad (8.11.3)$$

At avalanche is the collector current becomes very large ($I_C \rightarrow \infty$), which yields:

$$M\alpha_F = 1 \ or \ M = \frac{1}{\alpha_F} \quad (8.11.4)$$

Since V_{BE} is small compared to V_{CE} we can assume that $V_{CB} \cong V_{CE}$, such that Relationship 8.11.2 becomes:

$$M = \frac{1}{1 - \left(\dfrac{V_{CE}}{BV_{CB0}}\right)^n} \qquad (8.11.5)$$

Noting that the common-emitter breakdown voltage BV_{CE0} is equal to V_{CE} when $M = \frac{1}{\alpha_F}$ we can write:

$$BV_{CE0} = BV_{CB0} \sqrt[n]{1 - \alpha_F} \cong \frac{BV_{CB0}}{\sqrt[n]{\beta_F}} \qquad (8.11.6)$$

Relationship 8.11.6 shows that the collector breakdown voltage in the common-emitter configuration is lower than that in the common-base configuration by a factor 2 to 3, typically. When the transistor is used in the common-emitter and when the base is not actually open but connected to external circuitry, some of the base current generated by impact ionization can escape from the base. As a result the collector breakdown voltage will be higher than if the base was strictly open. In that case the breakdown voltage will have a value situated between BV_{CE0} and BV_{CB0}. Looking at the example of a resistor, R, connecting the base to ground (Figure 8.36), one easily concludes that $BV_{CE(R)} \rightarrow BV_{CB0}$ when $R \rightarrow 0$ and that $BV_{CE(R)} \rightarrow BV_{CE0}$ when $R \rightarrow \infty$.

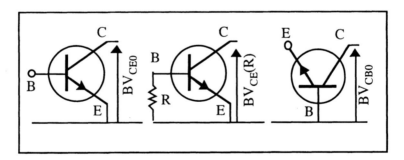

Figure 8.36: Collector breakdown voltage. Left: common emitter; center: common emitter with external circuitry between the base and ground; right: common base.

8.12. Charge-control model

The equations derived hitherto are time-independent, and while being satisfactory for solving many problems, they lack adequacy for analyzing the frequency response of a transistor or its switching behavior.

In the charge-control model the independent variables are no longer voltages or currents, but charges. The derivation of the charge-control model will be made assuming that the base doping is constant, *i.e.* that N_{aB} is independent of x.

8.12.1. Forward active mode

The excess minority carrier charge in the transistor base is given by:

$$Q_{nB} = - qA \int_0^{w_B} n'(x)dx \qquad (8.12.1)$$

As we have seen before, the currents in an NPN bipolar transistor are controlled by the base-emitter voltage, V_{BE}. This voltage influences not only the minority carrier charge in the base, Q_{nB}, but also other charges present in the transistor. These charges are described in Figure 8.37:

◊ a charge due to the holes injected from the base into the emitter, Q_{pE}, represented by the area under the excess hole concentration profile in the emitter, $p'(x)$:

$$Q_{pE} = qA \int_{emitter} p'(x)dx \qquad (8.12.2)$$

◊ a depletion charge Q_{VE} due to the variation of the emitter space-charge region caused by the application of a base-emitter bias, V_{BE}

◊ a depletion charge Q_{VC} due to the variation of the collector space-charge region caused by the application of a base-collector bias, V_{BC}:

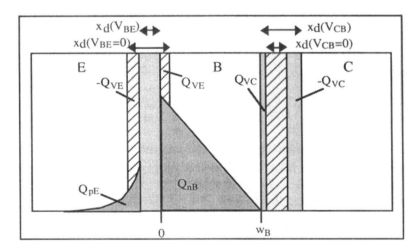

Figure 8.37: Charges used in the charge-control model in the forward active mode. [19]

In the case of a thin-base transistor operated in the forward-active mode the excess minority carrier concentration at the edges of the neutral base are:

$$n'(x=0) = \frac{n_i^2}{N_{aB}}\left(exp\left(\frac{qV_{BE}}{kT}\right) - 1\right) \quad \text{and} \quad n'(x=w_B) \cong 0 \qquad (8.12.3)$$

Assuming a linear distribution of the electron concentration in the base the total charge of excess minority carriers in the base, taken as a positive quantity, is therefore, equal to:

$$q_F \equiv -Q_{nB} = qA\int_0^{w_B} n'(x)dx = \frac{1}{2}A\,q\,w_B\frac{n_i^2}{N_{aB}}\left(exp\left(\frac{qV_{BE}}{kT}\right) - 1\right) \quad (8.12.4)$$

If we define

$$q_{F0} = \frac{1}{2}A\,q\,w_B\frac{n_i^2}{N_{aB}} \qquad (8.12.5)$$

we can write

$$q_F = q_{F0}\left(exp\left(\frac{qV_{BE}}{kT}\right) - 1\right) \qquad (8.12.6)$$

Assuming no recombination in the base, the collector current can be found using Relationship 8.12.3:

$$I_C = -AJ_{nB} = -A\,q\,D_{nB}\frac{dn'}{dx} = \frac{qA\,D_{nB}\,n_i^2\left(exp\left(\frac{qV_{BE}}{kT}\right) - 1\right)}{N_{aB}\,w_B} \qquad (8.12.7)$$

Using Equations 8.12.5, 8.12.6 and 8.12.7 the collector current can be rewritten in the following form:

$$I_C = \frac{2\,D_{nB}}{w_B^2}\,q_F \qquad (8.12.8)$$

which can be rewritten:

$$I_C = \frac{q_F}{\tau_F} \quad \text{with} \quad \tau_F = \frac{w_B^2}{2D_{nB}} \qquad (8.12.9)$$

τ_F is called the "transit time" of the minority carriers in the base. It represents the time it takes for the electrons injected from the emitter to reach the collector. It is proportional to the square of the width of the neutral base. Shrinking the base width is, therefore, an important design parameter for the improvement of bipolar transistor high-frequency performance.

The quasi-static base current has two components: the hole current injected by the base into the emitter, I_{pE}, and the hole current recombining with excess electrons in the base, I_{rB}. While I_{rB} can be neglected from the total collector current ($I_C >> I_{rB}$) it is, however, a substantial component of the base current, I_B.

The first component, I_{pE}, can be obtained from Relationship 8.3.10, which is valid if the doping concentration in the base is constant:

$$I_{pE} = A\, J_{pE} = -\frac{q\, A\, D_{pE}\, P_{oE}}{L_{pE}} \left[exp\left(\frac{q V_{BE}}{kT}\right) - 1 \right]$$

or, using Equations 8.12.5 and 8.12.6:

$$I_{pE} = -\frac{2\, D_{pE}\, P_{oE}}{L_{pE}\, w_B\, n_{oB}}\, q_F \qquad (8.12.10)$$

The second component, I_{rB}, can be found in Expression 8.6.16 and is equal to:

$$I_{rB} = -q\, A \int_0^{w_B} \frac{n'\, dx}{\tau_{nB}} = -\frac{q_F}{\tau_{nB}} \qquad (8.12.11)$$

Adding those two components we find the base current:

$$I_B = -I_{pE} - I_{rB} = \left(\frac{2\, D_{pE}\, P_{oE}}{L_{pE}\, w_B\, n_{oB}} + \frac{1}{\tau_{nB}} \right) q_F \qquad (8.12.12)$$

or:
$$I_B = \frac{q_F}{\tau_{BF}} \quad \text{with} \quad \tau_{BF} = \frac{1}{\dfrac{2\, D_{pE}\, P_{oE}}{L_{pE}\, w_B\, n_{oB}} + \dfrac{1}{\tau_{nB}}} \qquad (8.12.13)$$

Note that the common-emitter current gain is given by the following relationship:

$$\beta_F = \frac{I_C}{I_B} = \frac{\tau_{BF}}{\tau_F} \qquad (8.12.14)$$

The equations derived so far for I_B, I_E and I_C are quasi-static. To include time-dependent current components, the displacement currents due to the variation of charges in the device with emitter-base voltage have to be included in the model, which yields, for the base current:

$$I_B = \frac{q_F}{\tau_{BF}} + \frac{dq_F}{dt} + \frac{dQ_{VE}}{dt} + \frac{dQ_{VC}}{dt} \qquad (8.12.15)$$

Adding the quasi-static collector current to the displacement current flowing through the base-collector transition capacitance we obtain:

$$I_C = \frac{q_F}{\tau_F} - \frac{dQ_{VC}}{dt} \qquad (8.12.16)$$

The emitter current can readily be deduced from the base current and collector current expressions:

$$I_E = -(I_C + I_B) = -q_F\left(\frac{1}{\tau_F} + \frac{1}{\tau_{BF}}\right) - \frac{dq_F}{dt} - \frac{dQ_{VE}}{dt} \qquad (8.12.17)$$

We know that the quasi-static emitter current ($dI_E/dt = 0$) is given by I_E $= -I_{ES}\left[exp\left(\frac{qV_{BE}}{kT}\right)-1\right]$. Therefore, comparing the latter relationship with Expressions 8.12.6 and 8.12.17 we find the saturation current of the emitter-base junction:

$$I_{ES} = q_{F0}\left(\frac{1}{\tau_F} + \frac{1}{\tau_{BF}}\right) \qquad (8.12.18)$$

It is worthwhile noting that the capacitance $\dfrac{dq_F}{dt}$ is a diffusion capacitance due to the variation of minority carriers stored in the neutral base, while $\dfrac{dQ_{VE}}{dt}$ and $\dfrac{dQ_{VC}}{dt}$ are the transition capacitances of the emitter-base and collector-base junctions, respectively.

Expressions 8.12.15 to 8.12.18 corresponds the equivalent circuit shown in Figure 8.38.

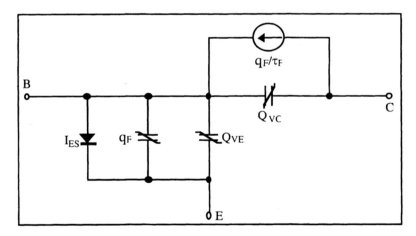

Figure 8.38: Equivalent circuit for the charge-control model of a bipolar transistor in the forward active mode. [20]

Example
  Consider the circuit shown in Figure 8.39, which represents an NPN bipolar transistor biased in the forward active mode. A $t=0$ the base current changes from a value I_{B1} to another value, I_{B2}. derive an expression for the change in collector current as a function of time.

Since the transistor is in the forward active mode the base-emitter voltage, V_{BE}, varies very little with base current since the base current is an exponential function of V_{BE}. As a result the variation of the charge stored in the emitter-base junction, dQ_{VE}/dt, is very small. In addition, we can assume V_{CB} is a constant, and, therefore, $dQ_{VC}/dt = 0$. Using these simplifications one can write from Expression 8.12.15:

$$I_B = + \frac{q_F}{\tau_{BF}} + \frac{dq_F}{dt}$$

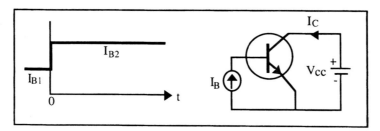

Figure 8.39: Example illustrating transient effects.

This is a first-order differential equation which can be solved for q_F using the following boundary conditions:

$$I_B(t=0) = I_{B1} \text{ and } I_B(t \to \infty) = I_{B2}$$
$$\Downarrow$$
$$q_F(t=0) = I_{B1} \tau_{BF} \text{ and } q_F(t \to \infty) = I_{B2} \tau_{BF}$$

The solution to the differential equation with the applied boundary conditions is thus:

$$q_F = \tau_{BF} [I_{B2} + (I_{B1} - I_{B2}) \exp(-t/\tau_{BF})]$$

from which we can derive the collector current:

$$I_C = q_F/\tau_F = \beta_F [I_{B2} + (I_{B1} - I_{B2}) \exp(-t/\tau_{BF})]$$

The evolution of the collector current with time is plotted in Figure 8.40.

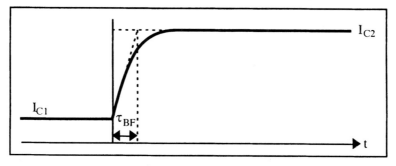

Figure 8.40: Evolution of the collector current with time. [21]

8.12.2. Large-signal model

The charge-control model is particularly useful when it comes to solving transient problems, *i.e.* when the transistor is switched from one mode of operation to another (forward active regime, saturation, cut-off, etc.)

As mentioned earlier (Figure 8.16) the distribution of minority carriers in a transistor in saturation is equal to the sum of the distributions in the forward active mode and in the reverse active mode. Superimposing those two distributions, and defining q_R as the charge injected by the collector into the base in the reverse active mode one obtains a set of three expressions that are applicable to any regime of operation of the transistor:

$$I_B = \frac{q_F}{\tau_{BF}} + \frac{dq_F}{dt} + \frac{q_R}{\tau_{BR}} + \frac{dq_R}{dt} + \frac{dQ_{VE}}{dt} + \frac{dQ_{VC}}{dt} \qquad (8.12.19)$$

$$I_C = \frac{q_F}{\tau_F} - \frac{dQ_{VC}}{dt} - q_R \left(\frac{1}{\tau_R} + \frac{1}{\tau_{BR}} \right) - \frac{dq_R}{dt} \qquad (8.12.20)$$

$$I_E = - q_F \left(\frac{1}{\tau_F} + \frac{1}{\tau_{BF}} \right) - \frac{dq_F}{dt} - \frac{dQ_{VE}}{dt} + \frac{q_R}{\tau_R} \qquad (8.12.21)$$

By analogy to the forward active mode one can define, in the reverse active mode:

$$q_R = q_{R0} \, [exp(qV_{BC}/kT)-1] \qquad (8.12.22)$$

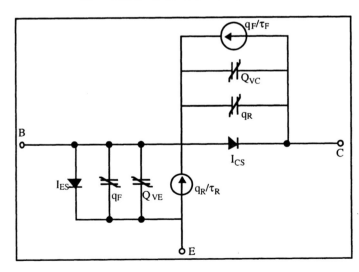

Figure 8.41: Equivalent circuit for the charge-control, large-signal model. [22]

From Equation 8.12.18 which is repeated here as Equation 8.12.23 we know that:

$$I_{ES} = q_{F0} \left(\frac{1}{\tau_F} + \frac{1}{\tau_{BF}} \right) \tag{8.12.23}$$

and, therefore,

$$I_{CS} = q_{R0} \left(\frac{1}{\tau_R} + \frac{1}{\tau_{BR}} \right) \tag{8.12.24}$$

Figure 8.41 shows the equivalent circuit corresponding to equations 8.12.19 to 8.12.24.

8.12.3. Small-signal model

In many instances the bipolar transistor is biased in the forward active mode by dc voltage supplies. A small ac signal applied to the circuit can then be amplified. If the amplitude of the small ac signal (a music signal, for example) is small compared to kT/q, it is possible to linearize the transistor equations around the dc operating point. This greatly simplifies the equations. The model obtained from this simplification is called a "small signal" model.

As we have seen previously the electron current in the base is given by (Relationships 8.6.12a and 8.6.12b):

$$J_{nE} = J_S \left[exp\left(\frac{qV_{BC}}{kT} \right) - exp\left(\frac{qV_{BE}}{kT} \right) \right] \quad \text{with} \quad J_S = \frac{q \, n_i^2 \, D_{nB}}{G_B}$$

In the forward active mode we obtain:

$$I_C \cong - A \, J_{nE} = I_S \, exp\left(\frac{qV_{BE}}{kT} \right) \tag{8.12.25}$$

If a small ac voltage variation is added to the dc bias the variation of collector current can be obtained:

$$\frac{dI_C}{dV_{BE}} = \frac{I_S}{kT/q} \, exp\left(\frac{qV_{BE}}{kT} \right) = \frac{I_C}{kT/q} \equiv g_m \tag{8.12.26}$$

The parameter $g_m = dI_C/dV_{BE}$ is called the transconductance of the transistor. Note that the transconductance is proportional to the collector current.

Using Equation 8.12.26 in conjunction with $I_C = \beta_F I_B$ one finds the base current:

$$\frac{dI_B}{dV_{BE}} = \frac{g_m}{\beta_F} = \frac{\tau_F}{\tau_{BF}} g_m \tag{8.12.27}$$

The variation of the electron charge in the base, q_F, with the emitter-base voltage is given by the following relationship:

$$\frac{dq_F}{dV_{BE}} = \frac{dI_C\ \tau_F}{dV_{BE}} = g_m\ \tau_F \equiv C_D \qquad (8.12.28)$$

where C_D represents the diffusion capacitance associated with the small-signal variation of the charge of the minority carriers injected by the emitter into the base and by the base into the emitter.

As far as small signals are concerned the Early effect influences the output conductance, which can be related to the transconductance as follows:

$$\frac{dI_C}{dV_{CB}} \equiv \frac{I_C}{-V_A} = \frac{g_m\ kT/q}{-V_A} \qquad (8.12.29)$$

If the Early effect is neglected ($V_A \to -\infty$) one can draw the small-signal equivalent circuit for the bipolar transistor shown in Figure 8.42. This equivalent circuit represents the hybrid-π model for the transistor. The word "hybrid" arises from the fact that the current source is controlled by a voltage, and the letter π comes from the fact that the circuit is shaped like the Greek letter π, upside down.

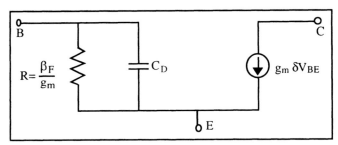

Figure 8.42: Small-signal equivalent circuit of the common-emitter configured transistor: the hybrid-π model. [23] δV_{BE} is a small ac variation of emitter-base voltage.

The hybrid-π model can directly be derived from the Ebers-Moll equations at low frequencies. Using Relationship 8.3.18 in the forward active mode, and neglecting I_{CS} one obtains:

$$I_E = -I_{ES}\ exp\left(\frac{qV_{BE}}{kT}\right) \text{ and } I_C = \alpha_F I_{ES}\ exp\left(\frac{qV_{BE}}{kT}\right)$$

from which we can derive:

$$\frac{dI_C}{dV_{BE}} = \frac{\alpha_F I_{ES}}{kT/q}\ exp\left(\frac{qV_{BE}}{kT}\right) = \frac{I_C}{kT/q} \equiv g_m \qquad (8.12.30)$$

and $\qquad I_C = \beta_F I_B \Rightarrow \frac{dI_B}{dV_{BE}} = \frac{g_m}{\beta_F} \qquad (8.12.31)$

Using these equations one can draw the equivalent circuit of Figure 8.43, directly from the Ebers-Moll equations.

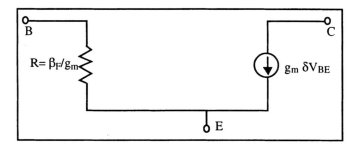

Figure 8.43: Hybrid-π model valid at low frequency.

Important Equations

Common-Emitter Current Gain

$$I_C = \beta_F I_B \quad \text{with} \quad \beta_F \equiv \frac{\alpha_F}{1 - \alpha_F} \tag{8.1.5}$$

Ebers-Moll Equation (Transport Model)

$$\begin{bmatrix} I_E \\ I_C \end{bmatrix} = I_S \begin{bmatrix} -\frac{1}{\beta_F} - 1 & 1 \\ 1 & -1 - \frac{1}{\beta_R} \end{bmatrix} \begin{bmatrix} exp(qV_{BE}/kT) - 1 \\ exp(qV_{BC}/kT) - 1 \end{bmatrix} \tag{8.5.4}$$

Emitter Efficiency

$$\gamma_F \cong \frac{1}{1 + \dfrac{D_{pE}\, p_{oE}\, L_{nB}\, w_B}{D_{nB}\, n_{oB}\, L_{pE}\, L_{nB}}} = \frac{1}{1 + \dfrac{\mu_{pE}\, p_{oE}\, w_B}{\mu_{nB}\, n_{oB}\, L_{pE}}} \tag{8.3.31}$$

Transport Factor in the Base

$$\alpha_T \cong 1 - \frac{(w_B/L_{nB})^2}{2} \tag{8.3.34}$$

Common-Base Current Gain

$$\alpha_F = \gamma_F\, \alpha_T \tag{8.3.35}$$

Problems

Problem 8.1:
Calculate the common-emitter current gain of an NPN bipolar transistor in the forward active mode. The doping concentrations in the base, emitter and collector are different constant (homogenous) values. The following data are given:

$$T = 300 \text{ K}$$
$$N_{aB} = 2 \times 10^{17} \text{ cm}^{-3}$$
$$N_{dE} = 5 \times 10^{19} \text{ cm}^{-3}$$
$$n_i = 1.45 \times 10^{10} \text{ cm}^{-3}$$
$$\mu_{pE} = 100 \text{ cm}^2\text{V}^{-1}\text{s}^{-1}$$
$$\mu_{nB} = 700 \text{ cm}^2\text{V}^{-1}\text{s}^{-1}$$
$$w_B = 0.2 \text{ }\mu\text{m}$$
$$L_{pE} = 0.02 \text{ }\mu\text{m}$$
$$L_{nB} = 4 \text{ }\mu\text{m}$$

 Problem 8.2

Using the Ebers-Moll model, plot the base current and the collector current of a silicon NPN bipolar transistor as a function of $V_{BE}\big|_{\text{at the contacts}}$. Plot the common-emitter current gain as a function of the collector current. Use log scales for all plots of currents. The following data are given:

T=300 K

Device area = 0.0001 cm^2

Dopant concentration in the emitter, base and collector =10^{19}, 10^{17} and 10^{16} cm^{-3}, respectively

Electron and hole mobility = 1000 and 300 cm^{2}V^{-1}s^{-1}, respectively

Neutral base width = 800 nm

Base resistance = 10 Ω

Lifetime of minority carriers in the base, emitter and collector = 1 μs, 100 ps and 300 ns, respectively

V_{CE} = 3 V

The voltage across the emitter-base transition region ranges from 0.01 to 0.85V

Use $V_{BE}\big|_{\text{across the junction}} = V_{BE}\big|_{\text{at the contacts}} - I_BR_B$ (Equation 8.9.1)

Problem 8.3:

This problem illustrates the Early effect. We have an NPN bipolar transistor with the following parameters:

N_{dE}	5.00E+19	cm-3
N_{aB}	1.00E+17	cm-3
N_{dC}	1.00E+16	cm-3
μ_n	8.00E+02	cm2/Vs
μ_p	3.00E+02	cm2/Vs
L_{pE}	1.00E-05	cm
L_{nB}	1.00E-03	cm

The width of the base region, defined as the distance between the two metallurgical junctions, is 1 μm. Assume the emitter junction width is much larger than L_{pE}.

Question: Calculate the common-emitter current gain when the transistor is in the forward active mode with V_{BE}=0.7V and V_{BC} is equal to 0 V and -5 V.

Problem 8.4:
A company manufactures NPN bipolar transistors. The β_F of these transistors is 100. One day the furnaces in the clean room of that company get contaminated by metallic impurities. As a result the lifetime of the minority carriers in the base of the devices drops by 50%. We will assume that the emitter efficiency of the devices is equal to unity. What value will the contaminated β_F have?

Problem 8.5:
Problem Figure illustrates a circuit fabricated using two NPN transistors connected at their bases, called a "current mirror".
1) Show that the current in resistor R is equal to the current in the 1000 Ω resistor if T_1 and T_2 are identical. What is the value of that current, assuming T_1 and T_2 are silicon devices operating at room temperature?
2) What is the current in the resistors if T_1 and T_2 are germanium transistors?

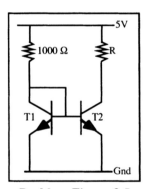

Problem Figure 8.5

 Problem 8.6
Using the Ebers-Moll model, plot the common-emitter current gain of a silicon NPN bipolar transistor as a function of neutral base width (0.1μm < w_B < 2μm). The following data are given:
  T=300 K
  Device area = 0.0001 cm^2
  Dopant concentration in the emitter, base and collector =10^{19}, 3×10^{17} and 10^{16} cm^{-3}, respectively
  Electron and hole mobility = 1000 and 300 cm^{2}V^{-1}s^{-1}, respectively
  Lifetime of minority carriers in the base, emitter and collector = 100 μs, 50 ps and 300 ns, respectively
  V_{CE} = 3 V
  V_{BE} =0.7 V

Problem 8.7

This problem illustrates the Early effect. We have an NPN bipolar transistor in the common-emitter configuration with the following parameters:

```
W=0.5e-4;                      % Metallurgical base width (cm)
NdE=5e19;NaB=1e17;NdC=1e16;    % Doping concentrations (cm-3)
mun=800;mup=300;               % Mobilities (cm2/Vs)
LpE=1e-5;LnB=1e-3;             % Diffusion lengths (cm)
VBE=0.7;                       % Base voltage (V)
A=0.0001;                      % Area (cm2)
```

The width of the neutral base is equal to the width of the metallurgical base minus the extension of the depletion regions from the emitter and collector junctions into the base.

Plot the collector current versus the collector-emitter voltage, V_{CE}, for $1V < V_{CE} < 5V$ and for $I_B = 100\mu A, 200\mu A, 300\mu A \ldots 1$ mA. Then draw a tangent to each $I_C(V_{BE})$ curve until it intercepts the x-axis ($I_C = 0$). This intercept gives us the Early voltage of the transistor (see Figure 8.23).

Problem 8.8
Derive equation 8.6.5 from equations 8.6.3 and 8.6.4.

References

1   J.M. Early, "Out to Murray Hill to play: an early history of transistors", *IEEE Transactions on Electron Devices*, Vol. 48, No. 11, p. 2468, 2001
2   A.S. Grove, *Physics and technology of semiconductor devices*, J. Wiley & Sons, p. 209, 1967
3   S.M. Sze, *Physics of semiconductor devices*, J. Wiley & Sons, p. 136, 1981
4   R.S. Muller and T.I. Kamins, *Device electronics for integrated circuits*, J. Wiley and Sons, p. 282, 1986

5   R.S. Muller and T.I. Kamins, *Device electronics for integrated circuits*, J. Wiley and Sons, p. 304, 1986
6   J.J. Ebers and J.L. Moll, "Large signal behavior of junction transistors", *Proceedings IRE*, Vol. 42, p. 1761, 1954
7   S.M. Sze, *Physics of semiconductor devices*, J. Wiley & Sons, p. 152, 1981
8   J.M. Feldman, *The physics and circuit properties of transistors*, J. Wiley & Sons, p. 371, 1972
9   D.J.Roulston, *Bipolar semiconductor devices*, Mc Graw Hill, p.195, 1990
10   J.L. Moll, *Physics of semiconductors*, McGraw-Hill, p. 151, 1964
11   R.S. Muller and T.I. Kamins, *Device electronics for integrated circuits*, J. Wiley and Sons, p. 299, 1986
12   H.K. Gummel and H.C. Poon, "An integral charge control model of bipolar transistors", *Bell System Technical Journal*, Vol. 49, No. 5, p. 827, 1970

13 S.M. Sze, *Physics of semiconductor devices*, J. Wiley & Sons, pp. 153-156, 1981
14 J.M. Early, "Effects of space-charge layer widening in junction transistors", *Proceedings IRE*, Vol. 40, p. 1401, 1952
15 R.S. Muller and T.I. Kamins, *Device electronics for integrated circuits*, J. Wiley and Sons, p. 321, 1986
16 C.T. Kirk, Jr., "A theory of transistor cut-off frequency (f_T) fall off at high current densities", *IRE Transactions on Electron Devices*, Vol. ED-9, p. 164, 1962
17 S.M. Sze, *Physics of semiconductor devices*, J. Wiley & Sons, p. 142, 1981
18 Device simulator MEDICI, AVANT! Corporation, Fremont, CA, USA.
19 R.S. Muller and T.I. Kamins, *Device electronics for integrated circuits*, J. Wiley and Sons, p.338, 1986
20 R.S. Muller and T.I. Kamins, *Device electronics for integrated circuits*, J. Wiley and Sons, p.342, 1986
21 R.S. Muller and T.I. Kamins, *Device electronics for integrated circuits*, J. Wiley and Sons, p.343, 1986
22 R.S. Muller and T.I. Kamins, *Device electronics for integrated circuits*, J. Wiley and Sons, p.345, 1986
23 J.M. Feldman, *The physics and circuit properties of transistors*, J. Wiley & Sons, p. 428, 1972

Chapter 9

HETEROJUNCTION DEVICES

9.1. Concept of a heterojunction

Silicon is not the only semiconductor used in the electronics industry. Beside elements from the fourth column of the periodic table and compounds thereof (Si, Ge, C, SiC and SiGe), a whole range of semiconductors can be synthesized using elements from columns III and V, such as GaAs, InP, $Ga_xAl_{1-x}As$, etc. In addition, it is also possible to fabricate semiconductors using elements from other columns of the periodic table, such as CdS and HgCdTe.

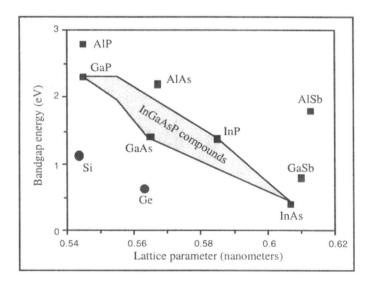

Figure 9.1: Energy bandgap of Si, Ge, and III-V compound semiconductors.[1]

The main parameter characterizing the electrical properties of these materials is the width of the bandgap. Figure 9.1 shows the bandgap energy for silicon, germanium, and different III-V compounds. Arbitrary values of the bandgap energy can be obtained using ternary or quaternary compounds, such as $Ga_xAl_{1-x}As$ and $Ga_xIn_{1-x}As_yP_{1-y}$. The desired bandgap energy can be reached by adjusting the x and y coefficients during the fabrication of the material.

A PN junction that encompasses two different semiconductors is called a *heterojunction*. The most distinctive feature of such junctions is that the P and the N region have different energy band gaps. A junction containing only one semiconductor, such as a classical silicon PN junction, is called a *homojunction*.

9.1.1. Energy band diagram

The presence of two materials with different bandgap energies introduces an additional level of difficulty in the energy band diagram of heterojunctions, when compared to homojunctions. Combining different semiconductor materials within a single device and the art of tailoring the shape of energy bands to achieve properties which could otherwise not be attained is often referred to as "bandgap engineering".

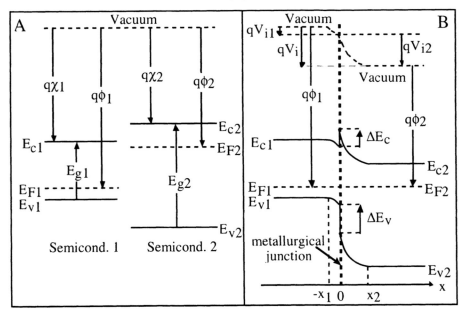

Figure 9.2: A: Energy band diagram of the two semiconductors taken separately; B: Energy band diagram of the materials connected (heterojunction).[2,3]

Consider the example of Figure 9.2 which illustrates how the energy band diagram of a heterojunction can be drawn. Two different semiconductor materials are combined. Let Semiconductor #1 be P-type and have an energy band gap, a work function, and an electron affinity equal to E_{g1}, $q\phi_1$, and $q\chi_1$, respectively. The work function is the energy difference between the vacuum level and the Fermi level; it represents the energy required to remove an electron of energy E_F from the semiconductor. The electron affinity is the energy needed to remove an electron in the conduction band to the vacuum level, as previously explained in Section 5.1.1. Similarly, we will suppose Semiconductor #2 is N-type, and its energy band gap, work function, and electron affinity are E_{g2}, $q\phi_2$, and $q\chi_2$, respectively.

The procedure for drawing the energy band diagram is the following:

1- Under equilibrium conditions the Fermi level in the two semiconductors is equal and constant. Far from the junction the semiconductor materials will be neutral and their energy band diagram will be the same as when the two materials are taken separately.

2- The work functions $q\phi_1$ and $q\phi_2$ remain unchanged in the neutral zones. This enables us to draw the vacuum levels, far from the junction.

3- The vacuum levels of the two semiconductors are connected by a smooth, continuous curve. The exact shape of the curve is at present unknown and will be calculated later. It is, however, a good idea to assume that it will have a shape similar to the band bending in the transition region of a homojunction. The vacuum level bends only within the transition region, thus between $-x_1$ and x_2.

4- During the junction formation electrons will diffuse from the N-type semiconductor into the P-type material since $q\phi_1 > q\phi_2$, and holes will diffuse in the opposite direction from the N-type into the P-type semiconductor. The resulting charge distribution gives rise to a depletion region, an internal junction potential, and therefore, to a curvature of the energy bands. This curvature is parallel to that of the vacuum level. Knowing that the electron affinities, $q\chi_1$ and $q\chi_2$ remain constant in the transition region enables us to draw E_{V1}, E_{V2}, E_{C1} and E_{C2} in the transition region.

5- Finally the valence (E_{V1} and E_{V2}) and conduction (E_{C1} and E_{C2}) levels are connected using vertical line segments, at the metallurgical junction ($x=0$). This feature constitutes what is called a "band discontinuity".

The junction potential, V_i, is given by:

$$V_i = V_{i1} + V_{i2} = \phi_1 - \phi_2 \qquad (9.1.1)$$

where qV_{i1} and qV_{i2} is the band curvature in semiconductors 1 and 2, respectively.

Since both E_{C1} and E_{C2} are parallel to the vacuum level there will be a discontinuity of the energy bands at the metallurgical junction. The discontinuity is equal to:

$$\Delta E_C = q(\chi_1 - \chi_2) \qquad (9.1.2)$$

and
$$\Delta E_v = (E_{g2} - E_{g1}) - q(\chi_1 - \chi_2) \qquad (9.1.3)$$

The sum of the two band discontinuities is equal to the bandgap difference between the two semiconductors:

$$\Delta E_C + \Delta E_v = E_{g2} - E_{g1} \qquad (9.1.4)$$

The exact curvature of the energy bands within the transition region can be obtained by solving Poisson's equation in both semiconductor materials and using the depletion approximation.

Semiconductor #1	**Semiconductor #2**
P-type, doping concentration = N_a	N-type, doping concentration = N_d
permittivity = ε_1	permittivity = ε_2
width of depletion region = $-x_1$	width of depletion region = x_2

Poisson's equation is integrated to calculate the electric field:

$$\frac{d^2\phi}{dx^2} = -\frac{\rho}{\varepsilon_1} = \frac{qN_a}{\varepsilon_1} \qquad\qquad \frac{d^2\phi}{dx^2} = -\frac{\rho}{\varepsilon_2} = -\frac{qN_d}{\varepsilon_2}$$

$$\mathcal{E}_1 = -\frac{d\phi}{dx} = -\frac{qN_a}{\varepsilon_1}(x+x_1) \qquad\qquad \mathcal{E}_2 = -\frac{d\phi}{dx} = -\frac{qN_d}{\varepsilon_2}(x_2-x)$$

Using Gauss' theorem at the metallurgical junction ($x=0$) yields:

$$\mathcal{E}_1\,\varepsilon_1 = \mathcal{E}_2\,\varepsilon_2 \;\Rightarrow\; N_a\,x_1 = N_d\,x_2 \qquad (9.1.5)$$

which expresses charge neutrality in the transition region ($-x_1 \leq x \leq x_2$).

A second integration of Poisson's equation yields the potential:

$$\phi = A + \frac{qN_a}{2\varepsilon_1}(x+x_1)^2 \qquad\qquad \phi = B - \frac{qN_d}{2\varepsilon_2}(x_2-x)^2$$

and the band curvature:

$$\phi(x=0) - \phi(-x_1) = \frac{qN_a}{2\varepsilon_1}x_1^2 \qquad\qquad \phi(x_2) - \phi(x=0) = \frac{qN_d}{2\varepsilon_2}x_2^2$$

The sum of the two latter equations is equal to the junction potential, V_i:

$$\phi(x_2) - \phi(-x_1) = \frac{qN_a}{2\varepsilon_1}x_1^2 + \frac{qN_d}{2\varepsilon_2}x_2^2 = V_i = V_{i1} + V_{i2} = \phi_1 - \phi_2 \qquad (9.1.6)$$

Eliminating x_2 between 9.1.5 and 9.1.6, we obtain the built-in potential in semiconductor:

$$\frac{qN_a}{2\varepsilon_1}x_1^2 + \frac{qN_d}{2\varepsilon_2}\frac{N_a^2}{N_d^2}x_1^2 = V_i \qquad (9.1.7)$$

from which the depletion width in semiconductor #1 can be extracted:

$$x_1 = \sqrt{\frac{2\varepsilon_1\varepsilon_2\,N_d\,V_i}{q\,N_a\,(\varepsilon_2 N_d + \varepsilon_1 N_a)}} \qquad (9.1.8a)$$

Using 9.1.5 and 9.1.8a we find the depletion width in semiconductor #2:

$$x_2 = \sqrt{\frac{2\varepsilon_1\varepsilon_2\,N_a\,V_i}{q\,N_d\,(\varepsilon_2 N_d + \varepsilon_1 N_a)}} \qquad (9.1.8b)$$

Knowing x_1, x_2, and $\phi(x)$ the energy band curvature can now be drawn with accuracy.

When an external bias, V_a, is applied to the diode, the electron and hole diffusion current densities injected respectively at $x = -x_1$ in the P-type and at $x = x_2$ in the N-type material are given by Equations 4.4.23 (where $l_n=x_2$) and 4.4.24 (where $-l_p=-x_1$) which, in the case of a heterojunction, becomes:

$$J_n(-x_1) = \frac{qD_n n_{i1}^2}{N_a L_n}\left[exp\left(\frac{qV_a}{kT}\right) - 1\right] \text{ and } J_p(x_2) = \frac{qD_p n_{i2}^2}{N_d L_p}\left[exp\left(\frac{qV_a}{kT}\right) - 1\right]$$

where n_{i1}^2 and n_{i2}^2 are the intrinsic carrier concentrations in semiconductor #1 and #2, respectively. The influence of the heterojunction on the diffusion currents is best illustrated by taking the J_n/J_p ratio at the edges of the depletion region:

$$\frac{J_n}{J_p} = \frac{D_n L_p N_d n_{i1}^2}{D_p L_n N_a n_{i2}^2} = \frac{D_n L_p N_d N_{v1} N_{c1} exp(-E_{g1}/kT)}{D_p L_n N_a N_{v2} N_{c2} exp(-E_{g2}/kT)}$$

$$= \frac{D_n L_p N_d\,(m_{n1}^* m_{p1}^*)^{3/2}}{D_p L_n N_a (m_{n2}^* m_{p2}^*)^{3/2}}\,exp\left(\frac{E_{g2}-E_{g1}}{kT}\right) \qquad (9.1.9)$$

where N_{v1}, N_{c1}, m_{n1}^*, m_{p1}^*, N_{v2}, N_{c2}, m_{n2}^*, m_{p2}^*, are the effective density of states in the valence and conduction band, and the effective electron and hole masses in semiconductor #1 and #2, respectively. An important conclusion can be drawn from Equation 9.1.9: the ratio of electron to hole current in the PN heterojunction is exponentially proportional to the difference of energy bandgaps between the two semiconductors.

9.2. Heterojunction bipolar transistor (HBT)

The Heterojunction Bipolar Transistor (HBT) was developed to overcome the limitations of conventional bipolar transistors. In a classical homojunction bipolar transistor the base width must be reduced to achieve high speed (the transit time of the minority carriers through the base is proportional to the square of the base width - Equation 8.12.9). However, if the width of the base is reduced, the base resistance is increased, which slows the device response time. The base resistance can be reduced by increasing the base doping concentration, but then the Gummel number in the base is increased, which decreases the current gain. It is, therefore, impossible to optimize the base thickness (width) and doping concentration for high-speed, high gain and low base resistance. The use of heterojunctions, however, permits improved transit time, current gain and base resistance simultaneously.

In a PN homojunction the ratio between the electron and hole current essentially depends on the ratio of impurity doping concentration between the N and the P regions (Relationship 8.1.1):

$$\frac{J_{nE}}{J_{pE}} = \sqrt{\frac{\mu_n \tau_p}{\mu_p \tau_n}} \frac{N_d}{N_a}$$

This is why a much larger doping concentration is used in the emitter of a bipolar transistor than in its base. This ensures that the emitter current is much larger than the base current, and as a consequence, that the emitter efficiency, γ_F, and the current gain, β_F, are large.

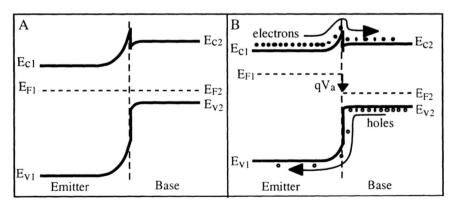

Figure 9.3: Energy band diagram in the emitter-base junction of an HBT. A: under equilibrium; B: under forward bias.

The use of a heterojunction for the emitter-base junction completely changes the ratio between electron and hole currents. Let us consider the example of Figure 9.3 where a wider bandgap material is used for the N-type emitter and a smaller bandgap semiconductor for the P-type base. It is easy to observe that when a forward bias, V_a, is applied to the junction, holes must overcome a much larger potential barrier than electrons. As a result the hole current injected into the emitter is much smaller than the electron current injected into the base, *even if the doping concentration in the base is higher than that of the emitter.*

One can take advantage of this strong asymmetry of carrier injection to achieve high gain. The emitter efficiency of a bipolar device is given by Relationship 8.6.33:

$$\gamma_F = \left.\frac{I_{nE}}{I_{nE}°+°I_{pE}}\right|_{V_{BC}=0} = \left.\frac{1}{°1°+\dfrac{°I_{pE}°}{I_{nE}}}\right|_{V_{BC}=0}$$

NPN HBTs provide values for γ_F which are very close to unity because $I_{nE} >> I_{pE}$ in the emitter-base junction, as demonstrated by Relationship 9.1.9. The use of silicon and silicon-germanium (SiGe) for such devices is quite popular. In that case the following structure is commonly used:

Emitter: N-type silicon (energy bandgap = 1.12 V)
Base: P-type Si (80%); Ge (20%) alloy (energy bandgap = 0.87 V)
Collector: N-type silicon (energy bandgap = 1.12 V)

The current gain of the transistor is directly proportional to the emitter efficiency. Using Equation 9.1.9 for the emitter junction, we find:

$$\frac{I_{nE}}{I_{pE}} \cong -\frac{I_C}{I_B} = \beta_F \ ^- exp(\Delta E_g/kT) \ where \ \ \Delta E_g = E_{g1} - E_{g2}$$

Using heterojunctions a high current gain can be achieved even if the doping concentration in the base is high. A thin, highly-doped base can be used, which satisfies the requirements for a low transit time for electrons and a low base resistance. This allows for the design of thin-base HBTs which have excellent high-frequency performances.[4]

9.2. High electron mobility transistor (HEMT)

The acronym HEMT stands for "High Electron Mobility Transistor". Sometimes this device is also called "modulation-doped field-effect transistor" (MODFET). HEMTs are usually realized on III-V semiconductor substrates, such as GaAs and InP.

The mobility of electrons in *lightly-doped* GaAs is very high, reaching values of 8,000, 200,000, and 1,500,000 cm^2/Vs at temperatures of 300, 77, and 4.2 K, respectively. Compared to the surface mobility of electrons in the channel of a silicon MOSFET, which is on the order of

650 cm^2/Vs, these numbers are quite impressive. However, if the impurity doping concentration is increased in GaAs, the electron mobility becomes significantly degraded because of impurity scattering.

The electron drift current in a semiconductor is given by $J_n = q \, \mu_n \, n \, \mathcal{E}$. Thus, for a given electric field, the current is proportional to both the electron concentration and the electron mobility. The operation of the HEMT is quite similar to that of a JFET. In both devices the current flows through a channel between source and drain, and the number of carriers in the channel is modulated by the gate voltage. The current in a JFET can be increased by increasing the doping concentration in the channel. Unfortunately, any increase of the doping concentration results in a decrease of mobility, which becomes a tradeoff: high mobility and high carrier concentration cannot be achieved at the same time. The use of a heterojunction structure allows one to circumvent that problem obtaining high electron concentrations in a lightly doped material which ensures high mobility. The energy band diagram of such a structure is presented in Figure 9.4.

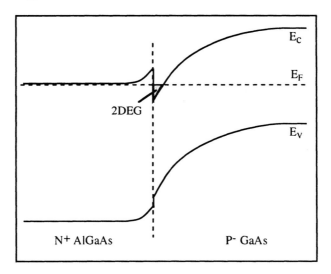

Figure 9.4: Formation of a two-dimensional electron gas in an N$^+$ AlGaAs / P$^-$ GaAs heterojunction.[5,6]

When an N$^+$ AlGaAs / P$^-$ GaAs heterojunction is used a particular band diagram is obtained, in which there is a region in the P$^-$ GaAs where the conduction band is located below the Fermi level. That region contains a very high concentration of electrons and additionally is located in lightly doped material where mobility is high. The region is very thin (5 - 10 nm) such that it has two-dimensional features, like the inversion layer in a MOSFET. Because of its small thickness, the electron layer is called a "**Two-D**imensional **E**lectron **G**as (2DEG)".

The electron concentration in the 2DEG can be modulated by applying a bias to the heterojunction, as shown in Figure 9.5.

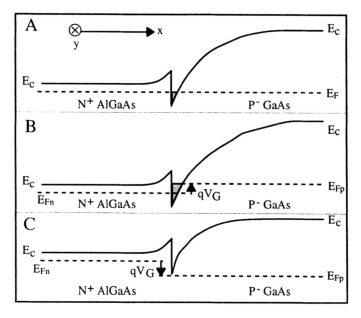

Figure 9.5: Modulation of the electron concentration in the 2DEG. The axes at the top indicate x in a positive direction from the N^+ AlGaAs to the P^- GaAs while y is pointed out of the page (from source to drain)A: $V_G=0$, B: $V_G>0$, C: $V_G<0$.

If a positive bias, V_G, is applied to the N^+ AlGaAs material the junction is reverse biased and E_C is further lowered below E_F in the 2DEG region, which increases the electron concentration. Conversely, if a negative bias is applied to the N^+ AlGaAs material the junction is forward biased and E_C is raised in the 2DEG region, resulting in an electron concentration decrease. If the bias is sufficiently negative, the 2DEG eventually disappears as shown in Figure 9.5.

The cross section of a HEMT is shown in Figure 9.6. The 2DEG forms a channel between the N^+ source and drain. A metallic Schottky contact to the N^+ AlGaAs forms the gate electrode. The application of a gate voltage changes the bias of the heterojunction, which, in turn, modulates the carrier concentration in the 2DEG channel. Note that there is also a parasitic MESFET structure in the AlGaAs layer, the conductivity of which is modulated by the variation of the Schottky gate potential as well. The output characteristics of the complete device are thus the parallel association of the HEMT and that of the parasitic MESFET. HEMTs are amongst the fastest solid-state transistors, owing to the high $\mu_n \times n$ product in the channel.

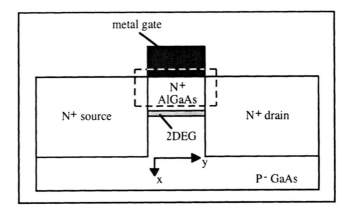

Figure 9.6: Cross section of a HEMT. The dotted line highlights the parasitic MESFET.

9.3. Photonic Devices

When a recombination event takes place in a direct-bandgap semiconductor a photon can be emitted. This phenomenon is called "radiative recombination". The wavelength of this photon depends on the bandgap energy of the semiconductor according to the relationship: $E_g = h\nu$. Radiative recombination is observed in many semiconductor materials such as SiC, GaN, GaAsP, AlInGaP and AlGaAs. Furthermore, the bandgap energy in semiconductor compounds can be tailored to produce devices capable of emitting photons with a specific desired color. There is a whole variety of solid-state devices that can emit and collect photons, but we will only focus here on the laser diode. However, it is necessary to understand the operation of the light-emitting diode before beginning the study of the laser diode.

9.3.1. Light-emitting diode (LED)

The Light-Emitting Diode, or LED, is a simple PN junction made in a semiconductor material which exhibits radiative recombination properties. This PN junction can either be a heterojunction or a homojunction. The energy bandgap of the semiconductor material determines the frequency of the emitted light, according to the relationship $E_g = h\nu$. Some examples of semiconductor materials used for the fabrication of LEDs, and the color of the emitted light are: GaN (blue), SiC (blue), GaP (green), $GaAs_{0.14}P_{0.86}$ (yellow), $GaAs_{0.35}P_{0.65}$ (orange), $GaAs_{0.6}P_{0.4}$ (red), and GaAs (infrared). In this section we will focus on the operation of a homojunction (single material) LED.

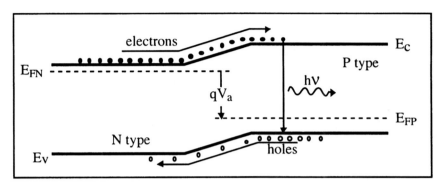

Figure 9.7: Operation of a light-emitting diode. [7]

Figure 9.7 illustrates a homojunction (single bandgap) PN$^+$ junction in the forward bias mode. Light emission is produced by the radiative recombination of electrons injected into the P-type material. Because $N_d >> N_a$ the electron current is much larger than the hole current. Electrons are injected when the PN junction is forward biased by V_a. The injection efficiency η_i relates the current of "useful" carriers (the electrons injected in the P-type region) to the total current in the junction:

$$\eta_i = \frac{I_n}{I_n + I_p + I_{rec}}$$

where I_n is the electron current injected into the P-type region, I_p is the hole current injected in the N-type region, and I_{rec} is the current of carriers recombining in a non-radiative way. Usually, η_i reaches values of 30 to 60%. As mentioned in Section 3.2 radiative recombination must satisfy conservation of momentum. This criterion is automatically met in direct-bandgap semiconductors since the momentum of electrons at the conduction band minimum is equal to that of holes at the valence band maximum. Light emission is, however, observed in indirect-bandgap semiconductors such as GaP and SiC. The only way radiative recombination can take place in these semiconductors is for the interaction to produce a particle, or something capable of acting like a particle, that can dissipate the initial electron momentum. Fortunately, an appropriate "particle" exists which is a quantum of vibrational energy in the crystal lattice, called a phonon. Phonons produce heat transfer to (or from) the lattice, which acts to reduce electron momentum and thereby enables radiative recombination. The interaction of concern is one in which an electron in the conduction band recombines with a hole in the valence band, and produces both a photon and a phonon. The combined energy of the photon and the phonon is equal to E_g and the sum of the initial electron momentum and the momentum of the phonon equals zero. This process is much more complex, and therefore, more unlikely to happen than radiative recombination as in direct bandgap semiconductors. As a result, the performance (in terms of brightness) of indirect bandgap LEDs is much lower than that of direct bandgap materials. The luminous intensity of indirect bandgap devices has, however, been substantially increased using the following "trick". The approach is to add an isoelectronic impurity, *i.e.* an impurity from the same column of the periodic table as the element it replaces. An example is nitrogen in GaP, designated GaP:N. Each nitrogen atom creates a localized strain in the crystal that can trap an electron. The electrons are bound so tightly to those traps that there is little uncertainty as to their position. But there is, according to the Heisenberg uncertainty principle, a large statistical uncertainty in their momentum. The uncertainty is large enough for each electron to have a significant probability of having zero momentum and undergoing radiative recombination. This quantum-mechanical "trick" raises the radiative recombination rate, but to date, not enough to rival the rate in direct bandgap semiconductors.[8]

9.3.2. Laser diode

The laser diode is a PN junction which can emit a laser beam. Laser light is coherent (*i.e.* the emitted photons are in all phase) and monochromatic (*i.e.* the emitted photons all have the same wavelength). Describing in detail how a laser works is beyond the scope of this book. It is, however, necessary to briefly describe the conditions required for a lasing effect to take place. The word "laser" means "**L**ight **A**mplification by **S**timulated **E**mission of **R**adiation". The key word in this definition is "stimulated emission". Stimulated emission is a phenomenon falling into the same category as generation and recombination, but in which an incident photon with an energy $E_g=h\nu$ triggers the recombination of an excited electron (an electron in the conduction band, in the case of a semiconductor laser). During the recombination event a new photon is emitted. This photon has the same wavelength as the incident photon and is in phase with it. This is why laser light is monochromatic (all photons have the same wavelength, fixed by the energy bandgap) and coherent (all photons have the same phase). This photon generation can of course be repeated, and the original photon can be amplified by 2, 4, 8, etc. as shown in Figure 9.8, resulting in a light amplification effect. If two parallel mirrors (which can reflect light) are placed at both sides of the semiconductor crystal light can travel back and forth inside the crystal and undergo significant amplification. Such a structure constitutes a Fabry-Pérot cavity. In practice, one of the mirrors is semitransparent, such that some of the laser light can escape from the crystal. Emitted photons which do not travel perpendicular to the mirrors exit the semiconductor and are lost (Figure 9.8).

A photon with energy $h\nu=E_g$ can not only stimulate the emission of another photon, but it can be absorbed by the semiconductor material and generate an electron-hole pair. This effect is highly undesirable in a laser diode since we do not want to see photons absorbed. Unfortunately, photon absorption is unavoidable. It is, however, possible to favor stimulated emission with respect to absorption. This can be achieved if the number of electrons in the excited state (*i.e.*, in the conduction band) is larger than the number of electrons in the ground state (*i.e.*, in the valence band). This condition is called "population inversion". It can be realized if an external source of energy "pumps" a large quantity of electrons from the fundamental state into the excited state. In a laser diode population inversion is obtained by injecting a large amount of electrons into a PN junction.

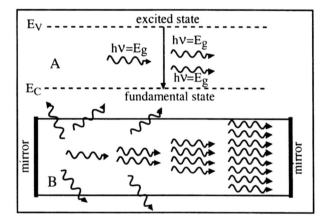

Figure 9.8: A: Principle of stimulated emission, B: Light amplification by stimulated emission.

Figure 9.9 shows a laser PN homojunction. The N^+ and P^+-type regions are degenerately doped and the Fermi level in the N^+ and P^+-type material is above the conduction band minimum and below the valence band maximum, respectively. When a forward bias is applied to the junction a thin region is formed which, instead of being depleted, is in population inversion. In that region there is a strong electron population in the conduction band and a high density of empty states (or holes) in the valence band. Under these conditions laser light is emitted through stimulated emission within the transition region.

A complete laser diode is presented in Figure 9.10. Two semi-transparent, parallel mirrors are obtained by cleaving the semiconductor along a natural crystal direction (*e.g.* (100)). Since the refractive index of the semiconductor material is larger than that of the surrounding air, the cleaved surfaces act as mirrors which reflect the light back into the crystal. These mirrors do not have a 100% reflectivity, however, which allows some of the laser light to be emitted from the device.

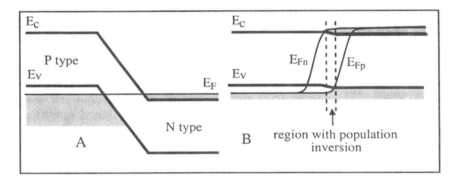

Figure 9.9: Laser PN junction. A: at equilibrium, B: under forward bias. [9]

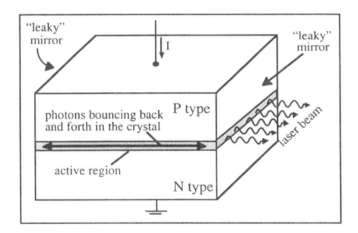

Figure 9.10: Laser diode in forward bias mode showing photon path within the transition region with population inversion.

The output light power of a laser diode is presented in Figure 9.11 as a function of the current injected into the diode. Below a given threshold, population inversion is not reached, however light is emitted because of radiative recombination. This light is incoherent and is similar to the light emitted by a LED. Above this threshold, population inversion takes place, and laser light is emitted. The light intensity then increases sharply as a function of the current in the diode. Because of the Fabry-Pérot cavity the spectrum of emitted light is compressed into one single spectral line. The emitted laser light is, therefore, monochromatic. Beside the "useful recombination" of electrons by stimulated emission in the population inversion region, a large quantity of electrons are injected into the P-type semiconductor where they can also recombine and emit either photons which do not take part in the lasing process, or phonons (heat). This renders homojunction laser diodes quite inefficient, and only a fraction of the electrical power supplied to the device is converted into laser light.

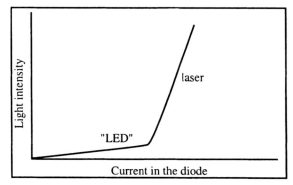

Figure 9.11: Emission of incoherent light (LED) and coherent light (laser) as a function of the current intensity. [10]

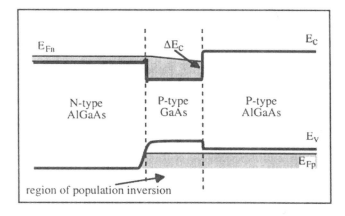

Figure 9.12: AlGaAs/GaAs/AlGaAs heterojunction laser diode.[11]

This problem can be solved by the use of a heterojunction structure. Let us take the example of the AlGaAs/GaAs/AlGaAs heterojunction laser diode shown in Figure 9.12: the electrons in the conduction band which are injected from a forward bias from the N-type AlGaAs into the P-type GaAs cannot spill over the potential barrier ΔE_C created by the P-type GaAs / P-type AlGaAs junction. These electrons are thus confined in the GaAs layer where the inversion population, and thus laser light emission, is produced. In addition, the refractive index of AlGaAs is lower than that of GaAs, which causes the junctions to act as mirrors. This helps confine the photons in the GaAs layer and limits the leakage of light into the AlGaAs layers. As a result the laser light emission efficiency is greatly enhanced and the current threshold for laser light emission is reduced.

Problems

Problem 9.1:

Using Matlab, plot the energy band diagram of a germanium-silicon heterojunction. The germanium is P-type with $N_a = 10^{17}$ cm^{-3}, and the silicon is N-type with $N_d = 10^{17}$ cm^{-3}. Use the following data:

$$\chi_{germanium} = 4.05 \text{ eV and } \chi_{silicon} = 4 \text{ eV}$$
$$E_g(\text{silicon}) = 1.12 \text{ eV and } E_g(\text{germanium}) = 0.66 \text{ eV}$$
$$n_i(\text{silicon}) = 1.45 \times 10^{10} \text{ cm}^{-3} \text{ and } n_i(\text{germanium}) = 2.43 \times 10^{13} \text{ cm}^{-3}$$
$$\varepsilon_o = 8.845 \times 10^{-14} \text{ F/cm}, \ \varepsilon_{silicon} = 11.7 \times \varepsilon_o \text{ and } \varepsilon_{germanium} = 16 \times \varepsilon_o$$

References

1   S.M. Sze, *Physics of semiconductor devices, 2nd edition*, J. Wiley & Sons, p. 706, 1981
2   S.M. Sze, *Physics of semiconductor devices, 2nd edition*, J. Wiley & Sons, p. 123, 1981
3   M. Shur, *GaAs devices and circuits*, Plenum Publishing Corporation, p. 515, 1987
4   M. Shur, *GaAs devices and circuits*, Plenum Publishing Corporation, pp. 615-633, 1987
5   H. Morkoç and H. Unlu, "Factors affecting the performances of (Al,Ga)As/GaAs and (Al,Ga)As/InGaAs modulation-doped field-effect transistors: microwave and digital applications", *Semiconductors and semimetals*, Academic Press, Vol. 24, *Applications of multiquantum wells, selective doping, and superlattices*, p. 135, 1987
6   N.T. Linh, "Two-dimensional electron gas FETs: microwave applications", *Semiconductors and semimetals*, Academic Press, Vol. 24, *Applications of multiquantum wells, selective doping, and superlattices*, p. 203, 1987
7   S.M. Sze, *Physics of semiconductor devices*, J. Wiley & Sons, pp. 681-703, 1981
8   K.I. Werner, "Higher visibility for LEDs", *IEEE Spectrum*, p. 30, July 1994
9   J. Wilson and J.F.B. Hawkes, *Optoelectronics: an introduction*, Prentice-Hall international series in optoelectronics, pp. 174-238, 1983
10 S.M. Sze, *Physics of semiconductor devices*, J. Wiley & Sons, p. 731, 1981
11 K.A. Jones, *Introduction to optical electronics*, Harper & Row Publishers, pp. 282-315, 1987

Chapter 10

QUANTUM-EFFECT DEVICES

10.1. Tunnel Diode

10.1.1. Tunnel effect

The tunnel diode was discovered by L. Esaki in 1958 for which he received the Nobel Prize for explaining the operation of the device.[1] In this section we will first briefly describe the physics underlying the tunnel effect and then explain how a tunnel diode works.

Tunneling of electrons through a potential barrier is an effect predicted by quantum mechanics that gives the electrons a finite probability of passing through the barrier, as opposed to the electrons needing an energy greater than the barrier potential energy to overcome it.

To illustrate this effect, let us take an infinite potential well and introduce a finite potential barrier in it (Figure 10.1A). The wave function of an electron in this potential well can be calculated using numerical simulations (see Problems 1.3 and 1.4). Let us focus on the lowest or ground-state energy level. In the absence of a potential barrier the lowest energy of an electron can be found using Equation 1.1.11: $E = \dfrac{\pi^2 \hbar^2}{2ma^2}$. For a well width of 50 nm the corresponding lowest energy value is approximately 0.15 meV. Let us introduce a potential barrier 40 mV in height and 2 nm in width inside the potential well. According to classical mechanics an electron confined in the left-hand side of the potential well does not possess enough energy to overcome the 40-mV potential barrier and venture into the right part of the well. If the calculation is made using quantum mechanics, on the other hand, one finds that there is a non-zero probability of finding the electron at the right of the potential barrier, as shown in Figure 10.1B.

In a more general sense, tunneling through a potential barrier can be characterized by a transmission coefficient which represents the probability of an electron passing through the barrier. The value of this transmission coefficient depends on the shape of the barrier (rectangular, triangular, etc.), on its width and its height. The thinner and the lower the barrier, the higher the transmission coefficient. In the particular case of a rectangular barrier, the transmission coefficient, T, is given by:

$$T = \cfrac{1}{1 + \cfrac{1}{4}\cfrac{V^2}{E(V-E)}\ sinh^2\left(\cfrac{a}{\hbar}\sqrt{2m(V-E)}\right)}$$

where a and V are the width and the height of the potential barrier, respectively, and E is the energy of the electron $(E<V)$.[2]

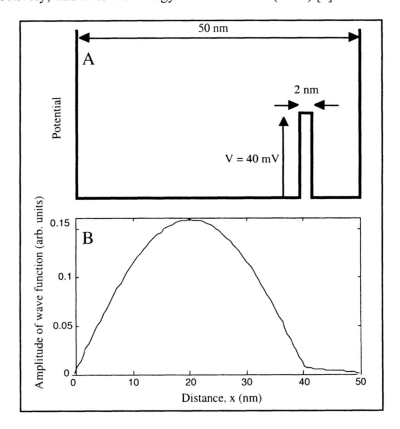

Figure 10.1: A: Infinite potential well with a potential barrier inside it; B: Corresponding lowest-energy wave function.

10.1.2. Tunnel diode

A tunnel diode is a PN junction where both P- and N-type regions are degenerately doped. As a result, the Fermi level in the N-type material is above the minimum of the conduction band and the Fermi level in the P-type material is below the maximum of the valence band. The doping concentrations are so high that the width of the space-charge region at the junction is extremely thin (Equation 4.2.12), and usually measures less than 10 nm.

As in any PN junction the existence of a space-charge region gives rise to a potential barrier. This barrier height is noted Φ_o which is a function of the doping concentrations according to Equation 4.2.9. The barrier prevents electrons from diffusing from the N-type region into the P-type material and vice-versa. Φ_o is relatively large because of the doping levels, but the width of the barrier is very small ($\leq$ 10 nm).

In order for electrons to tunnel through the potential barrier certain conditions must be met:

1- The energy of the electron must be conserved. In terms of an energy band diagram representation, this condition means that an electron tunneling from the N-type region into the P-type region must do so in a horizontal trajectory (Figure 10.2B).

2- There must be occupied states on the side of the junction that emits electrons.

3- There must be empty permitted states on the side of the junction which receives the electrons. Because of condition (1), these states must have the same energy as the states defined in (2).

4- The potential barrier height must be low enough and its width must be small enough for tunneling to take place.

The electron current from the N-type conduction band into the P-type valence band is given by:

$$I_{c \to v} = A \int F_C(E) \, n_c(E) \, T_t \, [1 - F_v(E)] \, n_v(E) \, dE \qquad (10.1.1)$$

where A is the area of the diode, $F_C(E)$ and $F_v(E)$ are the Fermi-Dirac distribution functions in the N-type conduction band and the P-type valence band, respectively, $n_c(E)$ and $n_v(E)$ are the density of states in the conduction and valence band, and T_t is the tunneling probability of an electron. This probability depends essentially on the width of the potential barrier, and it is independent of the direction of the electron (left to right or right to left). The positive sign of the current is due to the fact that electrons carry a negative charge and flow in the negative x-direction (Figure 10.2). The current due to the electron flow from the N-type conduction band into the P-type valence band is equal to:

$$I_{v \to c} = - A \int F_V(E) \, n_v(E) \, T_t \, [1 - F_c(E)] \, n_c(E) \, dE \qquad (10.1.2)$$

The total current is obtained by adding 10.1.1 and 10.1.2:

$$I_t = I_{c \to v} + I_{v \to c} = A \int T_t \, [F_c(E) - F_V(E)] \, n_v(E) \, n_c(E) \, dE \qquad (10.1.3)$$

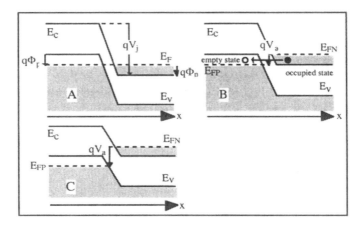

Figure 10.2: Energy band diagram for different increasing forward bias values. A: Zero applied bias; B: Maximum tunneling current; C: Tunneling current has vanished. The shaded areas represent states filled by electrons. [3] V_j is the built-in junction potential and V_a is the external applied voltage.

Calculating the tunnel current is relatively complex. We will only describe qualitatively what happens using the energy band diagrams of Figure 10.2.

A: Let us start with a zero applied bias. In that case $F_V(E)$ and $F_c(E)$ are equal because the Fermi level, E_F, is unique, and the tunneling current is equal to zero, according to Equation 10.1.3 - Figure 10.3.A.

B: If a forward bias, V_a, is applied the quasi-Fermi level and the energy bands in the N-type region move up with respect to the P-type region. As a result there are empty states in the P-side valence band which have the same energy as occupied states in the N-side conduction band. This condition allows for a tunneling current $I_{c \to v}$ to take place. This current increases with increased applied bias, V_a, until a maximum is reached. The maximum current occurs when the number of states in the N-conduction band having the same energy as empty states in the P-valence band is maximum (Figure 10.3.B).

C: If the applied bias, V_a, is further increased the number of empty valence states having the same energy as occupied conduction states decreases until the tunneling current eventually vanishes. A "valley" point of the I-V characteristics is reached when tunneling ceases (Figure 10.3.C).

D: In addition to the band-to-band tunneling current a "regular" PN junction current flows through the diode. As the forward bias is increased the current will increase again, as in a regular PN junction diode (Figure 10.3.D). In the part of the curve between the peak and the valley the tunnel diode has a negative resistance characteristics ($R = dV/dI < 0$).

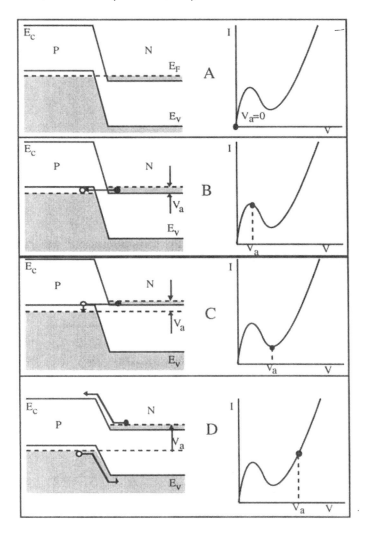

Figure 10.3: Energy band diagram I-V characteristics of a tunnel diode; A: V_a=0; B: peak tunneling current; C: valley current where tunneling ceases; D: "regular" PN junction diffusion current.

10.2. Low-dimensional devices

In a low-dimensional device carriers are no longer moving in a three-dimensional crystal, but they are confined within a two-, one- or zero-dimensional space. This is realized by fabricating devices where carriers are confined within a thin crystal, such as a quantum wire, or in a low-dimensional potential well, such as a quantum-well device.

In the case of a three-dimensional (3D) crystal the density of allowed states in an energy band is a square root function of the energy, as demonstrated in Section 1.1.8 and shown in Figure 10.4.

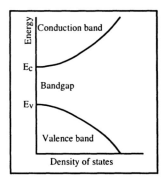

Figure 10.4: Density of states in the conduction and valence band near bandgap in a 3D semiconductor.

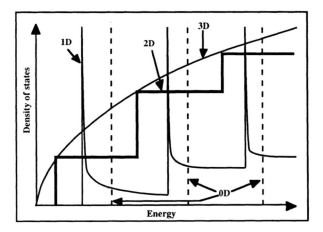

Figure 10.5: Density of states in zero- (0D), one- (1D), two- (2D) and three-dimensional (3D) crystals. [4]

In the case of low-dimensional structures the energy bands, and in particular the distribution of permitted states, is quite different from that of a 3D crystal (Figure 10.5). In a zero-dimensional (0D) crystal (also called "quantum dot") the permitted energy levels are discrete. In a one-dimensional (1D) crystal (also called "quantum wire") they are basically also discrete, but tend to spread out between the "quantized" levels. In a two-dimensional (2D) crystal the density of states is a staircase function of the energy. Figure 10.6 shows the different geometries (3, 2, 1 and 0-D samples) which correspond to the densities of stated in Figure 10.5.

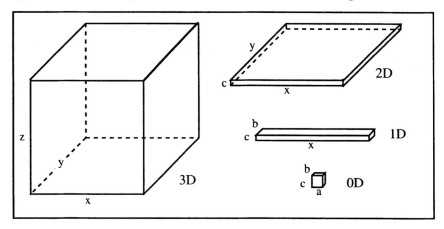

Figure 10.6: Geometry of 3D, 2D, 1D and 0D samples. x, y, and z represent spatial directions and a, b, and c represent small dimensions in the x, y, and z direction, respectively. [5,6,7]

10.2.1. Energy bands

The energy band calculations are based on the time-independent Schrödinger equation:

$$-\frac{\hbar^2}{2m} \nabla^2 \Psi(x,y,z) + V(x,y,z)\ \Psi(x,y,z) = E\ \Psi(x,y,z) \qquad (10.2.1)$$

which can be re-written, if $r = (x,y,z)$:

$$-\frac{\hbar^2}{2m} \nabla^2 \Psi(r) + V(r)\ \Psi(r)\ = E\ \Psi(r) \qquad (10.2.2)$$

We have solved this equation in Section 1.1.3 using the Krönig-Penney model. In the case of a *three-dimensional crystal* we have seen that near the bottom of the conduction band the energy of the electron as a function of the *k*-vector is parabolic, and behaves approximately as a free electron. In that case the periodic potential variation in the crystal can be neglected and one obtains:

$$-\frac{\hbar^2}{2m} \nabla^2 \Psi(r) = E\ \Psi(r)$$

The solution to the latter equation is $\Psi(r) = A\ exp(jkr)$, from which the energy can be found:

$$\frac{\hbar^2 k^2}{2m}\ \Psi(r) = E\ \Psi(r) \Rightarrow \frac{\hbar^2 k^2}{2m} = E_k \tag{10.2.3}$$

The k-vectors for a 3D sample can be found by imposing the Born-von Karman boundary conditions (Expression 1.1.13): $\Psi(x,y,z) = \Psi(x+NL,y,z)$, $\Psi(x,y,z) = \Psi(x,y+NL,z)$ and $\Psi(x,y,z) = \Psi(x,y,z+NL)$ where L is the size (length) of the crystal unit cell, and N is the number of such cells in each direction of space. If the crystal has a cubic lattice and has a cubic shape, each dimension of the crystal is equal to NL and one obtains:

$$k_x = \frac{2\pi}{NL}\ n_x,\ k_y = \frac{2\pi}{NL}\ n_y,\ k_z = \frac{2\pi}{NL}\ n_z \tag{10.2.4}$$

The unit volume in k-space corresponding to each permitted k value is:

$$3D\ unit\ volume = \left(\frac{2\pi}{NL}\right)^3 = \frac{8\pi^3}{V} \tag{10.2.5}$$

where V is the crystal volume.

Using equation (1.1.31) we obtain the values of the permitted wave number:

$$k = \frac{2\pi n}{NL}\quad (n=0,\pm1,\pm2,\pm3,...,\pm(N-1)/2,\ +N/2)$$

where N is the number of crystal cells (about 10^{22} per cubic centimeter). The number of permitted k values is, therefore, very large, and one can consider that k does not vary in a discrete manner, but in a continuous way. Finally the permitted energy levels in a three-dimensional crystal are given by:

$$E_k = \frac{\hbar^2}{2m}\left[\left(\frac{2\pi n_x}{NL}\right)^2 + \left(\frac{2\pi n_y}{NL}\right)^2 + \left(\frac{2\pi n_z}{NL}\right)^2\right] \tag{10.2.6}$$

If we now reduce the size of crystal in the in the z-direction to a very small value, c, we obtain a *two-dimensional crystal* (Figure 10.6). The wave functions in the z-direction are confined within an infinite potential well having a width, c, which is equal to the sample thickness. In the z-direction the wave function is finite inside the crystal and it is equal to zero outside it. Using the technique of separation of variables the wave function can be written as the product of two wave functions:

$$\Psi(x,y,z) = A\ exp(jkr)\ \Phi_c(z) \tag{10.2.7}$$

with $r = (x,y)$. In the z-direction the electron behaves like a "particle in a box" in an infinite potential well of width c. From Section 1.1.1.2 we know that the equation to be solved is:

$$- \frac{\hbar^2}{2m} \frac{d^2 \Phi_c(z)}{dz^2} = E \, \Phi_c(z) \qquad (10.2.8)$$

and that its solution is:

$$\Phi_c(z) = A \, exp(jkz) + B \, exp(-jkz) \qquad (10.2.9)$$

Using the boundary conditions of vanishing wave function at the sides of the crystal $\Phi_c(0) = 0$ and $\Phi_c(c) = 0$ we obtain:

$$\Phi_c(z) = C \, sin\left(\frac{\pi n_z z}{c}\right)$$
$$\text{with } n_z = 1, 2, 3,... \qquad (10.2.10)$$

The energy values in the z-direction can then be extracted:

$$- \frac{\hbar^2}{2m} \frac{d^2\left(C \, sin\left(\frac{\pi n_z z}{c}\right)\right)}{dz^2} = E_z \, C \, sin\left(\frac{\pi n_z z}{c}\right) \Rightarrow E_z = \frac{\hbar^2}{2m}\left(\frac{\pi n_z}{c}\right)^2 \quad (10.2.11)$$

The permitted energy levels (eigenvalues) for the electrons in the crystal can be obtained by summing the energy levels in the z-direction and the energy levels for $r = (x,y)$:

$$E_{ck} = \frac{\hbar^2}{2m} \left[\left(\frac{2\pi n_x}{NL}\right)^2 + \left(\frac{2\pi n_y}{NL}\right)^2 + \left(\frac{\pi n_z}{c}\right)^2\right]$$

which can also be written:

$$E_{ck} = E_c + \frac{\hbar^2}{2m} \left[\left(\frac{2\pi n_x}{NL}\right)^2 + \left(\frac{2\pi n_y}{NL}\right)^2\right] \qquad (10.2.12)$$

where $E_c = \frac{\hbar^2}{2m}\left(\frac{\pi n_z}{c}\right)^2$. The volume of the crystal is $V=c(NL)^2$. The 2D unit volume in k-space corresponding to each permitted k value in the sample is:

$$2D \text{ unit volume} = \left(\frac{2\pi}{NL}\right)^2 = \frac{4 \pi^2 c}{V} \qquad (10.2.13)$$

The permitted energy values are obtained by adding the energy levels which are a function of k_x and k_y and a series of discrete energy levels produced by the wave function confinement in the z-direction. For each discrete energy level resulting from the confinement, $E_c(n_z)$, there exists a 2D energy band corresponding to the possible k_x and k_y values. Such an energy band is called an energy subband (Figure 10.7). It is worth noting that the minimum energy of the electron, which was equal to zero in the three-dimensional case (when $n_x=n_y=n_z=0$ in (10.2.6)) is now equal to:

$$\frac{\hbar^2}{2m}\left(\frac{\pi}{c}\right)^2 \neq 0 \text{ (for } n_z = 1).$$

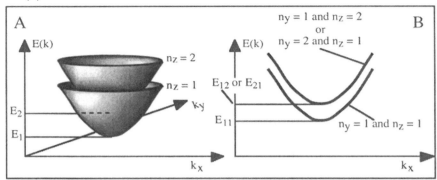

Figure 10.7: Energy vs. wave vector or wave number in a (A) two-dimensional and (B) one-dimensional semiconductor. The width and height of the 1D crystal are taken equal (*b=c*). Two subbands are shown for each sample.

In the case of a *one-dimensional crystal* the dimensions in both the y and z directions of the sample are very small as shown in Figure 10.7. The width of the crystal is noted "*b*" and its height is noted "*c*". The wave functions are now confined in both the y and z directions. Using the technique of separation of variables the wave function can be written as the product of two wave functions:

$$\Psi(x,y,z) = A \ exp(jkx) \ \Phi_{bc}(y,z) \qquad (10.2.14)$$

The wave function in the directions of confinement, $\Phi_{bc}(y,z)$, corresponds to that of a particle in two-dimensional infinite potential of width "*b*" and height "*c*". The wave function can be found using the Schrödinger equation adapted to this particular geometry:

$$-\frac{\hbar^2}{2m}\left(\frac{d^2\Phi_{bc}(y,z)}{dy^2} + \frac{d^2\Phi_{bc}(y,z)}{dz^2}\right) = E \ \Phi_{bc}(y,z) \qquad (10.2.15)$$

which has the solution:

$$\Phi_{bc}(y,z) = \Phi_b(y) \ \Phi_c(z)$$
$$\Phi_b(y) = A \ exp(jky) + B \ exp(-jky)$$
$$\Phi_c(z) = C \ exp(jkz) + D \ exp(-jkz) \qquad (10.2.16)$$

Using the boundary conditions $\Phi_b(0) = 0$, $\Phi_c(0) = 0$, $\Phi_b(b) = 0$ and $\Phi_c(c) = 0$, one obtains:

$$\Phi_b(y) = F \sin\left(\frac{\pi n_y y}{b}\right) \text{ and } \Phi_c(z) = G \sin\left(\frac{\pi n_z z}{c}\right)$$

with $n_y = 1, 2, 3,...$ and $n_z = 1, 2, 3,...$ (10.2.17)

The permitted energy levels can then be found:

$$-\frac{\hbar^2}{2m}\frac{d^2\left(F \sin\left(\frac{\pi n_y y}{b}\right)\right)}{dy^2} = E_y F \sin\left(\frac{\pi n_y y}{b}\right) \Rightarrow E_y = \frac{\hbar^2}{2m}\left(\frac{\pi n_y}{b}\right)^2$$

$$-\frac{\hbar^2}{2m}\frac{d^2\left(G \sin\left(\frac{\pi n_z z}{c}\right)\right)}{dz^2} = E_z G \sin\left(\frac{\pi n_z z}{c}\right) \Rightarrow E_z = \frac{\hbar^2}{2m}\left(\frac{\pi n_z}{c}\right)^2 \quad (10.2.18)$$

The permitted energy levels for the electrons in the crystal can therefore be obtained by summing the energy levels in the x, y and z directions:

$$E_{bck} = \frac{\hbar^2}{2m}\left[\left(\frac{2\pi n_x}{NL}\right)^2 + \left(\frac{\pi n_y}{b}\right)^2 + \left(\frac{\pi n_z}{c}\right)^2\right]$$

or:
$$E_{bck} = E_{bc} + \frac{\hbar^2}{2m}\left[\left(\frac{2\pi n_x}{NL}\right)^2\right] \quad (10.2.19)$$

where $E_{bc} = \frac{\hbar^2}{2m}\left[\left(\frac{\pi n_y}{b}\right)^2 + \left(\frac{\pi n_z}{c}\right)^2\right]$. The 1D unit volume in k-space corresponding to each permitted k value is:

$$1D \text{ unit volume} = \frac{2\pi}{NL} = \frac{2\pi b c}{V} \quad (10.2.20)$$

The permitted energy values are thus obtained by adding the energy levels which are a function of k_x (which vary in a continuous manner) and a series of discrete energy levels produced by the wave function confinement in the y and z directions. The discrete energy levels resulting from the confinement, $E_{bc}(n_y, n_z)$, are the minima of energy subbands. The other energy values in each subband are obtained by adding $E_{bc}(n_y, n_z)$ to the energies corresponding to k_x values (Figure 10.7.B). It is worth noting that the minimum energy of the electron, which was equal to zero in the three-dimensional case (when $n_x = n_y = n_z = 0$ in Equation 10.2.6) is now equal to $\frac{\hbar^2}{2m}\left(\left(\frac{\pi}{b}\right)^2 + \left(\frac{\pi}{c}\right)^2\right) \neq 0$ (for $n_y = n_z = 1$).

In the case of a *zero-dimensional crystal* the dimensions in all x, y and z directions are very small. The length, width and height of the crystal are noted "a", "b" and "c". The wave function is now confined in the x, y and z direction. Using the technique of separation of variables the wave function can be written as the product of separate wave functions:

$$\Psi(x,y,z) = \Phi_a(x)\ \Phi_b(y)\ \Phi_c(z) \qquad (10.2.21)$$

The wave function $\Psi(x,y,z)$ can be found by solving the Schrödinger equation in a three-dimensional potential well:

$$-\frac{\hbar^2}{2m}\left(\frac{d^2\Psi(x,y,z)}{dy^2} + \frac{d^2\Psi(x,y,z)}{dy^2} + \frac{d^2\Psi(x,y,z)}{dz^2}\right) = E\ \Psi(x,y,z) \quad (10.2.22)$$

which has the solution:

$$\Psi(x,y,z) = \Phi_a(x)\ \Phi_b(y)\ \Phi_c(z)$$
$$\Phi_a(x) = A\ exp(jkx) + B\ exp(-jkx)$$
$$\Phi_b(y) = C\ exp(jky) + D\ exp(-jky)$$
$$\Phi_c(z) = E\ exp(jkz) + F\ exp(-jkz) \qquad (10.2.23)$$

Applying the following boundary conditions $\Phi_a(0) = 0$, $\Phi_b(0) = 0$, $\Phi_c(0) = 0$, $\Phi_a(a) = 0$, $\Phi_b(b) = 0$ and $\Phi_c(c) = 0$, one obtains:

$$\Phi_a(x) = G\ sin\left(\frac{\pi n_x x}{a}\right),\ \Phi_b(y) = H\ sin\left(\frac{\pi n_y y}{b}\right)\ and\ \Phi_c(z) = I\ sin\left(\frac{\pi n_z z}{c}\right)$$

where n_x, n_y and n_z can take on values 1, 2, 3,... $\qquad (10.2.24)$

The energy eigenvalues in the different directions are:

$$-\frac{\hbar^2}{2m}\frac{d^2\left(G\ sin\left(\frac{\pi n_x x}{a}\right)\right)}{dx^2} = E_x\ G\ sin\left(\frac{\pi n_x x}{a}\right) \Rightarrow E_x = \frac{\hbar^2}{2m}\left(\frac{\pi n_x}{a}\right)^2$$

$$-\frac{\hbar^2}{2m}\frac{d^2\left(H\ sin\left(\frac{\pi n_y y}{b}\right)\right)}{dy^2} = E_y\ H\ sin\left(\frac{\pi n_y y}{b}\right) \Rightarrow E_y = \frac{\hbar^2}{2m}\left(\frac{\pi n_y}{b}\right)^2$$

$$-\frac{\hbar^2}{2m}\frac{d^2\left(I\ sin\left(\frac{\pi n_z z}{c}\right)\right)}{dz^2} = E_z\ I\ sin\left(\frac{\pi n_z z}{c}\right) \Rightarrow E_z = \frac{\hbar^2}{2m}\left(\frac{\pi n_z}{c}\right)^2 \quad (10.2.25)$$

where the constants G, H and I have been determined by applying the boundary conditions. The electron energy values are obtained by summing the three latter equations, which yields:

$$E_{abc} = \frac{\hbar^2}{2m}\left[\left(\frac{\pi n_x}{a}\right)^2 + \left(\frac{\pi n_y}{b}\right)^2 + \left(\frac{\pi n_z}{c}\right)^2\right] \qquad (10.2.26)$$

The permitted energy levels are thus a succession of discrete levels produced by the confinement in the three-dimensional potential well. The minimum energy value (when $n_x = n_y = n_z = 1$) is equal to

$$\frac{\hbar^2}{2m}\left(\left(\frac{\pi}{a}\right)^2 + \left(\frac{\pi}{b}\right)^2 + \left(\frac{\pi}{c}\right)^2\right) \neq 0 \, .$$

10.2.2. Density of states

In a *three-dimensional crystal* the volume of a lattice unit cell is equal to L^3 and the volume V of the crystal is equal to $V = (NL)^3$. The unit volume

corresponding to each permitted state (*i.e.* to each k value) is equal to

$$\left(\frac{2\pi}{NL}\right)^3 = \frac{8\pi^3}{V} \text{ (Relationship 10.2.5).}$$

Using a similar approach to that of Section 1.1.8 we will now consider a *sphere* in k-space which contains all the wave vectors corresponding to the electrons having an energy below a given maximum value. To each wave vector, $k \leq k_{max}$, correspond two electrons by virtue of the Pauli exclusion principle. The number of electrons is thus given by:

$$n = 2\left(\frac{4\pi}{3}k_{max}^3\right)\frac{V}{8\pi^3} \tag{10.2.27}$$

and, in a unit volume ($V=1$):

$$n = 2\left(\frac{4\pi}{3}k_{max}^3\right)\left(\frac{1}{2\pi}\right)^3 \tag{10.2.28}$$

The latter relationship enables us to link k_{max} to the electron concentration: $k_{max} = (3\pi^2 n)^{1/3}$.

The density of states is defined by $\rho = dn/dE$. We will use the symbol ρ for the density of states instead of $n(E)$, which was used in Equation 1.1.48 to avoid confusion between the number of electrons, n, and the density of states.

Using the following relationships we can relate the density of states, ρ, to energy values:

$$\frac{dn}{dk} = \frac{d}{dk}\left[2\left(\frac{4\pi}{3}k^3\right)\left(\frac{1}{2\pi}\right)^3\right] = 2\frac{4\pi k^2}{(2\pi)^3}$$

$$E = E_k = \frac{\hbar^2 k^2}{2m} \Rightarrow k = \left(\frac{2m}{\hbar^2}E_k\right)^{1/2}$$

$$\frac{dk}{dE} = \left(\frac{2m}{\hbar^2}\right)^{1/2} \frac{E_k^{-1/2}}{2} \tag{10.2.29}$$

Finally we obtain the density of states as a function of E:

$$\rho = \frac{dn}{dE} = \frac{dn}{dk}\frac{dk}{dE} = \frac{1}{2\pi^2}\left(\frac{2m}{\hbar^2}\right)^{3/2} E_k^{1/2} \tag{10.2.30}$$

Thus, the density of states near a band extremum, such as the minimum of the conduction band, varies as the square root of the energy.

———

In a *two-dimensional crystal* confined in the z-direction the 2D volume of a lattice unit cell is equal to L^2 and the volume of the crystal is equal to $V = c(NL)^2$. The unit volume corresponding to each permitted state (*i.e.* to each k value) is equal to $\left(\frac{2\pi}{NL}\right)^2 = \frac{4\pi^2 c}{V}$ (Relationship 10.2.13). Using a similar approach to that of Section 1.1.7 we now have to consider a *circle* in k-space which contains all the wave vectors corresponding to the electrons having an energy below a given maximum value. To each wave vector, $k \leq k_{max}$, correspond two electrons by virtue of the Pauli exclusion principle. The number of electrons is thus given by:

$$n = 2\left(\pi k_{max}^2\right)\frac{V}{4\pi^2 c}$$

and, in a unit volume ($V=1$): $n = 2\left(\pi k_{max}^2\right)\left(\frac{1}{2\pi}\right)^2\frac{1}{c} \tag{10.2.31}$

The latter relationship enables us to link k_{max} to the electron concentration: $k_{max} = (2\pi nc)^{1/2}$. The density of states in a subband is defined by: $\rho = dn/dE$. Thus we find:

$$\frac{dn}{dk} = \frac{d}{dk}\left[2\left(\pi k^2\right)\left(\frac{1}{2\pi}\right)^2\frac{1}{c}\right] = \frac{k}{\pi c}$$

$$E = E_{ck} = E_c + \frac{\hbar^2 k^2}{2m} \Rightarrow \frac{dE}{dk} = \frac{\hbar^2 k}{m} \Rightarrow \frac{dk}{dE} = \frac{m}{\hbar^2 k}$$

$$\rho = \frac{dn}{dE} = \frac{dn}{dk}\frac{dk}{dE} = \frac{m}{\hbar^2\pi}\frac{1}{c} \tag{10.2.32}$$

Thus, the density of states near a subband extremum, such as the minimum of the conduction band, is constant and independent of the energy. However, one has to take into account that there are several subbands. The total number of electrons is obtained by adding the number of electrons in the different subbands:

$$n(E) = \int \rho dE = \frac{1}{c} \frac{m}{\hbar^2 \pi} \sum_{c} \theta(E-E_c) \tag{10.2.33}$$

where the function θ is defined as:

$$\theta(E-E_c) = 0 \text{ if } E < E_c \text{ and } \theta(E-E_c) = E-E_z \text{ if } E \geq E_c.$$

In a *one-dimensional* crystal the 1D volume of a lattice unit cell is equal to L and the volume of the crystal is equal to $V = b\, c\, NL$. The unit volume corresponding to each permitted state (*i.e.* to each k value) is equal to $\left(\dfrac{2\pi}{NL}\right)$

$= \dfrac{2\pi\, b\, c}{V}$ (Relationship 10.2.20). Using a similar approach to that of Section 1.1.8 we now have to consider a *line segment* in k-space which contains all the wave vectors corresponding to the electrons having an energy below a given maximum value. The length of this segment is $2k_{max}$. To each wave vector, $k \leq k_{max}$, correspond two electrons by virtue of the Pauli exclusion principle. The number of electrons is thus given by:

$$n = 4\, k_{max}\, \frac{V}{2\pi\, b\, c}$$

and, in a unit volume ($V=1$):

$$n = 2\, \frac{k_{max}}{\pi}\, \frac{1}{b\, c} \tag{10.2.34}$$

The latter relationship enables us to link k_{max} to the electron concentration: $k_{max} = \dfrac{\pi n b c}{2}$.

The density of states in a subband is defined by: $\rho = dn/dE$. Thus we find:

$$\frac{dn}{dk} = \frac{d}{dk}\left[\frac{2k}{\pi} \frac{1}{b\, c}\right] = \frac{2}{\pi} \frac{1}{b\, c}$$

Using $E = E_{bck} = E_{bc} + \dfrac{\hbar^2 k^2}{2m} \Rightarrow k = \left(\dfrac{2m}{\hbar^2}(E_{bck} - E_{bc})\right)^{1/2}$

we find

$$\frac{dk}{dE} = \left(\frac{2m}{\hbar^2}\right)^{1/2} \frac{(E_{bck} - E_{bc})^{-1/2}}{2}$$

and thus:

$$\rho = \frac{dn}{dE} = \frac{dn}{dk}\frac{dk}{dE} = \frac{1}{\pi}\left(\frac{2m}{\hbar^2}\right)^{1/2}(E_{bck} - E_{bc})^{-1/2}\frac{1}{b\, c} \tag{10.2.35}$$

where E_{bck} is a continuous function of k in the x direction and a discrete function in the y and z directions.

Thus, the density of states near a subband extremum, such as the minimum of the conduction band, now varies as an inverse square root function of the energy as a function of k. Again, one has to take into account that there are several subbands corresponding to the discretization in the y and z directions. The total number of electrons is obtained by adding the number of electrons in the different subbands:

$$n(E) = \int \rho dE = \frac{2}{\pi} \left(\frac{2m}{\hbar^2}\right)^{1/2} \frac{1}{b\,c} \sum \theta(E_{bck} - E_{bc}) (E_{bck} - E_{bc})^{1/2} \qquad (10.2.36)$$

where the function θ is defined as:

$\theta(E_{bck} - E_{bc}) = 0$ if $E_{bck} - E_{bc} < 0$ and $\theta(E_{bck} - E_{bc}) = 1$ if $(E_{bck} - E_{bc}) \geq 0$.

The density of states for a 1D and 2D and 3D semiconductor sample with specified dimensions is shown in Figures 10.8 to 10.11.

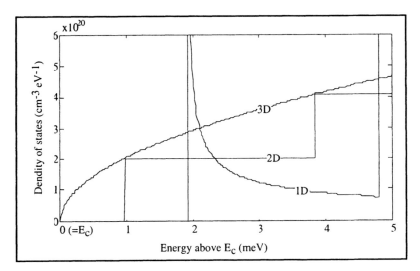

Figure 10.8: Density of states in the conduction band, as a function of energy, in silicon 1D, 2D, and 3D crystals. In the 2D sample the crystal height, c, is 20 nm, and in the 1D crystal the height, c, and the width, b, of the sample, are both equal to 20 nm.

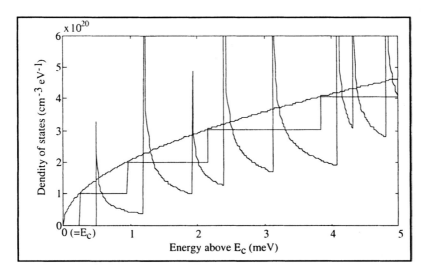

Figure 10.9: Density of states in the conduction band, as a function of energy, in silicon 1D, 2D, and 3D crystals. In the 2D sample the crystal height, c, is 40 nm, and in the 1D crystal the height, c, and the width, b, of the sample, are both equal to 40 nm.

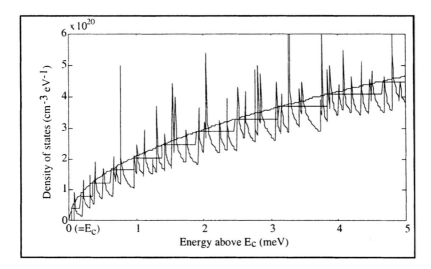

Figure 10.10: Density of states in the conduction band, as a function of energy, in silicon 1D, 2D, and 3D crystals. In the 2D sample the crystal height, c, is 100 nm, and in the 1D crystal the height, c, and the width, b, of the sample, are both equal to 100 nm. The dimensions b and c are now large enough for both the 1D and 2D distributions to "follow" the 3D curve.

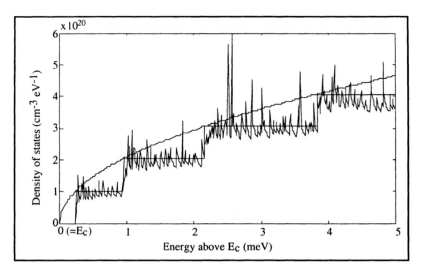

Figure 10.11: Density of states in the conduction band, as a function of energy, in silicon 1D, 2D, and 3D crystals. In the 2D sample the crystal height, c, is 40 nm, and in the 1D crystal the height, c, and the width, b, of the sample, are equal to 40 and 400 nm, respectively. The width, b, is now large enough for the 1D distribution to "follow" the 2D curve.

10.2.3. Conductance of a 1D semiconductor sample

Consider a one-dimensional semiconductor sample. We will assume that the electrons move without interacting with the crystal lattice. Such electrons are called *ballistic electrons* and can be found in very short MOS devices.

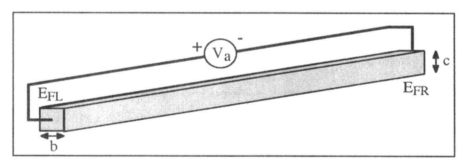

Figure 10.12: One-dimensional sample with applied bias.

When a potential difference, V_a, is applied to the ends of a 1D semiconductor sample, a difference in the Fermi levels, $E_{FR} - E_{FL} = qV$

appears between the right and the left of the sample (Figure 10.12). The current density in the sample is given by:

$$I = q \, v \, n \quad (10.2.37)$$

where v is the electron group velocity, and n is the electron concentration. If the applied bias is such that $E_{FR} > E_{FL}$, the electrons with an energy lower than E_{FL} will not contribute to any current. The density of electrons contributing to a current flow is given by:

$$n_e(E) = \frac{1}{2} \int_{E_{FL}}^{E_{FR}} \rho_n(E) \, dE = \frac{1}{2} \rho_n(E) \, q \, V_a$$

where ρ_n is the density of states in the n-th subband. The factor 1/2 accounts for only one direction of electron motion (from left to right).[8,9] The total current density is obtained by adding the current in the various subbands where there are n subbands:

$$I = \sum_{all \; subbands} I_n = \frac{1}{2} \sum_n q \, v_{nk} \, \rho_n(E) \, q \, V_a \quad (10.2.38)$$

and v_n is the group velocity in the n-th subband. The units for v_n and ρ_n are ms^{-1} and m^{-1} eV^{-1}, respectively.

The electron group velocity, v_n, is given by (see Table A.1 in the Annex):

$$v_k = \frac{1}{\hbar} \frac{dE}{dk} \quad (10.2.39)$$

Using (10.2.35) we find:

$$\rho_n = \frac{dn}{dE} = \frac{dn}{dk} \frac{dk}{dE} = \frac{2}{\pi} \left(\frac{dE}{dk} \right)^{-1} \frac{1}{b \, c} \quad (10.2.40)$$

and thus:

$$v_k \rho_n = \frac{2}{\pi \hbar} \frac{1}{b \, c} = \frac{4}{h} \frac{1}{b \, c} \quad (10.2.41)$$

Introducing the latter result into 10.2.38 the current can be obtained:

$$I = \sum_n \left(2 \frac{q^2}{h} \right) V \frac{1}{b \, c} \quad (10.2.42)$$

from which the conductance of the quantum wire can be extracted:

$$G = \frac{I}{V} = 2 \, n \frac{q^2}{h} \frac{1}{b \, c} \quad (10.2.43)$$

The latter expression is known as the Landauer formula.[10] It describes the conductance of a one-dimensional sample, which varies in a staircase manner as a function of the Fermi level. The height of each step is equal to $2q^2/h$ ohm^{-1} = $(26 \; k\Omega)^{-1}$ per electron spin.

10.2.4. 2D and 1D MOS transistors

2D MOS transistor
If a MOS transistor is made in a thin silicon film electron transport can become two-dimensional. A positive gate voltage is applied such that band bending occurs. We assume that the temperature is equal to 0 K and that the drain voltage is small, for simplicity. When the gate voltage is such that $E_F > E_C$ free electrons occur. The electron current is given by Expression 10.2.37:

$$I = q \, v_k \, n$$

with
$$n = \int_{E_C}^{E_F} \rho(E) \, dE \quad \text{if} \quad E_F > E_C$$

and
$$n = 0 \quad \text{if} \quad E_C \leq E_F$$

where the density of states in the conduction band, $\rho(E)$, is given by Equation 10.2.32. The current is, therefore, proportional to the shaded areas of Figure 10.13. The relative position between the Fermi level and the minimum of the conduction band depends on the applied gate voltage.

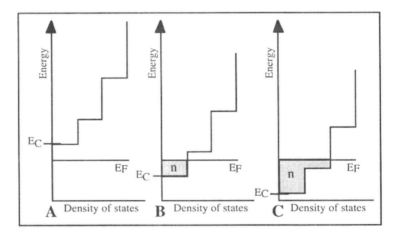

Figure 10.13: Density of states and free electron concentration in a 2D MOS transistor. A: Below threshold (no electrons in the conduction subbands); B: V_G is increased and there are electrons in the first subband; C: V_G is further increased and electrons occupy the first and second subband.

Figure 10.14 shows the transconductance, dI_D/dV_G, of a thin, double-gate SOI MOSFET (see Figure 7.41) measured at a temperature of 0.3 K with a silicon film thickness of 40 nm. For gate voltages below -0.18V there are no electrons in the conduction subbands and the current is equal to zero. When V_G is increased to -0.18V the lowest energy subband becomes populated with electrons (for -0.18V $< V_G <$ -0.02 V for this particular device). At higher gate voltages ($V_G >$ -0.02 V) electrons populate the second subband as well. The drop in transconductance around V_G = -0.1 V is due to mobility reduction by scattering between electrons in the first and the second subband, called "inter-subband scattering". The transconductance decrease for $V_G > 0.1$ V is attributed to classical surface mobility reduction (see Section 7.5).

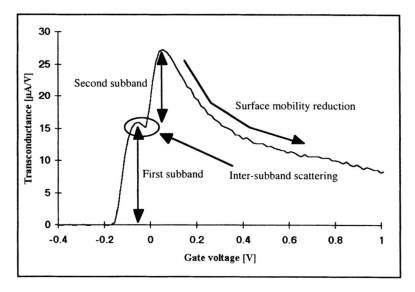

Figure 10.14: Transconductance of a double-gate, two-dimensional SOI MOSFET.[11]

1D MOS transistor

If a MOS transistor is made in a thin silicon and narrow silicon wire, electron transport can become one-dimensional. A positive gate voltage is applied such that band bending occurs. We assume that the temperature is equal to 0 K and that the drain voltage is small, for simplicity. When the gate voltage is such that $E_F > E_C$ free electrons occur. The electron current is given by Expression 10.2.37:

$$I = q \, v_k \, n$$

with
$$n = \int_{E_C}^{E_F} \rho(E) \, dE \quad \text{if} \quad E_F > E_C$$

and
$$n = 0 \quad \text{if} \quad E_C \leq E_F$$

where the density of states in the conduction band, $\rho(E)$, is given by Equation 10.2.35. The current is, therefore, proportional to the gray areas of Figure 10.15. The relative position between the Fermi level and the minimum of the conduction band depends on the applied gate voltage.

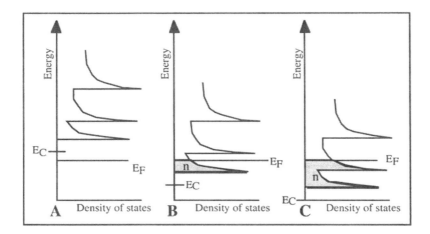

Figure 10.15: Density of states and free electron concentration in a 1D MOS transistor. A: Below threshold (no electrons in the conduction subbands); B: V_G is increased and there are electrons in the first subband; C: V_G is further increased and electrons occupy the first and second subband.

Figure 10.16 shows the current of a 1D MOSFET sample measured at low temperature. The silicon wire width and thickness is 80 nm ($b=c=80$ nm). Let us focus on the curve measured at $T=4.2$ K (liquid helium). For gate voltages below 0.3 V there are no electrons in the conduction subbands and the current is equal to zero. When the gate voltage is increased energy subbands become populated with electrons and current oscillations are observed. These are due to the "spiky" nature of the density of states, and to some extent, to inter-subband scattering. Note that the oscillations disappear at higher temperatures. The separation between the subband energy levels must be larger than the thermal voltage kT/q for the quantum oscillations to be observable; in this sample the energy difference between the different subbands is relatively small such that a temperature above approximately 35K is sufficient to cause the measurement to look continuous.

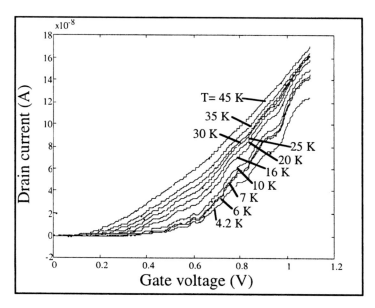

Figure 10.16: Current in a 1D SOI MOSFET at different temperatures.[12]

10.3. Single-electron transistor

10.3.1. Tunnel junction

The single-electron transistor makes use of tunnel junctions. Such a junction is made of a thin insulator sandwiched between two pieces of semiconductor or metal, as shown in Figure 10.17A. If a voltage is applied at the terminals the structure behaves like a capacitor. If the applied voltage is large enough it may be energetically favorable for an electron to tunnel through the insulator, giving rise to a brief current spike. If a constant voltage is applied to the structure, periodic current oscillations, called "Coulomb oscillations", are produced (Figure 10.17B). If the average current through the structure is I, the frequency of the Coulomb oscillations is equal to $f=I/q$, and are each caused by the tunneling of a single electron through the insulator.[13,14]

Consider the circuit shown in Figure 10.18. Electrons are injected through a tunnel junction into a small piece of semiconductor or metal, called a "dot", where the dot is capacitively coupled to ground. As the external applied voltage, $V_a<0$, is ramped up in absolute value, the potential of the dot, V_{dot}, will increase in a staircase manner if the displacement current is neglected.

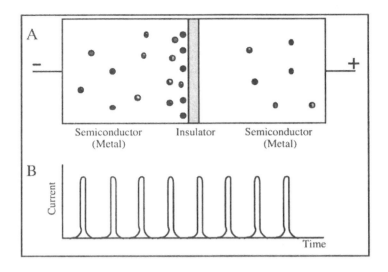

Figure 10.17: A: Tunnel junction. The small spheres represent electrons; B: Coulomb oscillations.

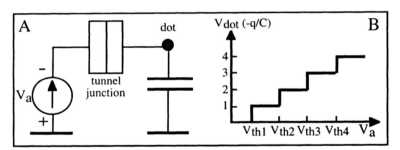

Figure 10.18: Charging of a small material dot through a tunnel junction. A: Circuit schematics; B: Dot potential vs. applied bias.

Each increase of V_{dot} corresponds to the injection of an electron into the dot. If the capacitance between the dot and ground is small the injection of a single electron into the dot will give rise to a measurable increase of the dot potential since $dQ = C\, dV$. For instance, if the dot capacitance is equal to 1.6 aF (1 aF $= 10^{-18}$F), each electron injected into the dot increases its potential by 100 mV. If we consider the dot to be an isolated sphere embedded in silicon dioxide, electrostatics tells us that its capacitance is given by $C = 4\pi\varepsilon_{ox}R$, where R is the radius of the sphere. For example, a dot with a radius of 3.7 nm has a capacitance of 1.6 aF.

Electrons can be transferred into the dot by ramping up V_a. To transfer an electron into the dot, a coulombic energy $E_C = q^2/2C$ is required. No electron is injected into the dot as long as the applied voltage is smaller than $V_{th1}=q/C$, since there is not enough coulombic energy available for electron tunneling into the dot. This behavior is called the "Coulomb blockade" and is repeated for applied voltages smaller than $V_{th2} = 2q/C$, $V_{th3}=3q/C$, etc. (Figure 10.18B). The Coulomb blockade effect can only be observed if the thermal energy, kT/q, is lower than the electrostatic energy in the dot. This condition imposes C to be lower than 12 and 3 aF to observe Coulomb blockade effects at temperatures of 77 and 300K, respectively.

In order to understand how a single-electron transistor works, it is necessary to analyze the energies stored in different parts of the device. The energy supplied over a period of time by all the voltage sources in a circuit, E_s, may be written as the time integral of the power delivered to the system by each source:

$$E_S = \sum_{sources} \left(\int V(t)\ I(t)\ dt \right) \tag{10.3.1}$$

Following any tunneling event, charges flow to and from the contacts until equilibrium is reached. It is assumed that the duration of this charge relaxation caused by tunneling or changing voltage sources, which typically take place in 10 femtoseconds, is much shorter than the time between two tunneling events. Voltage sources are considered to be ideal, that is their internal resistance is zero, and for constant voltage sources, the change in energy due to storing or removing an electron from the dot may be written as:

$$\Delta E_s = \pm qV_{dot} + \sum_i V_i\ \Delta Q_i \tag{10.3.2}$$

where first the term is the work variation due to storing or removing an electron from the dot, and the second term is the work accomplished to account for possible voltage variations at each node of the circuit.

The Helmholtz free energy of the device, F, is defined as the difference between the total energy stored in the device, E_T, and the work done by the power sources:

$$F = E_T - E_S$$

10.3.2. Double tunnel junction

Consider the circuit of Figure 10.19A, which contains two tunnel junctions, one dot, and a single ideal voltage source, V_a. The voltage drops across junction 1 and junction 2 are noted V_1 and V_2, and the charges on the tunnel junctions and their capacitances are $Q_1, Q_2, C1$ and $C2$, respectively; n_1 and n_2 are the number of electrons that tunnel through junctions 1 and 2, respectively. One can write: [9]

$$Q_1 = C_1 V_1$$
$$Q_2 = C_2 V_2$$
$$Q_{dot} = Q_1 - Q_2 = -n\,q + Q_0$$

(10.3.3)

where Q_o is an initial charge that might have been present on the dot before biasing the circuit, and $n = n_1 - n_2$.

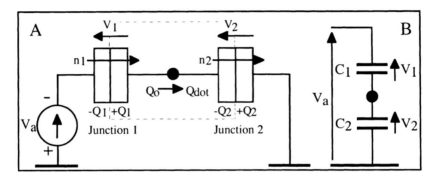

Figure 10.19: A: Circuit with double tunnel junction; B: Equivalent capacitor circuit.

Since $V_a = V_1 + V_2$, one finds, using 10.3.3:

$$V_1 = \frac{C_2 V_a - n\,q + Q_o}{C_1 + C_2} \quad \text{and} \quad V_2 = \frac{C_1 V_a + n\,q - Q_o}{C_1 + C_2}$$

(10.3.4)

The electrostatic energies stored in the junctions are:

$$E_T = \frac{Q_1^2}{2C_1} + \frac{Q_2^2}{2C_2} = \frac{C_1 C_2 V_a^2 + (n\,q - Q_o)^2}{2(C_1 + C_2)}$$

(10.3.5)

We can now calculate the energy supplied by the voltage source. If one electron tunnels through junction 1 the voltage drop variation across junction 1 is equal to $\Delta V_1 = -q/C_1$. To this variation corresponds, according to the capacitive divider of Figure 10.19B, a charge equal to $-q$ $C_2/(C_1+C_2)$ supplied by the voltage source V_a. Thus, for n_1 electrons tunneling through junction 1 the energy supplied by the voltage source according to 10.3.2, is equal to:

$$E_{s1} = -\frac{n_1\,q\,V_a\,C_2}{C_1 + C_2}$$

(10.3.6)

A similar calculation yields the energy supplied by the voltage source for n_2 electrons tunneling through junction 2:

$$E_{s2} = - \frac{n_2 \, q \, V_a \, C_1}{C_1 + C_2} \qquad (10.3.7)$$

The Helmholtz free energy of the complete system is given by:

$$F(n_1,n_2) = E_T - E_{s1} - E_{s2}$$

$$= \frac{\frac{1}{2}\left(C_1 C_2 V_a^2 + (n q - Q_o)^2 \right) + q \, V_a \, (C_1 \, n_2 + C_2 \, n_1)}{C_1 + C_2} \qquad (10.3.8)$$

If one electron is added to, or removed from, the dot through junctions 1 or 2, the variation of free energy is given by:

$$\Delta F_1 = F(n_1 \pm 1, \, n_2) - F(n_1,n_2) = \frac{q}{C_1 + C_2}\left(\frac{q}{2} \pm (V_a C_2 + n q - Q_o) \right) \qquad (10.3.9)$$

and

$$\Delta F_2 = F(n_1, \, n_2 \pm 1) - F(n_1,n_2) = \frac{q}{C_1 + C_2}\left(\frac{q}{2} \pm (V_a C_1 - n q + Q_o) \right) \qquad (10.3.10)$$

Tunneling will be possible if the Helmholtz free energy is reduced in that process. Remembering that $V_a < 0$, if we assume equal values for the two junction capacitances $C_1 = C_2 \equiv C$ and if we start with an uncharged dot ($n=0$ and $Q_o = 0$), the condition for tunneling ($\Delta F < 0$) becomes:

$$|V_a| > \frac{q}{2C} \qquad (10.3.11)$$

The inhibition of tunneling for low bias voltage is a manifestation of the Coulomb blockade effect. The current-voltage characteristics of the double tunnel junction is shown in Figure 10.20. The voltage span over which no current flows through the device is called the "Coulomb gap" and its width is equal to q/C.

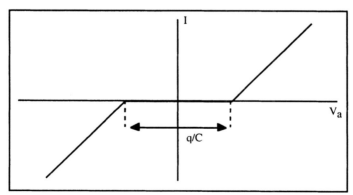

Figure 10.20: Current-voltage characteristics of a double tunnel junction.

If an electron enters the dot via junction 1, it can flow to ground through tunnel through junction 2. After a small period of time a new electron can then tunnel through junction 1, etc. and current flows through the device.

10.3.3. Single-electron transistor

If we add to the double tunnel junction a gate electrode that is capacitively coupled with the dot we obtain a single-electron transistor (SET), which is schematically represented in Figure 10.21. The dot potential, and thus the current flow, can now be controlled by the gate voltage. The charge on the dot (Equation 10.3.3) now becomes:

$$Q_{dot} = -n\,q + Q_o + C_G(V_G - V_2) \tag{10.3.12}$$

The expressions derived for the double junctions can, therefore, be used for the SET, provided that Q_o is replaced by $Q_o + C_G(V_G - V_2)$ and C_G is introduced in the capacitive network of the device. [9]

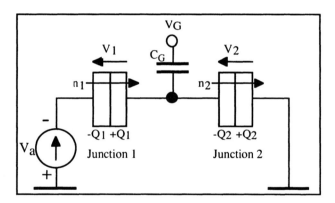

Figure 10.21: Equivalent circuit for the single-electron transistor (SET) with a gate voltage V_G.

In particular, the voltage drops across the tunnel junctions are now given by:

$$V_1 = \frac{(C_2 + C_G)V_a + C_G V_G - n\,q + Q_o}{C_1 + C_2 + C_G} \tag{10.3.13}$$

and

$$V_2 = \frac{(C_1 + C_G)V_a - C_G V_G + n\,q - Q_o}{C_1 + C_2 + C_G} \tag{10.3.14}$$

The change in free energy after a tunneling event in junctions 1 and 2 becomes:

$$\Delta F_1 = F(n_1 \pm 1, n_2) - F(n_1, n_2) = \frac{q}{C_1 + C_2 + C_G}\left(\frac{q}{2} \pm (C_2 + C_G)V_a - C_G V_G + nq - Q_o\right)$$

$$(10.3.15)$$

and

$$\Delta F_2 = F(n_1, n_2 \pm 1) - F(n_1, n_2) = \frac{q}{C_1 + C_2 + C_G}\left(\frac{q}{2} \pm (C_1 + C_G)V_a + C_G V_G - nq + Q_o\right)$$

$$(10.3.16)$$

At low temperature, only transitions producing a negative change in free energy are permitted: $\Delta F_1 < 0$ or $\Delta F_2 < 0$. This condition can be used along with Equations 10.3.15 and 10.3.16 to plot the conditions for current flow in the V_a-V_G plane (Figure 10.22). In such a plot, domains where Coulomb blockade prevents current from flowing through the device can be identified. These have a characteristic rhombus shape and appear periodically along the V_G axis as the number of electrons, n, injected into the dot increases or decreases. Figure 10.23 shows lines of equal current in the V_a-V_G plane, measured on an actual SET. One can easily identify the rhombus-shaped domains where no current flows because of the Coulomb blockade effect.

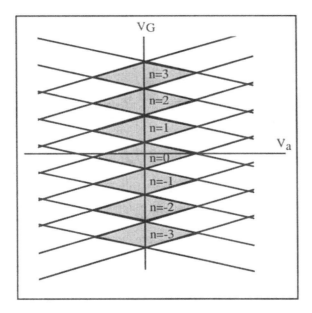

Figure 10.22: Condition for current flow in an SET. The shadowed areas indicate regions of the V_a-V_G plane where Coulomb blockade prevents current flow and n=0,1,2, etc. indicate the number of electrons stored in the dot.

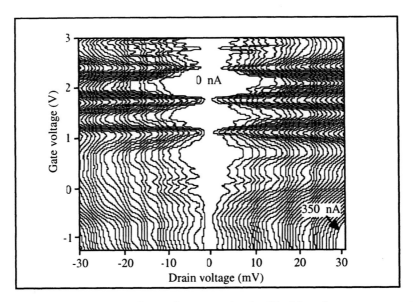

Figure 10.23: Lines of equal current in the V_a-V_G plane, measured on an actual SET. The rhombus-shaped domains where no current flows reveal the Coulomb blockade effect.[15]

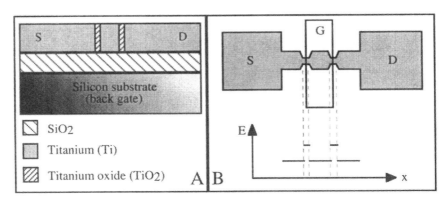

Figure 10.24: A: Metal - oxide SET; B: Silicon SET. B also shows the minimum energy levels as a function of position.

Figure 10.24 presents two practical implementations of single-electron transistors. In Figure 10.24A a metal (titanium) SET is presented, where insulating TiO_2 forms the tunneling junctions and where the silicon substrate is used as a the gate electrode. [16] In Figure 10.24B a silicon-on-insulator SET is shown. Here, there is no actual tunnel insulators on both sides of the dot.[17] Instead, constrictions of the silicon island introduce potential barriers for the electrons and act as tunnel barriers.

Problems

Problem 10.1:
Using Matlab, calculate and plot the density of states above the minimum of the conduction band in a 3D silicon sample, in a 2D silicon sample with a height of 40 nm, and in a 1D silicon sample having a 40nm x 40 nm cross section (Figure 10.9). Assume that the electron mass in silicon is 0.98 times the mass of a free electron.

Problem 10.2:
Tunnel effect: Using Matlab and the finite-difference numerical method described in Problem 1.3, calculate the first (ground-state) wave function of an electron in the potential well shown in Problem Figure 10.1, for ΔV = -10 mV, +3.8 mV, +3.95 mV and +5 mV.

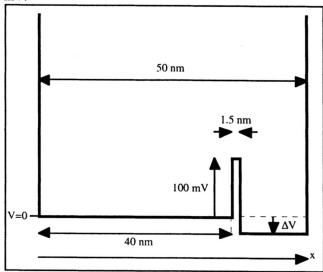

Problem Figure 10.1

References

1   L. Esaki, "New phenomenon in narrow germanium p-n junctions", *Phys. Rev.*, Vol. 109, p. 603, 1958
2   R.L. Liboff, *Introductory Quantum Mechanics*, Addison-Wesley, p. 221, 1988
3   S.M. Sze, *Physics of semiconductor devices*, J. Wiley & Sons, p. 513, 1981
4   L. Esaki, "The evolution of semiconductor quantum structures in reduced dimensionality - Do-it-yourself quantum mechanics", *Electronic properties of multilayers and low-dimensional semiconductor structures*, NATO ASI Series, Plenum Press, Series B: Physics Vol. 321, p. 1, 1990

5   T. Ando, A.B. Fowler, and F. Stern, "Electronic properties of two-dimensional systems", *Review of Modern Physics*, Vol. 54, p. 437, 1982

6   P.N. Butcher, *Physics of low-dimensional semiconductor structures*, Edited by P.N. Butcher, N.H. March andM.P. Tosi, Plenum Press, p. 95, 1993

7   F. Stern, *Physics of low-dimensional semiconductor structures*, Edited by P.N. Butcher, N.H. March and M.P. Tosi, Plenum Press, p. 177, 1993

8   J.H. Davies and G. Timp, "The smallest electronic device: An electron waveguide", in *Heterostructures and quantum devices*, Ed. by. N.G. Einspruch aand W.R. Frensley, Academic Press, Vol. 24 in *VLSI Electronics: Microstructure Science*, p. 385, 1994

9   J.H. Davies, *The physics of low-dimensional devices*, Cambridge University Press, p. 163, 1998

10 R. Landauer, "Spatial variation of currents and firlds due to localization scatterers in metallc conduction", *IBM. J. Res. Develop.*, Vol. 1, No. 3, p. 233, 1957

11 X. Baie and J.P. Colinge, "Two-dimensional confinement effects in gate-all-around (GAA) MOSFETs", *Solid-State Electronics*, Vol. 42, No. 4, p. 499, 1998

12 X. Baie, J.P. Colinge, V. Bayot and E. Grivei, "A silicon-on-insulator quantum wire", *Solid-State Electronics*, Vol. 39, p. 49, 1996

13 Ch. Wasshuber, *About Single-Electron Devices and Circuits*, Österreichischer Kunst- und Kulturverlag, Postfach 17, A-1016 Wien, Austria (ISBN 3-85437-159-4)

14 Ch. Wasshuber, H. Kosina, and S. Selberherr, "SIMON-A simulator for single-electron tunnel devices and circuits", *IEEE Transactions on Computer-Aided Design of Integrated Circuits ans Systems*, Vol. 16, No. 9, p. 937, 1997

15 X. Tang, X. Baie, V. Bayot, F. Van de Wiele and J.P. Colinge, "An SOI single-electron transistor", *Proceedings of the IEEE International SOI Conference*, p.46, 1999

16 K. Matsumoto, M. Ishii, K. Segawa, Y. Oka, B.J. Vartanian, and J.S. Harris, "Room temperature operation of a single electron transistor made by the scanning tunneling microscope nanooxidation process for the TiO_x/Ti system", *Applied Physics Letters*, Vol.68, No. 1, p.34, 1996

17 Y. Takahashi, H. Namatsu, K. Kurihara, K. Iwadate, M. Nagase, and K. Murase, "Size dependence of the characteristics of Si single-electron transistors on SIMOX substrates", *IEEE Transactions on Electron Devices*, Vol.43, No.8, p.1213, 1996

Chapter 11

SEMICONDUCTOR PROCESSING

11.1. Semiconductor materials

There exist many different semiconductor materials. The most important parameter distinguishing these materials is the width of the energy bandgap. The energy bandgap of the most common semiconductors is: 1.12 eV (silicon), 0.67 eV (germanium), and 1.42 eV (gallium arsenide). The main elements used in the semiconductor industry are shown in Figure 11.1.

III	IV	V
B	C	N
Al	**Si**	P
Ga	Ge	As
In		Sb

Figure 11.1: Main elements used in the semiconductor industry, either as elemental semiconductors (e.g. Si), III-V compounds (e.g. GaAs), or doping impurities (e.g. B).

Beside elemental semiconductors such as silicon and germanium, compound semiconductors can be synthesized by combining elements from column IV of the periodic table (SiC and SiGe), by combining elements from columns III and V (GaAs, GaN, InP, AlGaAs, AlSb, GaP, AlP and AlAs). Elements from other columns can sometimes be used as well (HgCdTe, CdS,...). Diamond exhibits semiconducting properties at high temperature, and tin (right below germanium in column IV of the periodic table) becomes a semiconductor at low temperatures. About 98% of all semiconductor devices are silicon based (integrated circuits, microprocessors, memory chips,...). The two remaining percents make use of III-V compounds (light-emitting diodes, laser diodes, RF components,...).

11.2. Silicon crystal growth and refining

Silicon is obtained by the chemical reduction of SiO_2 commonly produced from sand. The SiO_2 is reduced by carbon in an arc furnace equipped with graphite electrodes, according to the following reaction: $SiO_2 + C \rightarrow Si + CO_2$. The silicon produced by this method has a low purity and is called "metallurgical-quality silicon". The silicon then reacts with hydrochloric acid to produce trichlorosilane, according to the reaction $Si + 3 HCl \rightarrow SiHCl_3 + H_2$. The obtained trichlorosilane is then filtered and purified by distillation. Finally, trichlorosilane is decomposed at high temperature into HCl and silicon. Once pure silicon has been obtained it must be converted into a single crystal.

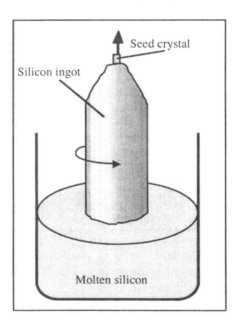

Figure 11.2: Czochralski growth.[1]

This is achieved using a technique called Czochralski growth, (CZ growth) which allows one to grow single-crystal rods up to 40 cm in diameter and over a meter long. During CZ growth the silicon is melted in a quartz crucible, and a small seed silicon crystal is dipped into the molten bath. The seed crystal is spun and slowly pulled out of the molten silicon. As the crystal is pulled upwards the temperature differential between the molten silicon bath and the gas ambient above it causes the silicon to crystallize. A rod-like silicon ingot is thus produced, as illustrated in Figure 11.2. Impurities such as boron or phosphorus can be added to the molten silicon to dope the silicon P- or N-type and give it the desired resistivity. The pulling of a silicon crystal by the Czochralski technique typically takes 24 hours. Once cooled, the ingot is cut like a salami into 0.5 to 2 mm-thick slices called silicon wafers. The wafers are then mirror polished using a combination of mechanical and chemical polishing agents.

The silicon ingot can be further refined using the float-zone technique, in which a section of the ingot is melted by induction using an RF coil. The molten zone is then swept from one side of the ingot to the other side (Figure 11.3). Several passes can be applied to further improve crystal purity.

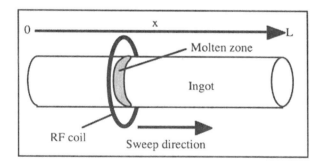

Figure 11.3: Float-zone purification technique.

The principle behind float-zone (FZ) refining is the following. The segregation coefficient for an impurity at the solid-liquid interface of the molten zone, k, is defined as the ratio of the concentrations of the impurity in the solid and the liquid: $k = \dfrac{C_S}{C_L}$. As an illustration, here are some segregation coefficients of some impurities in silicon: P: 0.32, As: 0.27, Sb: 0.02, B: 0.72, Ga: 0.0072, Au: 0.0000225. Since k is smaller than unity impurities are extracted from the solid and trapped in the molten zone during the float-zone refining process. To analyze the physics of the float-zone refining process let us define the following parameters: L is the length of the ingot, x is the position along the ingot, s is the amount of impurities dissolved in the molten zone, A is the cross-sectional area of the ingot, C_M is the original impurity concentration (per gram of silicon), and ρ is the silicon volumic mass. When the molten zone moves a distance dx the quantity of impurities introduced in the molten zone is equal to $C_M A \rho dx$, and the quantity of impurities left behind the trailing edge of the molten zone is equal to $k s \dfrac{dx}{L}$. The variation of the quantity of impurities in the molten zone is, therefore, equal to $ds = C_M A \rho dx - k s \dfrac{dx}{L}$. Using the initial condition $s = C_M A \rho L$ at $x = 0$ we can calculate the variation of the quantity of impurities in the molten zone as a function of position:

$$s = \frac{C_M A \rho L}{k}\left(1 - (1-k)\, e^{-kx/L}\right)$$

Since the impurity concentration left behind the molten zone, C_S, is equal to $\dfrac{k s}{A \rho L}$ we obtain:

$$C_S(x) = C_M \left(1 - (1-k)\, e^{-kx/L}\right) \tag{11.2.1}$$

Relationship 11.2.1 shows that the smaller the value of x, the more purified the crystal. The drawback of this refining method is, of course, that the impurity concentration is not constant along the ingot length. However, the use of several float-zone passes can produce higher material purity (Figure 11.4). Although Equation 11.2.1 is valid for the first pass only, the concentration versus position in the crystal can be calculated using a numerical calculation technique. The result is sketched in Figure 11.4.

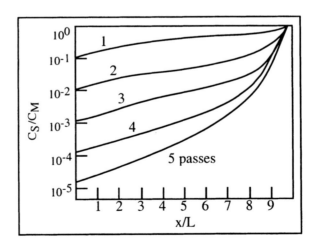

Figure 11.4: Decrease of impurity concentration obtained by float-zone refining as a function of the position in the ingot and number of passes. The segregation coefficient is 0.1 for this example. [2]

11.3. Doping techniques

The goal of doping a semiconductor is to introduce impurity atoms of a desired species into substitutional sites in the semiconductor crystal. Impurity atoms in interstitial (*i.e.*, non-substitutional) sites are electrically inactive and degrade the semiconductor properties such as carrier mobility. If an arsenic atom (column V of the periodic table) is substituted for a silicon atom the covalent bonds with neighboring atoms will be satisfied, and an extra electron will be released into the conduction band. Similarly, if a boron atom (column III of the periodic table) is substituted for a silicon atom all but one covalent bond with neighboring silicon atoms will be satisfied, and a hole will be released into the valence band.

The introduction of doping impurities can be carried out by different techniques: doping of molten silicon before Czochralski growth, diffusion from a gaseous doping substance into the silicon, ion implantation, growth of a doped silicon layer on an existing substrate (epitaxy), and neutron doping. In the latter technique silicon is submitted to a neutron (v) flux in a nuclear reactor. Some of the silicon atoms are transmuted into phosphorus atoms according to the following nuclear reaction:

$$^{30}_{14}Si + v \Rightarrow {}^{31}_{14}Si + \gamma \Rightarrow {}^{31}_{15}P + \beta$$

Neutron doping is only used in the fabrication of devices such as power thyristors where a low doping concentration and a high uniformity of doping concentration are needed.

11.3.1. Ion implantation

The ion implantation technique allows for the introduction of doping impurities in silicon with a high level of accuracy. An ion implanter is basically a particle accelerator composed of an ionization chamber called the source, an acceleration stage, a mass separation stage, an electrostatic deflection system, and a target chamber where the silicon wafers are placed for implant. A substance containing the doping element (e.g. gaseous BF_3 for boron implantation, or solid arsenic for arsenic implantation) is introduced in the source, where filament heating and microwave energy produce a plasma containing ions of the desired implant species. The ions are accelerated by an electric field to energies usually ranging from 5 keV to 200 keV, although some implanters are capable of producing ions with MeV energy. The ions are then deflected by an electromagnet depending on their mass. The current in the electromagnet is chosen so that only the desired ions continue their flight toward the silicon target. Other ions such as F^+, BF_2^+ in the case of boron implantation from a BF_3 source, are sent into a dead-end region of the implanter called a "beam trap". Further down the line an x-y electrostatic deflection system is used to uniformly distribute the ions across the silicon wafer in the target chamber by a raster scan method.

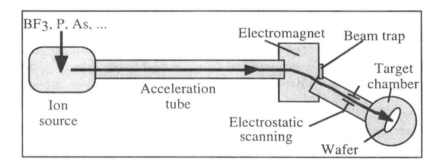

Figure 11.5: Ion implanter.

Ions accelerated by the implanter penetrate into the silicon and stop at a given depth in the crystal depending on the chosen implant energy. The higher the ion energy, the deeper the ions are implanted. The deceleration of the ions is due to interactions and collisions with the crystal atoms and electrons. These interactions are analyzed in the LSS (Lindhard, Scharff and Schiøtt) theory which predicts the stopping depth and the statistical distribution of the implanted atoms.[3]

The profile of the implanted atoms can be described within reasonable accuracy by a Gaussian distribution. The peak of the Gaussian distribution is located at a depth beneath the silicon surface called the "projected range", noted R_p. The width of the distribution is characterized by a standard deviation called the "straggle" and noted ΔR_p. Both the projected range and the straggle are expressed in centimeters. The concentration of implanted impurities, is therefore, described by the following relationship:

$$C(x) = C_p \, exp\left(-\frac{(x-R_p)^2}{2\Delta R_p^2}\right) \qquad (11.3.1)$$

where C_p is the concentration at the peak of the Gaussian distribution. By integrating the doping concentration over the entire Gaussian distribution one obtains the total implanted dose, which yields a relationship between the peak concentration, C_p (cm^{-3}) and the implanted dose N' (cm^{-2}):

$$C_p = \frac{N'}{\sqrt{2\pi} \, \Delta R_p} \qquad (11.3.2)$$

In a strict sense the above equations can be applied only to the implantation of atoms into an amorphous material. When implantation is performed in a single-crystal material such as silicon the impurities can penetrate deeper than predicted by theory. This phenomenon is called "ion channeling" and is due to the regularity of the position of the silicon atoms in the crystal. If the ion penetrates the material in the direction of preferential crystallographic directions, it will "see" rows of atoms separated by tunnels -or channels- along which it can penetrate much deeper into the crystal than predicted by the LSS theory. This effect is undesirable, and in practice, silicon wafers are tilted by an angle of 7 degrees with respect to the ion beam such that they present no preferential crystalline directions to the ion beam, and channeling is avoided.

The projected range R_p and the standard deviation ΔR_p can be expressed empirically as a function of the implantation energy, E. These expressions are found in Table 11.1 and graphically illustrated by Figures 11.6 and 11.7.

Table 11.1: Empirical expressions for R_p (nm) and ΔR_p (nm) as a function of implant energy, E (keV), in silicon.

Boron:	$R_p = 5.2629 \, E^{0.8909}$	$\Delta R_p = 5.34216 \, E^{0.5610}$
Phosphorus:	$R_p = 1.23612 \, E^{1.0000}$	$\Delta R_p = 0.76046 \, E^{0.8287}$
Arsenic:	$R_p = 1.09590 \, E^{0.8638}$	$\Delta R_p = 0.30303 \, E^{0.8038}$

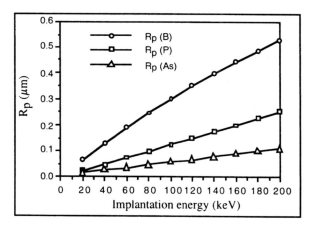

Figure 11.6: R_p for B, P and As in silicon as a function of energy. [4]

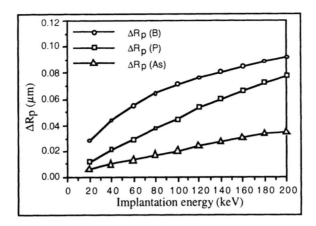

Figure 11.7: ΔR_p for B, P and As in silicon as a function of energy.

The energy lost by the implanted atoms during collisions with silicon atoms causes the formation of defects (vacancies, interstitials, etc.) in the silicon substrate, such that a certain degree of amorphization is created in the crystal. The distribution profile of these defects is Gaussian with a peak concentration located at a depth of 50%-80% of R_p. After an ion implantation step it is, therefore, necessary to restore the silicon crystal integrity. This is achieved by a thermal annealing step. A second function of the annealing step is to allow the impurities to diffuse into substitutional sites in the silicon lattice, since doping atoms occupying interstitial (non-substitutional) positions are electrically inactive.

Ion implantation can also be used to synthesize materials. By implanting a high dose of oxygen ions a buried SiO_2 layer can be created below the silicon surface. This process, called SIMOX (separation by implantation of oxygen) is used to fabricate silicon-on-insulator (SOI) substrates. SIMOX material consists of a thin (20 nm, typically) top layer of single-crystal silicon sitting on a buried oxide layer which is mechanically supported by the thick silicon wafer substrate.[5]

11.3.2. Doping impurity diffusion

Impurity diffusion is generally carried out at high temperature (800°-1100°C) in a furnace. At those temperatures the impurity atoms can diffuse throughout the crystal lattice through interactions with point defects (interstitials and vacancies). The equations governing the diffusion of impurities can be derived using Fick's first law of diffusion. Accordingly the variation of impurity concentration in an elementary volume in the crystal is given by:

$$\frac{\partial C}{\partial t} \, dx = F(x) - F(x+dx)$$

where C is the average impurity concentration in the volume under consideration and F is the flux of impurity atoms (Figure 11.8). Letting dx approach 0 we obtain $\dfrac{\partial C}{\partial t} = -\dfrac{dF}{dx}$.

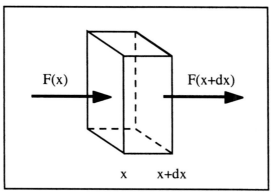

$$F(x) \qquad F(x+dx)$$

$$x \qquad x+dx$$

Figure 11.8: Elementary volume and flux of impurity atoms.

The flux is proportional to the impurity concentration gradient: $F = -D\,\dfrac{dC}{dx}$ where D is the diffusion constant of the impurity in silicon. Combining the previous equations we obtain the following relationship, known as Fick's second law of diffusion:

$$\frac{\partial C}{\partial t} = D\frac{d^2 C}{dx^2} \tag{11.3.3}$$

Strictly speaking the diffusion constant is not a real constant since it depends on both the temperature and the concentration of impurities in the silicon (the diffusing impurity or other impurities). The solution of Equation 11.3.3 yields a value called the "characteristic diffusion length", L, which is a function of temperature and time of diffusion: $L(T,t) = 2\sqrt{Dt}$. If the impurity concentration distribution before diffusion is Gaussian, it remains Gaussian after diffusion. The depth of the peak concentration, R_p, remains unchanged, but the peak concentration, C_p, decreases, and the distribution spreads out, such that its standard deviation increases. The new standard deviation, noted L' is a function of the pre-diffusion standard deviation, ΔR_p, and the diffusion length, $2\sqrt{Dt}$. It is equal to:

$$L' = \sqrt{2\Delta R_p{}^2 + 4Dt} \tag{11.3.4}$$

and the impurity concentration profile after diffusion is given by:

$$C(x,t) = \frac{N'}{L'\sqrt{\pi}}\,exp\left(-\left(\frac{x - R_p}{L'}\right)^2\right) \tag{11.3.5}$$

It is worthwhile noting that the concentration profiles of implanted impurities before and after diffusion are described by the same equation,

$C(x) = \dfrac{N'}{\sqrt{\pi}\,L}\, exp\left(-\left(\dfrac{x-R_p}{L}\right)^2\right)$, in which $L = \sqrt{2}\,\Delta R_p$ before diffusion and

$L = \sqrt{2\Delta R_p^2 + 4Dt}$ after diffusion.

The following computer code and Figure 11.9 illustrate the implantation followed by a diffusion of boron and phosphorus into an N-type substrate to form an NPN bipolar transistor.

```
TITLE BIPOLAR TRANSISTOR SUPREM SIMULATION
INIT SILICON <100> PHOS=5E15 THICK=1 DX=.005 SPACES=100
IMPLANT BORON ENERGY=20 DOSE=2E13
DIFFUSION TEMP=850 TIME=30 INERT
IMPLANT ARSENIC ENERGY=30 DOSE=5E15
DIFFUSION TEMP=900 TIME=30 INERT
COMMENT FIRST GRAPH
PLOT CHEMICAL ARSENIC TOP=1E21 BOTTOM=1E15 Y.LOGAR RIGHT=0.5
PLOT CHEMICAL BORON ADD
PLOT CHEMICAL PHOSPHOR ADD PAUSE
COMMENT SECOND GRAPH
PLOT NET ACTIVE TOP=1E21 BOTTOM=1E15 Y.LOGAR RIGHT=0.5 PAUSE
STOP
```

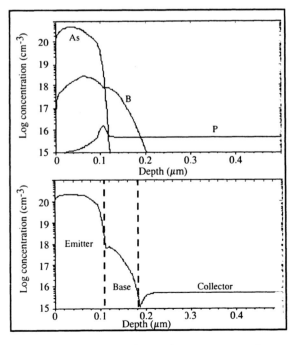

Figure 11.9: Simulation of the impurity concentration profiles in an NPN bipolar transistor. [6]

11.3.3. Gas-phase diffusion

Silicon can be doped in a high-temperature furnace in which an ambient gas containing atoms of doping impurities, such as BBr_3 (for boron doping) or $POCl_3$ (for phosphorus doping) is introduced.

Mass transport limited case
If the impurity atoms from the gas phase are present at the surface of the silicon in a relatively short time, such that no diffusion into the crystal takes place, the doping concentration profile can be approximated by a delta function. The impurity concentration is, therefore, given by $C(x=0) = N'$ (cm^{-2}) and $C(x>0) = 0$. If the sample is then submitted to a thermal annealing step called "drive-in" the impurity profile can be found by solving the diffusion equation (Relationship 11.3.3). The impurity concentration follows a Gaussian distribution and the peak concentration is located at the silicon surface ($x=0$):

$$C(x,t) = \frac{N'}{L'\sqrt{\pi}} exp\left(-\left(\frac{x-R_p}{L'}\right)^2\right) \tag{11.3.6}$$

The standard deviation of the Gaussian profile is equal to $L' = 2\sqrt{Dt}$ and depends on temperature (through the diffusion coefficient D) and the annealing time. If several annealing steps are carried out (the number of annealing steps being n) the final standard deviation is equal to:

$$L' = \sqrt{\sum_{i=1}^{n} L_i^2} = \sqrt{\sum_{i=1}^{n} 4D_i t_i} \tag{11.3.7}$$

Diffusion limited case
If the doping impurity gas is fed into the furnace continually the impurity concentration at the silicon surface, C_s, is maintained constant. In such a case the impurity concentration profile is no longer Gaussian, and is rather given by a complementary error function, $erfc(x)$:

$$C(x,t) = C_s\, erfc\left(\frac{x}{2\sqrt{Dt}}\right) = \frac{2C_s}{\sqrt{\pi}} \int_{x/2\sqrt{Dt}}^{\infty} e^{-v^2}\, dv \tag{11.3.8}$$

The complementary error function is defined by:

$$erfc(\eta) = 1 - erf(\eta) = 1 - \frac{2}{\sqrt{\pi}} \int_0^{\eta} exp(-v^2)\, dv$$

One of the properties of Relationship 11.3.8 is: $C(0,t) = C_s$

The total concentrations of impurities in the silicon is equal to:

$$N'(t) = \int_0^\infty C(x,t)dx = 2\sqrt{\frac{D\,t}{\pi}}\,C_s \qquad (11.3.9)$$

The maximum impurity concentration that can be introduced in silicon is fixed by the solid solubility of each doping species in silicon. Solid solubility is a function of temperature and is equal to 3.2×10^{20}, 2.1×10^{20} and 10^{20} cm^{-3} for P, As and B in silicon at 1000°C, respectively. Any attempt to introduce more doping atoms would result in the formation of impurity precipitates in the silicon crystal.

Modern silicon technology requires the formation of very shallow junctions. For that reason annealing techniques have been developed to activate implanted impurities with negligible diffusion. The rapid thermal annealing (RTA) technique achieves this goal: the silicon wafer is rapidly heated up to a high temperature (e.g.: 1100°C) for a few seconds, and then is quickly cooled down. Such an annealing step allows for the restoring of the silicon crystal and the placement of doping atoms into substitutional sites without causing diffusion over an appreciable distance.

11.4. Oxidation

Silicon is oxidized by oxygen or steam at high temperature according to the following chemical reactions:

$$Si(solid) + O_2(gas) \rightarrow SiO_2(solid) \text{ (dry oxidation)}$$

or

$$Si(solid) + 2H_2O(gas) \rightarrow SiO_2(solid) + 2H_2(gas) \text{ (wet oxidation)}$$

Two mechanisms influence the growth rate of the oxide. The first one is the actual chemical reaction rate between silicon and oxygen. The second one is the diffusion rate of the oxidizing species through an already grown oxide layer. When there is no or little oxide on the silicon the oxidizing agent easily reaches the silicon surface and the factor determining the growth rate is the kinetics of the silicon-oxide chemical reaction. In that case the oxidation process is reaction limited and the oxide thickness increases linearly as a function of time. If, on the other hand, the silicon is already covered by a sufficiently thick layer of oxide the oxidation process is mass-transport limited and the factor limiting the growth rate is the diffusion rate of O_2 or H_2O through the oxide, in which case oxide growth increases as a square root function of time. A steam ambient is usually preferred to a dry oxygen ambient for the growth of thick oxides: H_2O molecules are smaller than O_2 molecules, and as a result, they can diffuse more readily through SiO_2, which gives rise to higher oxidation rates.

To derive the equation describing silicon oxidation we will consider the mass transport of oxygen molecules from the gas ambient towards the silicon through a layer of already grown oxide (Figure 11.10). The flux of oxygen molecules is proportional to the differential in oxygen concentration between the ambient, C^*, and the oxide surface, C_o. The oxygen flux towards the oxide, F_1, is thus given by the following equation: $F_1 = h(C^* - C_o)$ where h is the mass transport coefficient for oxygen in the gas phase.

The diffusion of oxygen through the oxide is proportional to the difference of oxygen concentration between the oxide surface and the silicon/SiO_2 interface. The flux of oxygen through the oxide, F_2, is given by: $F_2 = D \dfrac{C_o - C_i}{t_{ox}}$ where C_i is the oxygen concentration at the silicon/SiO_2 interface, D is the diffusion coefficient of either O_2 or H_2O in oxide, and t_{ox} is the oxide thickness. Finally, the kinetics of the chemical reaction between silicon and oxygen is characterized by a reaction constant k, such that we have: $F_3 = k_s C_i$.

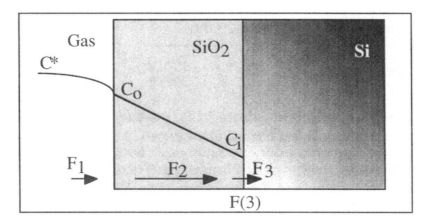

Figure 11.10: Oxygen concentration profile during oxidation. [7]

In steady state all flux terms are equal: $F_1 = F_2 = F_3 \equiv F$. Eliminating C_o from the flux equations we obtain:

$$C_i = \frac{C^*}{1 + \dfrac{k_s}{h} + \dfrac{k_s t_{ox}}{D}}$$ (11.4.1)

If N_{ox} is a constant representing the number of oxidizing gas molecules necessary to grow a unit thickness of oxide one can write:

$$\frac{dt_{ox}}{dt} = N_{ox} F = N_{ox} k_s C_i = \frac{N_{ox} k_s C^*}{1 + \frac{k_s}{h} + \frac{k_s t_{ox}}{D}}$$ (11.4.2)

The solution to this differential equation is:

$$\int_0^{t_{ox}} \frac{1 + \frac{k_s}{h} + \frac{k_s t_{ox}}{D}}{N_{ox} k_s C^*} \, dt_{ox} = \int_0^t dt$$

If $t_{ox}=0$ when $t=0$, the integration yields:

$$\frac{t_{ox}^2}{2} + \left(\frac{D}{k_s} + \frac{D}{h}\right) t_{ox} - N_{ox} C^* D t = 0$$

or: $$t_{ox}^2 + 2D\left(\frac{1}{k_s} + \frac{1}{h}\right) t_{ox} = 2DN_{ox}C^*t$$

Defining new constants A and B in terms of D, k_s, N_{ox} and C^*:

$$A = 2D\left(\frac{1}{k_s} + \frac{1}{h}\right) \quad \text{and} \quad B = 2DC^*N_{ox}$$

we obtain: $$t_{ox}^2 + A t_{ox} = B t$$

from which we find t_{ox}: $$t_{ox} = \frac{A}{2}\left(\sqrt{1 + \frac{(t + \tau)}{A^2/4B}} - 1\right)$$ (11.4.3)

Parameter τ is introduced to account for the possible presence of an oxide layer on the silicon before thermal oxide growth is performed. This pre-existing oxide layer can either be a native oxide layer due to the oxidation of bare silicon by ambient air or a thermally grown oxide produced during a prior oxidation step. $\tau=0$ if the thickness of the initial oxide is equal to zero. Equation 11.4.3 is referred to as the Deal-Grove model of oxidation. [8]

When thin oxides are formed the growth rate is limited by the kinetics of the chemical reaction between silicon and oxygen. In that case Equation 11.4.3 can be approximated by: $t_{ox} = \frac{B}{A} (t+\tau)$ which is linear with time.

The $\frac{B}{A}$ ratio is called the "linear growth coefficient" and is dependent on the crystal orientation of silicon.[9] When thick oxides are formed the growth rate is limited by the diffusion rate of the oxygen through the oxide. In that case Relationship 11.4.3 can be approximated by: $t_{ox} = \sqrt{B(t+\tau)} \cong \sqrt{Bt}$. The coefficient B is called the "parabolic growth coefficient" and is independent of the crystal orientation of silicon. The parabolic growth coefficient can be increased by increasing the pressure of the ambient oxygen up to 10 to 20 atmospheres (high-pressure oxidation, HIPOX).[10] The linear growth coefficient can be increased if the silicon contains a high concentration of impurities such as phosphorus. These impurities increase the concentration of point defects in the crystal which increase the oxidation reaction rate at the silicon/silicon dioxide interface. Similarly, the oxidation process generates point defects in silicon, which accelerates the diffusion of doping impurities (oxidation-enhanced diffusion, OED).[11] Therefore, some doping impurities diffuse faster when annealing is performed in an oxidizing ambient than when it is carried out in a neutral gas such as nitrogen.

Oxide growth consumes silicon. As a rule of thumb one can consider that the thickness of silicon consumed is 44% of that of the grown oxide. When an oxide is grown the doping atoms in the silicon redistribute between the silicon and the oxide and there exists a constant ratio between the impurity concentrations on both sides of the Si/SiO$_2$ interface called the "segregation coefficient" defined by: $m = \dfrac{C_{Si}}{C_{SiO2}}$. The segregation coefficient of arsenic and phosphorus in silicon is larger than unity, while that of boron is smaller than unity. In other words, the concentration of arsenic and phosphorus in the oxide is less than in silicon, and the concentration of boron is larger.[12] This effect results in a "pile-up" of arsenic or phosphorus in the silicon at a Si/SiO$_2$ interface or a depletion of boron in the silicon at a Si/SiO$_2$ interface. Impurity segregation and OED are illustrated in Figures 11.11 and 11.12 where annealing in both neutral (nitrogen) and oxidizing (dry oxygen) ambient have been simulated.

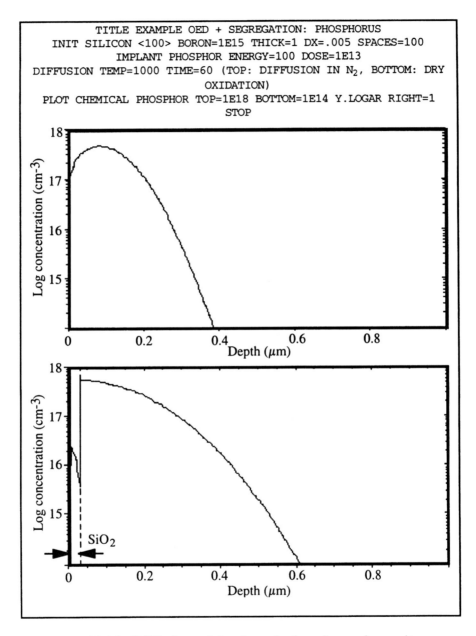

Figure 11.11: Diffusion of implanted phosphorus in a nitrogen ambient (top) and in dry oxygen (bottom) at 1000°C for one hour. The bottom plot shows a lower phosphorous concentration in SiO_2 than in silicon at the Si/SiO_2 interface.[13]

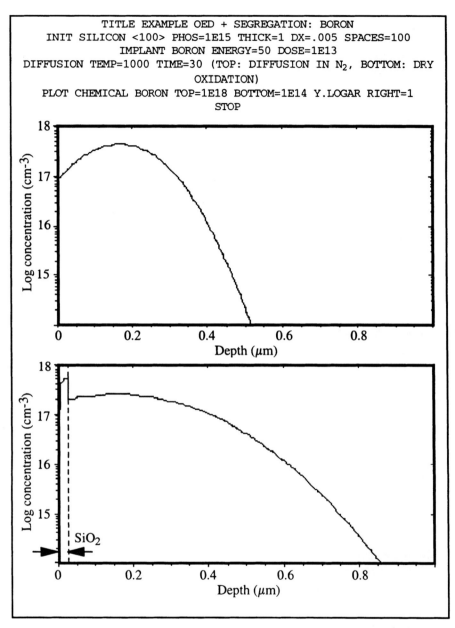

TITLE EXAMPLE OED + SEGREGATION: BORON
INIT SILICON <100> PHOS=1E15 THICK=1 DX=.005 SPACES=100
IMPLANT BORON ENERGY=50 DOSE=1E13
DIFFUSION TEMP=1000 TIME=30 (TOP: DIFFUSION IN N$_2$, BOTTOM: DRY OXIDATION)
PLOT CHEMICAL BORON TOP=1E18 BOTTOM=1E14 Y.LOGAR RIGHT=1
STOP

Figure 11.12: Diffusion of implanted boron in a nitrogen ambient (top) and in dry oxygen (bottom) at 1000°C for 30 minutes. The bottom plot shows boron segregating into the oxide layer.[14]

Sometime oxide must be grown over selected areas of a silicon wafer. The LOCOS (local oxidation of silicon) technique has been widely used in MOS and bipolar integrated circuit manufacturing processes to laterally isolate devices from one another.[15] This oxide between devices is also called "field oxide". The LOCOS process is illustrated in Figure 11.13. It is based on the fact that oxygen does not diffuse through silicon nitride. A layer of silicon nitride is deposited and patterned using photolithography and etching. Usually a thin thermal oxide layer called the "pad oxide" is grown prior to nitride deposition. This layer acts as a buffer between the nitride and the silicon to avoid build-up of mechanical stress between those two materials during thermal cycles. Such a stress would generate crystal defects in the silicon. Both silicon and silicon nitride are hard materials, while SiO_2 is rather soft at high temperature.

If the substrate is P-type, boron is then implanted after the nitride has been patterned to create a P^+ region under the LOCOS oxide and prevent the formation of an inversion layer under the field oxide. Field inversion can be caused by the presence of positive charges commonly found in oxide or by the presence of a positively biased metal line running over the oxide. If field inversion occurs, the N-type diffusions of the circuit can be short-circuited under the field oxide. Field inversion can occur in P-type silicon only and no field implantation is needed in N-type silicon.

The thick field oxide is then grown, usually in a wet oxygen ambient. The nitride and the pad oxide are chemically removed in hot phosphoric acid (H_3PO_4) and hydrofluoric acid (HF), respectively. Because of its shape the side of the field oxide, which grew underneath the nitride, has been nicknamed "bird's beak".

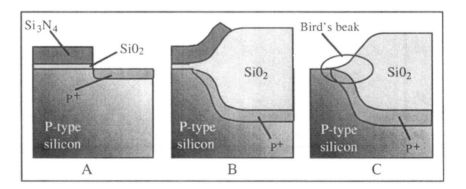

Figure 11.13: Local growth of field oxide (LOCOS). A: Patterning of nitride layer and pad oxide and P^+ field implantation; B: Field oxide growth; C: etching of the nitride and pad oxide.

The lateral extension of the bird's beak is on the order of 0.1 to 0.2 μm, which is too large for modern, deep-sumicron devices. Therefore, a new type of field isolation called "shallow trench isolation" (STI) has been developed. In this process a shallow trench is etched in the silicon and a nitride cap layer (Figure 11.14A). Then a thin thermal oxide is grown and CVD oxide is deposited to fill the trench (Figure 11.14B). Chemical-mechanical polishing (CMP) is then used to remove the excess oxide at the surface of the wafer. The nitride mask conveniently acts as an etch stop for the CMP step. The nitride is then removed by selective chemical etching.

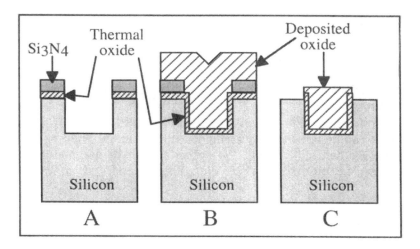

Figure 11.14: Shallow trench isolation (STI). A: Trench etching; B: Growth of thin thermal oxide and oxide deposition; C: Chemical-mechanical polishing and nitride etch.[16]

11.5. Chemical vapor deposition (CVD)

In chemical vapor deposition (CVD) a chemical reaction between gas-phase reactants is used to deposit layers of solid material. Such a technique can be employed to deposit silicon, insulating materials such as SiO_2 and Si_3N_4 or metals (e.g. tungsten). Variations on the CVD technique include LPCVD (low-pressure chemical vapor deposition) where the deposition is carried out under reduced gas pressure for better uniformity and step coverage and PECVD (plasma-enhanced chemical vapor deposition) where plasma excitation of the gas phase is used to deposit materials at low temperature.

11.5.1. Silicon deposition and epitaxy

Epitaxy is a processing technique in which a single-crystal layer of silicon is grown on silicon. Epitaxial growth is carried out at high temperature in a reactor where pyrolysis of gases such as silane (SiH_4) or dichlorosilane (SiH_2Cl_2) is used to deposit the silicon layer. The chemical reactions involved in silicon epitaxial growth are either

$$SiH_4(gas) \rightarrow Si(solid) + 2H_2(gas)$$

or

$$SiH_2Cl_2(gas) + 2\ H_2(gas) \rightarrow Si(solid) + 4HCl(gas)$$

If the deposition temperature is high enough (900°-1250°C) the silicon atoms generated by the pyrolysis reaction are not deposited in a random order. Rather they position themselves in alignment with the silicon atoms at the substrate surface such that a single-crystal silicon layer, having the same crystal orientation as the substrate, is grown. The epitaxial layer can be doped *in situ* by introducing small amounts of gases such as phosphine (PH_3) or diborane (B_2H_6) in the reaction chamber. Silicon can be epitaxially grown selectively in certain areas of the silicon wafer. This can be achieved by etching openings in an oxide layer grown on silicon. By carefully tuning the epitaxial growth parameters, silicon can be grown in those areas where the silicon is exposed, and not on the oxide.

If the deposition is carried out at low temperature (<600°C) the silicon layer is amorphous. Amorphous silicon is usually not employed in the fabrication of integrated circuits, but it is widely used in the fabrication of thin-film transistors (TFTs) for flat-panel displays and the fabrication of amorphous solar cells used to power some "solar" pocket calculators and wristwatches. In those applications the amorphous silicon is usually deposited on a glass substrate.

LPCVD is also used to deposit polycrystalline silicon -or polysilicon- layers. Polysilicon is commonly used as the gate electrode material of MOS transistors. Polysilicon deposition is obtained by pyrolysis of silane under low pressure and at a temperature of 620°C. The deposited film is composed of silicon crystallites separated by grain boundaries. The crystallites have a diameter of approximately 100 nm and a height equal to the film thickness.[17] Lightly doped polysilicon can be used to form high-value resistors in integrated circuits, while heavily doped polysilicon is used to form MOSFET gate electrodes and local interconnections.

11.5.2. Dielectric layer deposition

The CVD, LPCVD and PECVD techniques are widely used for the deposition of insulating dielectric layers. For example, the following reactions are used to produce silicon dioxide and silicon nitride layers:

$$SiH_4(gas)+2O_2(gas) \rightarrow SiO_2(solid)+2H_2(gas)$$

and

$$3SiH_2Cl_2(gas)+4NH_3(gas) \rightarrow Si_3N_4(solid)+6\ HCl(gas)+6H_2(gas)$$

The reactant gases are usually mixed right before their introduction in the reaction chamber and flow continuously over the silicon wafer. The flow is laminar but slows down at the surface of the wafers, such that the velocity of the gas mixture varies from its nominal value far from the sample to zero at the sample surface. The region where the gas velocity varies near the wafer is called the "boundary layer". Before reaching the silicon surface the gas reactants must diffuse through the boundary layer, after which the chemical reaction and thus the dielectric material deposition can take place.

The gas flux through the boundary layer, F_1, is obtained using the following relationship: $F_1 = D \dfrac{C_g - C_s}{\delta}$, where δ is the thickness of the boundary layer, D is the diffusivity of the gases in the boundary layer, and C_g and C_s are the concentrations of the gas reactants in the ambient and at the silicon surface, respectively. The diffusivity of the gases in the boundary layer is virtually temperature independent.

At the silicon surface the gas flux, F_2, is given by $F_2 = k_s C_s$, where k_s is the rate of the chemical reaction, which depends exponentially on temperature: $k_s = k_{so} \exp(-E_a/kT)$, where E_a is the activation energy of the reaction (Figure 11.15).

In steady state $F_1 = F_2 \equiv F$ such that the deposition rate can be calculated: $R_d = \dfrac{F}{N} = \dfrac{C_g/N}{\delta/D + 1/k_s}$, where N is the number of molecules per unit volume in the layer being deposited. The term δ/D represents the diffusion rate of the gaseous reactants through the boundary layer and $1/k_s$ represents the chemical reaction rate at the silicon surface.

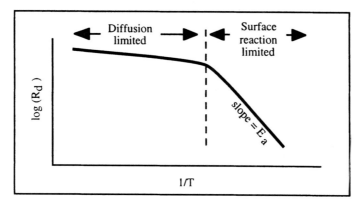

Figure 11.15: Deposition rate (log scale) as a function of $1/T$. [18]

As can be seen on Figure 11.15 the deposition rate is limited at lower temperatures by the reaction rate at the silicon surface, and at higher temperatures, by the gas diffusion through the boundary layer. To obtain good control of the deposition rate it is, therefore, suitable to operate at high temperatures, where the deposition rate is less sensitive to temperature fluctuations.

11.6. Photolithography

Photolithography is used in device fabrication process every time a pattern must be transferred to the silicon surface. It also allows one to perform ion implantation or etch a material in selected areas on the wafer. Photoresist is a photosensitive organic substance which is a sticky liquid with a high viscosity. After having been spun onto a wafer it is thermally hardened in an oven. There are two types of photoresists: positive and negative. In a positive resist exposure to light breaks down long-chain organic molecules into shorter chain molecules which can be dissolved by an appropriate chemical solution called a developer. In a negative resist exposure to light induces the cross-linking of organic molecules such that a higher atomic mass is achieved, *i.e.*, longer-chain molecules are produced. An appropriate developer solution is then used to remove the resist that has not been exposed to light.

The transfer of the desired patterns onto the resist is made using ultraviolet light exposure through a mask. The mask is a quartz plate which contains the patterns corresponding to a given processing operation, such as gate material etching or metal interconnection etching. The mask basically plays the role of the negative in conventional photography. In the simplest photolithography tools the mask and the wafer are either placed in contact or at close proximity to one another (Figure 11.16). In the case of contact exposure the mask actually touches the photoresist, which allows for an excellent printing resolution, however defects in the quartz mask or the wafer can occur such as scratches. In a proximity aligner the mask is held at a small, but finite distance from the wafer surface (Figure 11.16). As a result the light and dark patterns projected onto the photoresist are less sharp than in contact mode. The minimum feature size that can be printed using a proximity aligner is on the order of $l_m \cong \sqrt{\lambda g}$ where g is the distance (gap) between the mask and the resist on the wafer, and λ is the wavelength of the light used for exposure. For example, if $\lambda = 0.4 \mu m$ and $g = 10 \mu m$, then the minimum printable feature size is $l_m = 2 \mu m$.

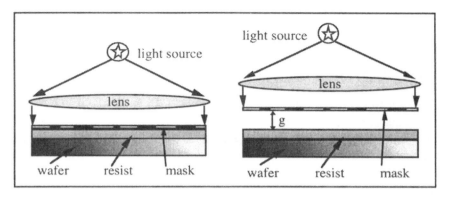

Figure 11.16: Contact (left) and proximity (right) photolithography.

The poor resolution of proximity systems can be overcome by using a complicated optics system between the mask and the wafer. This is used in projection systems called wafer steppers in which UV light is shone through a mask called a "reticle" in which the patterns are usually 5 or 10 times larger than the features to be printed onto the photoresist. The system optics reduces the size of the features and projects them on the wafer. In such a system the minimum printable feature size is equal to:

$$l_m \cong \lambda/NA \tag{11.6.1}$$

where λ is the wavelength of the light and NA is the numerical aperture of the optics. The operation of such an optical system is illustrated in Figure 11.17. The numerical aperture NA is defined as the sine of the angle formed by the light beam reaching a point in the focal plane. Simple geometric analysis shows that in order to print a minimum feature size equal to l_m the distance between the wafer surface and the focal plane must be smaller than the depth of focus Δz:

$$\Delta z = \frac{\pm l_m/2}{tan\Theta} \cong \frac{\pm l_m/2}{sin\Theta} = \pm \frac{\lambda}{2(NA)^2} \tag{11.6.2}$$

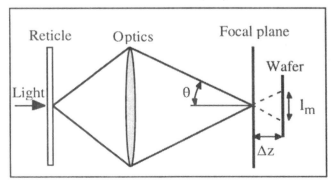

Figure 11.17: Projection photolithography; light shines through the patterned reticle, optics reduces the pattern size, and the pattern is projected onto the wafer.

The resolution of a projection system can be improved by using shorter light wavelengths, such as the deep-UV light produced by excimer lasers. It can also be improved by planarizing the surface of the wafer and placing it close to the focal plane with the highest possible accuracy.

High-resolution lithography can also be obtained by writing the patterns into a resist-covered wafer using a focused electron beam or a focused ion beam. The drawback of such systems is the lengthy exposure time necessary to expose an entire silicon wafer. Fine-line lithography is also possible when X-rays are used instead of UV light. Since X-rays cannot be bent or focused by optical systems the X-ray source must emit a non-divergent beam that will go through the mask onto the wafer. This requirement imposes the use of sophisticated X-ray sources such as synchrotrons.

To illustrate how photolithography is used, let us take the example of Figures 11.18a to 11.18f, where a metal layer must be etched to form a contact to the silicon surface. The metal is first deposited over the entire silicon wafer. Etched holes in the oxide allow for the metal to contact the desired diffusions (Figure 11.18a). Positive photoresist is the spun (deposited) onto the metal (Figure 11.18b). A mask is used to expose the photoresist in the areas the metal will be removed (Figure 11.18c). The exposed photoresist is then removed in a developer solution (Figure 11.18d). The metal is then etched, for example, by a chemical that does not react with either the photoresist or silicon dioxide (Figure 11.18e). Those parts of the metal film which are covered by the resist are not etched. Finally the resist is stripped off the metal in an appropriate chemical bath (Figure 11.18f).

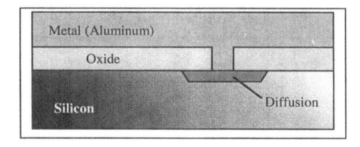

Figure 11.18a: The metal layer is deposited over the entire wafer.

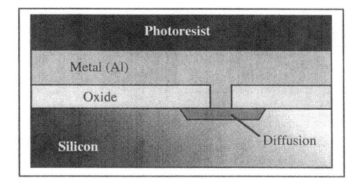

Figure 11.18b: Photoresist is deposited by spinning on the liquid photoresist.

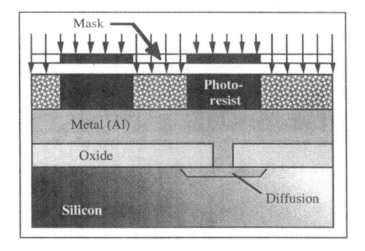

Figure 11.18c: Photoresist is exposed, using light, through a mask.

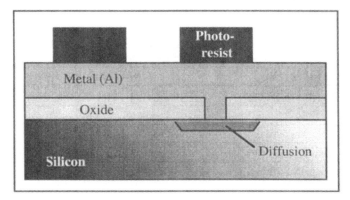

Figure 11.18d: Photoresist developing using a chemical solution (developer).

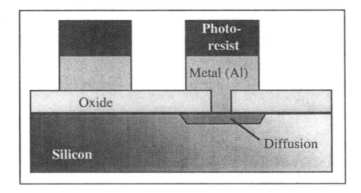

Figure 11.18e: Metal etch removes the "unwanted" aluminum.

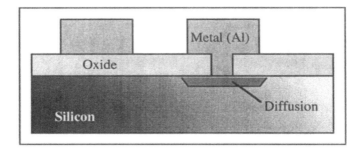

Figure 11.18f: Photoresist strip in a chemical solution.

11.7. Etching

Photolithography steps are usually followed by either an ion implantation or an etching step. Etching can take place by placing the wafer in a chemical bath (wet etch) or in a plasma (dry etch). The most commonly used chemicals used in wet etching are listed in Table 11.2.

Table 11.2: Chemicals commonly used for wet etching.

Material etched	Chemical etchant
Silicon, polysilicon	$HF + HNO_3$, hot KOH, hot TMAH*
SiO_2	HF
Si_3N_4	hot H_3PO_4
Aluminum	HCl, H_3PO_4
Photoresist (strip)	$H_2SO_4 + H_2O_2$, fuming HNO_3, acetone

* Tetramethyl Ammonium Hydroxide

Chemical etchants are usually very selective: for instance, an hydrofluoric acid solution (buffered HF) etches SiO_2 while it virtually does not react with Si, Si_3N_4, or photoresist. The selectivity of an etching agent is defined as the ratio of the etching rates produced in different materials. For example, if a buffered HF solution etches 60 nm of SiO_2 per minute and 0.1 nm of silicon per minute, the selectivity is 600:1 (read "six hundred to one"). Chemical etching is usually isotropic, meaning it has the same etch rate in every direction (x or y). As a result etch profiles in the shape of an arc of a circle are obtained (Figure 11.19) and the etched region extends underneath the masking material (the photoresist in Figure 11.9), thereby giving rise to overetching in the y direction.

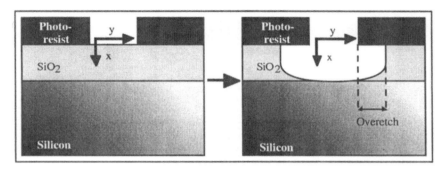

Figure 11.19: Isotropic etch of SiO_2 by an HF solution.

Wet etching is very selective and easy to use. However, it becomes obsolete when feature size becomes smaller than 1 or 2 micrometers because of overetching problems. For etching small patterns, dry etching in a plasma is generally used instead.

The degree of anisotropy of an etch is defined as $A_f = 1 - \frac{v_l}{v_v}$ where v_l and v_v are the lateral (y direction) and vertical (x direction) etch rates, respectively. If $A_f = 1$ the etching is completely anisotropic (vertical), while it is perfectly isotropic if $A_f = 0$ (Figure 11.20).

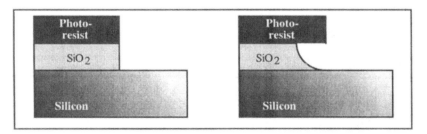

Figure 11.20: Anisotropic (left) and isotropic (right) SiO_2 etch profiles of SiO_2.

Plasmas used for dry etching are produced using low-pressure (10^{-2}-10^{-3} Torr) gas mixtures. A plasma is formed by applying either a high dc voltage or a radio-frequency (RF) bias to the low-pressure gas. Fresh gaseous reactants are continuously fed into the reactor while the reaction byproducts are pumped out to maintain a controlled pressure value in the reactor chamber (Figure 11.21).

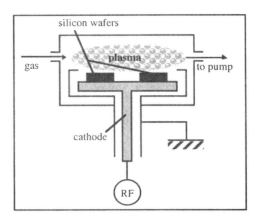

Figure 11.21: Plasma etch reactor.

The process by which a plasma etches a material has two components: a chemical component due to the presence of chemically active species such as F^- or Cl^-, and a physical erosion component due to the bombardment of the sample surface by ions. This erosion mechanism is, in a way, similar to sand blasting. The chemical component gives rise to isotropic, selective etching characteristics, while the physical erosion mechanism is anisotropic and non-selective. The art of plasma etching consists of fine-tuning the gas mixture composition, chamber pressure, and the RF power delivered to the plasma to maximize both the anisotropy and the selectivity of the etch. Plasma etching machines that combine physical erosion and chemical reaction are called "reactive ion etching" (RIE) reactors. The gas mixtures most frequently used in silicon dry processing are listed in Table 11.3.

Table 11.3: Some gases used in dry etching.

Material etched	Gas used in the plasma
Si, polysilicon	CF_4+O_2, SF_6+O_2, CCl_4, HBr
SiO_2, Si_3N_4	CF_4+H_2, C_2F_6
Aluminum	CCl_4, Cl_2
Photoresist	O_2

11.8. Metallization

Metal layers separated by layers of dielectric material such as SiO_2 are used to interconnect devices, supply electrical power, and clock signals, etc. Each metal layer consists of a series of metal lines running to selected destinations. In early integrated circuits the use of a single metal layer would provide enough connectivity to an entire chip. In today's complex very large scale integrated circuits (VLSI circuits), however, up to 10 metallization levels can be used.

11.8.2. Metal deposition

The deposition of a metal layer is usually carried out in a vacuum. The metal is evaporated from a solid source, and re-deposited onto the silicon wafers. Metal evaporation can be achieved by hitting the metal source with an electron beam or by an argon plasma that sputters fine metal particles into the vacuum of the deposition chamber. Sometimes chemical vapor deposition is also used. For instance, tungsten can be deposited by CVD according to the following chemical reaction: $WF_6 + 3H_2 \rightarrow W + 6HF$.

Aluminum and copper are the most widely used metal for the fabrication of integrated circuits. Another metal, tungsten, is frequently used to form "plugs" through vias etched in dielectric layers. The plugs allow for the passage of electric signals between different interconnection levels, *i.e.*, different metal layers. As devices are scaled down their speed is increased, such that the clock speed of microprocessors has increased from a few megahertz in the early eighties to over a gigahertz in year 2000. To cope with these high frequencies the RC delay of interconnection lines must be reduced as much as possible. This means using low-resistivity metals and low-permittivity dielectric materials between the metal lines. With the exception of silver, copper is the metal that has the lowest resistivity, and therefore, is a good candidate for making interconnections in integrated circuits. Unfortunately, it is virtually impossible to etch copper using conventional dry etching techniques. Instead, a patterning process called the "damascene process" is used (Figure 11.22). A dielectric layer, such as silicon dioxide is first etched in order to form trenches. Copper is deposited and finally chemical-mechanical polishing (CMP) is used to remove the excess copper and the interconnections are formed.

Another way of reducing the RC delay of metal lines is to use dielectric materials with a low electrical permittivity between the metal lines. The permittivity of a material is given by: $\varepsilon_{material} = \kappa_{material} \times \varepsilon_0$, where ε_0 is the permittivity of vacuum. The values of κ for air and SiO_2 are 1.001

and 3.9, respectively. Dielectric materials with low κ values, called "low-K dielectrics" are increasingly used as a replacement for SiO$_2$. Such materials include polymers and polymers with air bubbles, and even air itself. In the latter case, a "dummy" or "sacrificial" organic layer is deposited, upon which metal is deposited and patterned. Exposure to a solvent or an oxygen plasma strips the polymer and leaves "free-standing" metal lines at the surface of the integrated circuit.

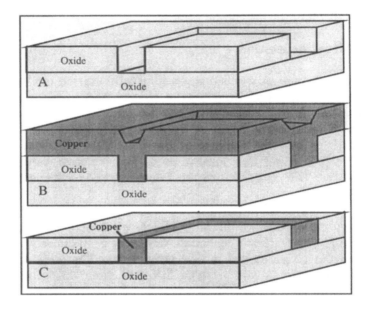

Figure 11.22: Damascene process: A: Trenches are plasma etched in the dielectric; B: Copper is deposited; C: Chemical-mechanical polishing removes the excess copper.

11.8.3. Metal silicides

As devices shrink in size, the source and drain diffusions also become shallower. This, in turn, increases the parasitic resistance of the source/drain and reduces the device speed. Refractory metal silicides are used to decrease the source, drain and gate resistance of MOSFETs. The most widely used silicides are titanium, cobalt and tungsten silicides (TiSi$_2$, CoSi$_2$ and WSi$_2$). Silicide layers can be deposited by co-sputtering of metal and silicon or can be formed by chemical reaction between a metal and silicon at relatively high temperature (600-850°C). Silicides have a higher resistivity than metals, but they can withstand relatively high thermal budgets which makes them attractive as local interconnection materials.

In the SALICIDE process (self-aligned silicide) the silicide is formed by chemical reaction between the silicon of the source, drain and gate electrodes and a deposited metal such as titanium. The process sequence used is described in Figure 11.23. Oxide spacers are formed at the edges of the polysilicon gate using CVD oxide deposition. Next isotropic RIE etching is used to remove the oxide from the source, the drain and the top of the drain electrode, thereby creating "spacers" on the gate electrode sidewalls. Titanium is then deposited. Upon annealing $TiSi_2$ is formed on the source, drain and gate of the MOSFET. The unreacted titanium over the oxide regions is then stripped in a mixture of sulfuric acid and hydrogen peroxide.

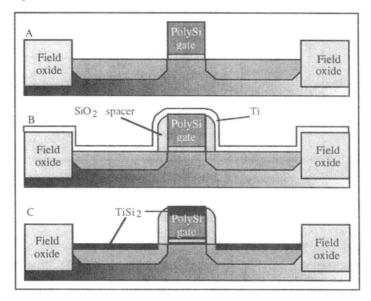

Figure 11.23: SALICIDE process: A: Initial MOS transistor; B: Formation of spacers by CVD oxide deposition and RIE etch of the oxide, and titanium deposition; C: $TiSi_2$ formation and stripping of the unreacted titanium.

11.9. CMOS process

CMOS technology is by far the dominant fabrication technology in the semiconductor industry. The word CMOS means "complementary MOS" and arises from the fact that both n-channel and p-channel transistors can be formed side by side on a same chip. Historically the first MOS integrated circuits were fabricated using a pMOS process, where only p-channel transistors could be fabricated. pMOS was then replaced by nMOS technology where only n-channel devices were used due to their superior mobility over pMOS devices. Finally, CMOS was introduced and quickly replaced all other MOS fabrication processes because CMOS circuits consume virtually no power when on standby, unlike nMOS or pMOS circuits. In addition, circuit design is more efficient in CMOS than in the other MOS families. The CMOS fabrication process can be fully described by the sequence of operations used to fabricate the simplest CMOS logic gate: the CMOS inverter (Figure 11.24).

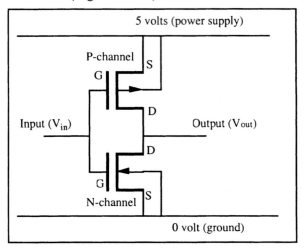

Figure 11.24: CMOS inverter. [19]

N-channel transistors must be made on a P-type substrate, while p-channel MOSFETs require the use of an N-type substrate. To integrate both types of devices on a single silicon wafer the need arises to form both N-type and P-types regions in the substrate. If an N-type wafer is used P-type regions called "P-wells" must be created to host n-channel transistors. If a P-type wafer is the starting substrate, N-wells will be formed for the P-channel devices. An N-well CMOS process is described below in Figures 11.25a to 11.25p. The fabrication sequence yields a CMOS inverter.

Figure 11.25a: Starting P-type substrate.

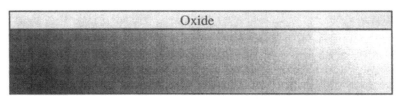

Figure 11.25b: Growth of a thick wet oxide layer.

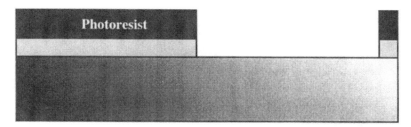

Figure 11.25c: Opening of N-well area using lithography and oxide etch.

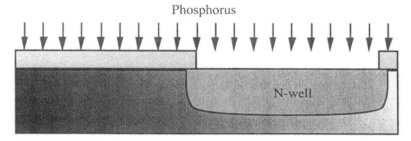

Figure 11.25d: Phosphorus implantation and N-well formation.

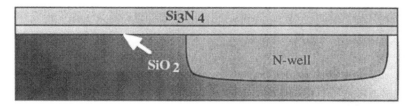

Figure 11.25e: Oxide strip, pad oxide growth and silicon nitride deposition.

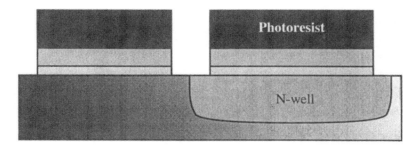

Figure 11.25f: Photolithography: definition of the active device areas.

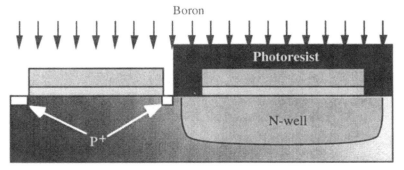

Figure 11.25g: Photolithography and field implantation around the future n-channel devices.

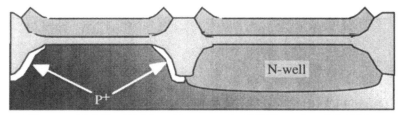

Figure 11.25h: Field oxide growth (FOX) using LOCOS.

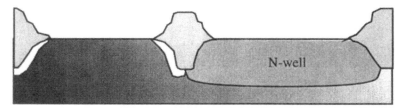

Figure 11.25i: Nitride and pad oxide strip.

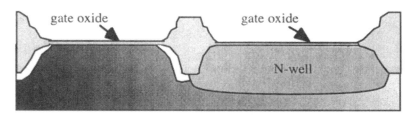

Figure 11.25j: Thin gate oxide growth over the active regions.

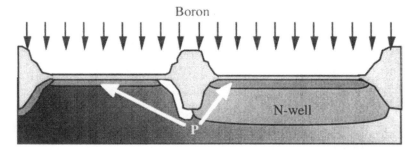

Figure 11.25k: Boron implantation to adjust the p-channel threshold voltage.

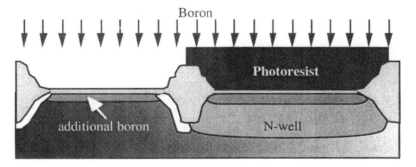

Figure 11.25l: Photolithography and additional boron implantation to adjust the n-channel threshold voltage.

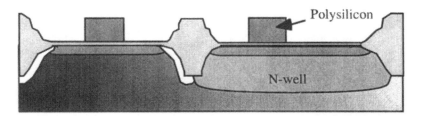

Figure 11.25m: Deposition and N^+ doping of polysilicon; photolithography and polysilicon etch to form the gate electrodes.

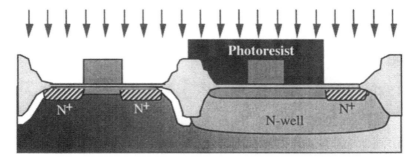

Figure 11.25n: Photolithography and arsenic implantation to form N-channel source and drain and N-well contact.

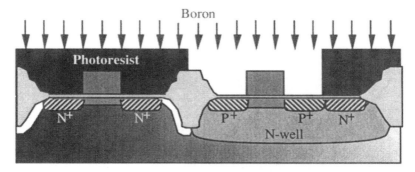

Figure 11.25o: Photolithography and formation of the p-channel source and drain.

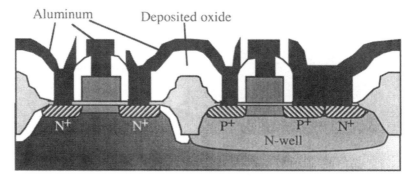

Figure 11.25p: Oxide deposition, lithography and contact via opening (*i.e.* oxide removal at source, drain, gate and N-well contacts). Metal deposition, lithography and metal etch; metal sintering anneal.

A plane view of the CMOS inverter is shown in Figure 11.26. The well contact is connected to V_{DD} such that the N-well/substrate junction is always reverse biased. The dotted line represents the region where the cross-section of Figure 11.25p is taken.

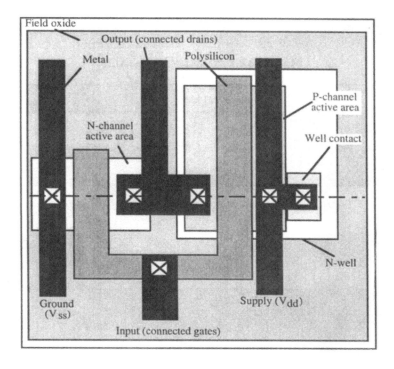

Figure 11.26: Plane view of a CMOS inverter.

11.10. NPN bipolar process

One characteristic feature of bipolar transistor processing is the use of epitaxy, although epitaxial growth can also be used in the fabrication of CMOS integrated circuits. To reduce the collector resistance a highly doped buried collector diffusion is made below the active area of the device. Epitaxy is then used to grow a lightly n-doped silicon film on top of the buried collector. A fabrication process that yields both NPN bipolar transistors and CMOS devices is called a BiCMOS process. If, in addition, it allows for the formation of PNP bipolar transistors, it is called a CBiCMOS process. A typical processing sequence used in the fabrication of NPN bipolar integrated circuits is shown in Figures 11.27a to 11.27o.

Figure 11.27a: Starting substrate.

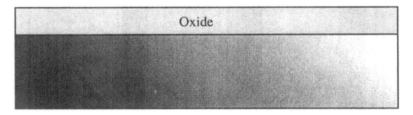

Figure 11.27b: Growth of a thick wet oxide.

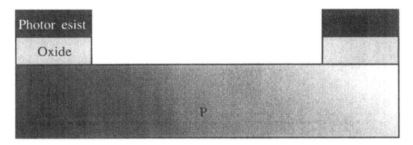

Figure 11.27c: Photolithography and oxide etch to define the buried collector.

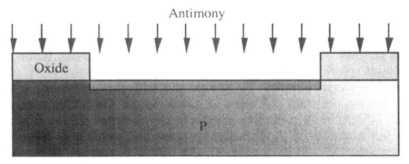

Figure 11.27d: Buried collector implantation (antimony).

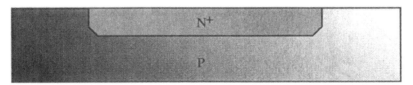

Figure 11.27e: Buried collector diffusion.

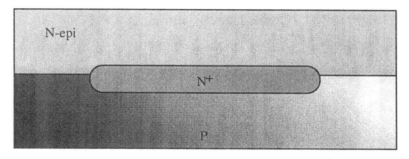

Figure 11.27f: Epitaxial growth of a lightly-doped N-type silicon layer.

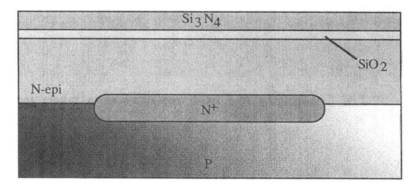

Figure 11.27g: Pad oxide growth and silicon nitride deposition.

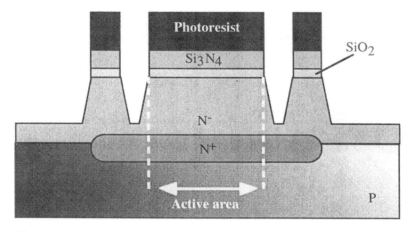

Figure 11.27h: Active area lithography, nitride etch, pad oxide etch and silicon etch.

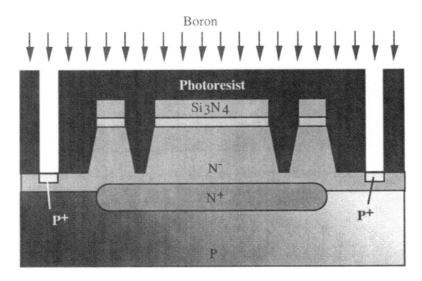

Figure 11.27i: Lithography and field implantation to create junction isolation areas between transistors.

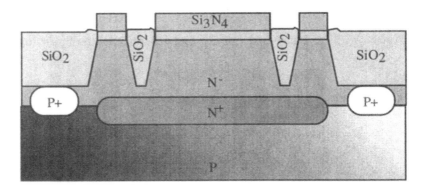

Figure 11.27j: LOCOS field oxide growth. All trenches are filled.

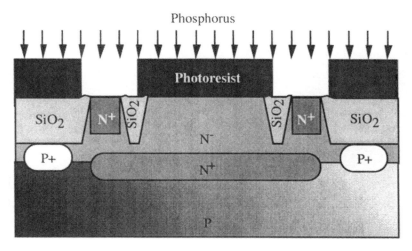

Figure 11.27k: Nitride strip and pad oxide strip. Lithography and implantation to form the low-resistivity (N^+) collector contacts.

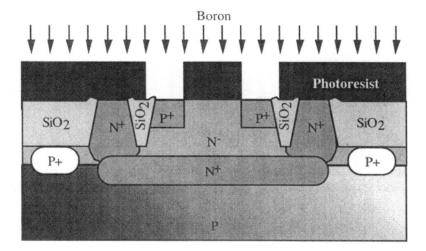

Figure 11.27l: Collector contact diffusion, lithography and base contact implantation (P^+ regions).

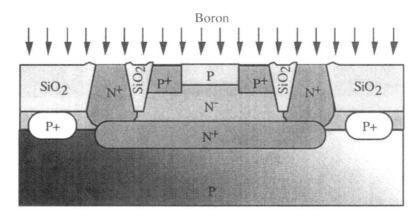

Figure 11.27m: Active base implantation (P-type).

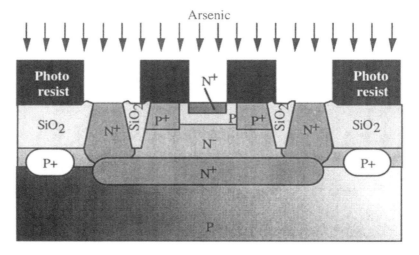

Figure 11.27n: Lithography to define the emitter region. N^+ collector contacts are also opened. Emitter implant and annealing.

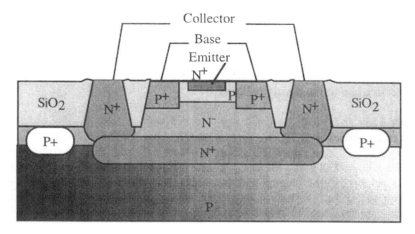

Figure 11.27o: Final device with no metallization process shown, *i.e.* contacts to collector, base and emitter.

Problems

 Problem 11.1:

A dose of boron of 5×10^{14} cm^{-2} is implanted into N-type silicon ($N_d = 10^{15}$ cm^{-3}) with an energy of 80 keV.

After implantation the sample is annealed for 30 minutes at 1000°C. The diffusion constant for boron at 1000°C is $D_{boron} = 1.53\times10^{-14}$ cm^{2}s^{-1}.

a)   Plot C(x) for $0 < x < 1$ µm before and after annealing. C(x) should be plotted on a log scale.
     - What is the maximum (peak) concentration of boron right after implantation? (unit: cm^{-3})
     - Determine the junction depth after implantation from the plot. (unit: micrometer)
     - What is the maximum (peak) concentration of boron after annealing? (unit=cm^{-3})
     - Determine the junction depth after annealing from the plot. (unit: micrometer)

b)   How much longer should the sample be annealed at 1000°C to obtain a junction depth of 0.8 µm?

 Problem 11.2:

An NPN bipolar transistor is fabricated using the following process steps:
The starting material is phosphorus-doped silicon; the phosphorus atom concentration is $N_d=10^{15}$ cm^{-3}. The collector contact is at the back of the silicon

wafer. Boron is implanted to form the base. The implantation energy and dose are 20 keV and 2×10^{12} cm^{-2}, respectively. To form the emitter arsenic is implanted at an energy of 30 keV and a dose of 10^{14} cm^{-2}. The wafer is then annealed for 2 hours at a temperature of 900°C. The diffusion coefficient of boron and arsenic at T=900°C are 2.5×10^{-15} and 1.18×10^{-16} cm^{2}s^{-1}, respectively.
What is the width of the transistor base?

Problem 11.3:
A wet oxide is grown at 900°C. The target thickness for the oxide is 500 nm. Not knowing the oxidation rate it is decided to proceed by trial and error and grow the oxide for 1 hour. The oxide thickness is then measured. It is equal to 135.7 nm. The wafer is put back in the furnace and the oxidation is continued for an extra 2 hours. The oxide thickness is then measured again and equals 348.7 nm. How much longer do we have to continue the oxidation to reach the target thickness (500 nm)?

 Problem 11.4:
The doping concentration in the channel region of an actual MOSFET is not constant as a function of depth. If we take the example of an n-channel MOSFET, the P-type substrate doping concentration is in the order of 10^{15} cm-3, and boron is implanted underneath the gate oxide to minimize short-channel effects and to adjust the value of the threshold voltage (this implantation step is called a "threshold implant"). In Problem 7.2 we have developed a technique to measure a uniform doping concentration using a MOS capacitor. Here we will use a similar technique to measure a non-uniform doping profile.

1) If the doping concentration in a MOS capacitor is $N_a(x)$, show that $N_a(x)$ can be obtained from a capacitance-voltage measurement $C(V_G)$, according to the following equation (valid when the capacitor is in depletion):

$$N_a(x) = \left(\frac{q \varepsilon_{Si}}{2} \frac{d}{dV_G} \left(\frac{1}{C^2} \right) \right)^{-1}$$

2) Using Matlab, simulate the fabrication of the capacitor and its C-V characteristics: A dose of boron of 5×10^{11} cm^{-2} is implanted into P-type silicon ($N_d = 10^{15}$ cm^{-3}) with an energy of 20 keV. After implantation the sample is annealed for 30 minutes at 1000°C. The diffusion constant for boron at 1000°C is $D_{boron} = 1.53 \times 10^{-14}$ cm^{2}s^{-1}. A 100-nm gate oxide is then deposited (assume no thermal step for the oxide deposition). Metal is evaporated on the structure and etched to produce a capacitor with an area of 1 cm^2. We will assume that the flat-band voltage, V_{FB}, is equal to 0 volt and the measurement is taken at room temperature (T=300K). Plot the capacitance of the MOS capacitor versus gate voltage for depletion depths ranging from 0 to 0.5 µm.

3) Using the equation obtained in Part 1 of this problem and the C-V data from Part 2, plot the doping concentration as a function of depth (0 µm < x < 0.5 µm). Compare the profiles.

Problem 11.5:

MOSFETs are fabricated in two P-type silicon wafers. The doping concentration in the wafers is uniform and equal to 5×10^{15} cm^{-3}. The gate material is degenerately doped N-type polysilicon, and the gate oxide thickness is 20 nm.

Wafer one is implanted with boron prior to gate oxide growth with a dose of 1.8×10^{12} cm^{-2} at an energy of 20 keV. Wafer two receives no implantation. The gate oxide growth is carried out at 1000°C for 30 minutes. The diffusion constant for boron at 1000°C is $D_{boron} = 1.53 \times 10^{-14}$ cm^{2}s^{-1}. Neglect any oxidation-enhanced diffusion effects.

Using a finite-difference numerical technique (see Problem 2.4), calculate the threshold voltage in the two types of transistors (wafer one, implanted and wafer two, non-implanted) and plot the doping concentration as a function of depth, as well as the depletion charge as a function of depth at threshold (C/cm^2) for both wafers. There are no charges in the oxide and no interface states. Linearization of the Poisson equation is recommended (see Problem 7.16).

References

1   D.V. Morgan and K. Board, *An introduction to semiconductor microtechnology*, John Wiley and Sons, p. 18, 1991

2   S.K. Ghandhi, *The theory and practice of microelectronics*, John Wiley and Sons, p. 38, 1968

3   J. Lindhard, M. Scharff, and H.E. Schiøtt, "Range Concepts and Heavy Ion Ranges", *Matematisk-fysiske Meddelelser Det Kongelige Danske Videnskabernes Selskab*, Vol. 33, No. 14, p.1, 1963

4   R.S. Muller and T.I. Kamins, *Device electronics for integrated circuits*, J. Wiley and Sons, pp. 81-83, 1986

5   J.P. Colinge, *Silicon-on-Insulator Technology: Materials to VLSI*, 2nd Edition, Kluwer Academic Publishers, 1997

6   SUPREM-IV, Avant! Corporation, Fremont, CA, USA

7   A.S. Grove, *Physics and technology of semiconductor devices*, J. Wiley & Sons, p. 24, 1967

8   B.E. Deal and A.S. Grove, "General relationships for the thermal oxidation of silicon", *Journal of Applied Physics*, Vol. 36, p. 3770, 1965

9   E.A. Irene, H.Z. Massoud, E. Tierney, "Silicon oxidation studies: silicon orientation effects on thermal oxidation", *Journal of the Electrochemical Society*, Vol. 133, No. 6, p. 1253, 1986

10 S. Marshall, R.J. Zeto, C.G. Thornton, "Dry pressure local oxidation of silicon for IC isolation", *Journal of the Electrochemical Society*, Vol. 122, No. 19, p. 1411, 1975

11 M. Miyake, "Oxidation-enhanced diffusion of ion-implanted boron in silicon in extrinsic conditions", *Journal of Applied Physics*, Vol. 57, No. 6, p. 1861, 1985

12 A.S. Grove, *Physics and technology of semiconductor devices*, J. Wiley & Sons, p. 70, 1967

13 SUPREM-IV, Avant! Corporation, Fremont, CA, USA

14 SUPREM-IV, Avant! Corporation, Fremont, CA, USA

15 E. Kooi and J.A. Appels, "Selective oxidation of silicon and its device applications", *Journal of the Electrochemical Society*, Vol. 120, No. 3, Abstracts of the Electrochemical Society Meeting, Chicago, IL, USA, p. 101C, 1973

16 B. Davarik, C.W. Koburger, R. Schulz, J.D. Warnock, T. Furukawa, M. Jost, Y. Taur, W.G. Schwittek, J.K. DeBrosse, M.L. Kerbaugh, and J.L. Mauer, "A new planarization technique, using a combination of RIE and chemical mechanical polish (CMP)", *Technical Digest of the International Electron Devices Meeting* , p. 61, 1989

17 T. Kamins, *Polycrystalline silicon for integrated circuit applications*, Kluwer Academic Publishers, 1988

18 A.S. Grove, *Physics and technology of semiconductor devices*, J. Wiley & Sons, p. 12, 1967

19 H. Taub and D. Schilling, *Digital integrated electronics*, Mc Graw-Hill, p. 263, 1977

ANNEX

A1. Physical Quantities and Units

QUANTITY	USUAL SYMBOL	UNIT
Capacitance	C	Farad (F)
Charge	Q	Coulomb (C)
Conductance	G	Siemens (S) = Ω^{-1}
Conductivity	σ	S cm^{-1} = Ω^{-1} cm^{-1}
Current	I	Ampere (A)
Current density	J	A cm^{-2}
Distance	d, l, w, x, y, z	Centimeter (cm) 1 micrometer (μm) = 10^{-4} cm 1 nanometer (nm) = 10^{-7} cm 1 angström (Å) = 10^{-8} cm
Electric field	$\mathcal{E} = -\nabla\Phi$ $\mathcal{E} = -d\Phi/dx$ *(1 dimension)*	V cm^{-1}
Energy	E	Joule (J) Electron-volt (1 eV = 1.6×10^{-19} J)
Frequency	f	s^{-1}
Potential	V, Φ	Volt (V)
Resistance	R, r	Ohm (Ω)
Resistivity	ρ	Ω cm
Temperature	T	Kelvin (K) 0°C = 273.15 K
Time	t	Second (s)

A2. Physical Constants

SYMBOL	MEANING	VALUE	UNIT
E_g (GaAs)	GaAs bandgap energy	1.42	eV
E_g (Ge)	Ge bandgap energy	0.67	eV
E_g (Si)	Si bandgap energy	1.124	eV
ε_o	Permittivity of vacuum	8.854×10^{-14}	F cm^{-1}
h	Planck constant	6.63×10^{-34}	J s
$\hbar$	Reduced Planck constant	$h/2\pi$	J s
k	Boltzmann constant	1.3805×10^{-23}	J K^{-1}
κ (GaAs)	Dielectric constant of GaAs	13.1	dimensionless
κ (Ge)	Dielectric constant of Ge	16	dimensionless
κ (Si)	Dielectric constant of Si	11.7	dimensionless
κ (SiO$_2$)	Dielectric constant of SiO$_2$	3.9	dimensionless
kT/q	Thermal voltage (at T=300K)	0.02586	V
L (GaAs)	Lattice parameter (GaAs)	5.6533×10^{-8}	cm
L (Ge)	Lattice parameter (Ge)	5.64613×10^{-8}	cm
L (Si)	Lattice parameter (Si)	5.43095×10^{-8}	cm
μ_n (GaAs)	Electron mobility (intrinsic GaAs)	8800	cm^2 V^{-1} s^{-1}
μ_n (Ge)	Electron mobility (intrinsic Ge)	3900	cm^2 V^{-1} s^{-1}
μ_n (Si)	Electron mobility (intrinsic Si)	1417	cm^2 V^{-1} s^{-1}
m_o	Free electron mass	9.11×10^{-31}	kg
μ_p (GaAs)	Hole mobility (intrinsic GaAs)	400	cm^2 V^{-1} s^{-1}
μ_p (Ge)	Hole mobility (intrinsic Ge)	1900	cm^2 V^{-1} s^{-1}
μ_p (Si)	Hole mobility (intrinsic Si)	471	cm^2 V^{-1} s^{-1}
N_c (GaAs)	Effective density of states in cond. band (GaAs)	4.7×10^{17}	cm^{-3}
N_c (Ge)	Effective density of states in cond. band (Ge)	1.04×10^{19}	cm^{-3}
N_c (Si)	Effective density of states in cond. band (Si)	2.8×10^{19}	cm^{-3}
N_v (GaAs)	Effective density of states in valence band (GaAs)	7×10^{18}	cm^{-3}
N_v (Ge)	Effective density of states in valence band (Ge)	6×10^{18}	cm^{-3}
N_v (Si)	Effective density of states in valence band (Si)	1.04×10^{19}	cm^{-3}
n_i (GaAs)	Intrinsic carrier concentration (GaAs)	1.1×10^7	cm^{-3}
n_i (Ge)	Intrinsic carrier concentration (Ge)	2.5×10^{12}	cm^{-3}
n_i (Si)	Intrinsic carrier concentration (Si)	1.45×10^{10}	cm^{-3}
q	Electron charge (absolute value)	1.6×10^{-19}	C

All values are given for T = 300K.

A3. Concepts of Quantum Mechanics

In this Annex the Reader is reminded of some concepts from quantum mechanics that will be used in this book.

1) A particle can be fully described by a function, called wave function. The wave function is noted $\Psi(x,y,z,t)$ and it contains all measurable information about the particle.

2) To each dynamic variable corresponds a quantum-mechanic operator:

   ◊ To the position x corresponds the operator $\hat{x} \equiv x$ (A3.1)

   ◊ To momentum p_x corresponds the operator $p_x \equiv \dfrac{\hbar}{j} \dfrac{\partial}{\partial x}$ (A3.2)

   ◊ To the total energy E corresponds the operator $\hat{E} \equiv -\dfrac{\hbar}{j} \dfrac{\partial}{\partial t}$ (A3.3)

   ◊ To the potential energy $V(x,y,z)$ corresponds the operator
   $$\hat{V} \equiv V(x,y,z)$$ (A3.4)

   where $j = \sqrt{-1}$ and where $\hbar = h/2\pi$, h being Planck's constant.

3) The wave function also gives the probability of finding the particle in a given region of space. If the wave function is real (*i.e.*, not complex) the probability of finding the particle between positions a and b in one dimension (x) is given by:

   $$probability = \int_a^b \Psi^* \Psi \, dx \quad (= \int_a^b \Psi^2 \, dx \text{ if } \Psi \text{ is a real function})$$

   For all space in one dimension the particle must be *somewhere* between $x = -\infty$ and $x = +\infty$ and therefore, we obtain the normalization condition:

   $$\int_{-\infty}^{+\infty} \Psi^* \Psi \, dx = 1 \quad (\int_{-\infty}^{+\infty} \Psi^2 \, dx = 1 \text{ if } \Psi \text{ is a real function}) \quad (A3.5)$$

Consider the total energy of a particle in a classical Newtonian physics approach. If the particle has a momentum p and a potential energy V, its total energy is given by:

$$E = \frac{p^2}{2m} + V \qquad (A3.6)$$

Note that $p=p(x,y,z)$, $p^2 = p_x^2 + p_y^2 + p_z^2$ and $V=V(x,y,z)$

Applying these concepts to an electron having a mass m for the one-dimensional case one obtains Table A.1:

Table A.1: Physical variables and operators.

Quantity	Classical mechanics	Quantum mechanics
Momentum	$p = mv$	$\dfrac{\hbar}{j}\dfrac{d}{dx}$
Kinetic energy	$\dfrac{p^2}{2m}$	$\dfrac{1}{2m}\dfrac{\hbar}{j}\dfrac{d}{dx}\left(\dfrac{\hbar}{j}\dfrac{d}{dx}\right) = -\dfrac{\hbar^2}{2m}\dfrac{d^2}{dx^2}$
Potential energy	V	$\hat{V}$
Total energy	$E = \dfrac{p^2}{2m} + V$	$-\dfrac{\hbar}{j}\dfrac{\partial}{\partial t}$
Mass	$m = \dfrac{1}{d^2 E/dp^2}$	$m = \dfrac{\hbar^2}{d^2 E/dk^2}$
Velocity, group velocity	$v = \dfrac{dE}{dp}$	$v_k = \dfrac{1}{\hbar}\dfrac{dE}{dk}$

In this Table, k is a wave vector or a wave number that corresponds to the momentum of the particle.

The Schrödinger equation is basically the quantum mechanical equivalent of classical mechanics $E = \dfrac{p^2}{2m} + V$. For the one-dimensional case the quantum mechanical equivalent of total energy is:

$$-\frac{\hbar^2}{2m}\frac{\partial^2\Psi}{\partial x^2} + V(x,t)\Psi = -\frac{\hbar}{j}\frac{\partial\Psi}{\partial t} \qquad (A3.7)$$

and, in three dimensions:

$$-\frac{\hbar^2}{2m}\nabla^2\Psi + V(x,y,z,t)\Psi = -\frac{\hbar}{j}\frac{\partial\Psi}{\partial t} \qquad (A3.8)$$

where ∇^2 is the Laplacian operator defined by:

$$\nabla^2\Psi(x,y,z,t) \equiv \frac{\partial^2\Psi}{\partial x^2} + \frac{\partial^2\Psi}{\partial y^2} + \frac{\partial^2\Psi}{\partial z^2}$$

If the potential energy function is time independent ($\partial V/\partial t = 0$) one is able to construct a solution to the Schr dinger equation through the technique of separation of variables where the wave function is written as the product of a time-independent term, $\psi(x,y,z)$ and a space-independent term, $T(t)$, such that $\Psi(x,y,z,t) = \psi(x,y,z)\,T(t)$. The introduction of these terms into (A3.8) yields:

$$T(t)\left(-\frac{h^2}{2m}\cdot\nabla^2\psi(x,y,z)\right) + V(x,y,z)\,\psi(x,y,z)\,T(t)$$

$$= \psi(x,y,z)\left(-\frac{h}{j}\cdot\frac{\partial T(t)}{\partial t}\right)$$

or

$$\frac{1}{\psi(x,y,z)}\left(-\frac{h^2}{2m}\cdot\nabla^2\psi(x,y,z)\cdot+{}^\circ V(x,y,z)\cdot\psi(x,y,z)\right) = \frac{1}{T(t)}\left(-\frac{h}{j}\cdot\frac{\partial T(t)}{\partial t}\right) \qquad (A3.9)$$

The left-hand term of this equation depends only on space, while the right-hand term depends only on time, which indicates that the separation of Ψ into the product of ψ and T was successful. We can now solve the Schr dinger equation for the variables ψ and T separately, and with this solution find $\Psi = \psi T$. Equation A3.9 makes sense only if both terms are equal to a constant which we shall call E, therefore, we can write:

$$E\,T(t) = -\frac{h}{j}\cdot\frac{\partial T(t)}{\partial t} \quad\Rightarrow\quad T(t) = exp\left(\frac{-jEt}{h}\right) \qquad (A3.10)$$

and therefore:

$$\Psi(x,y,z,t) = \psi(x,y,z)\,exp\left(\frac{-jEt}{h}\right) \qquad (A3.11)$$

Introducing Expression A3.11 into A3.8 one obtains the time-independent Schr dinger equation:

Time-independent Schr dinger equation

$$\boxed{-\frac{h^2}{2m}\,\nabla^2\psi(x,y,z) + [V(x,y,z) - E]\,\psi(x,y,z) = 0} \qquad (A3.12)$$

where E is the (constant) energy of the particle, where the energy of the particle is given by:

$$-\frac{h}{j}\cdot\frac{\partial\Psi(x,y,z,t)}{\partial t} = \psi(x,y,z)\left(-\frac{h}{j}\cdot\frac{\partial T(t)}{\partial t}\right) = \psi(x,y,z)\,E\,T(t) = E\,\Psi(x,y,z,t)$$

A4. Crystallography – Reciprocal Space

Most semiconductors are crystalline materials. Elemental semiconductor atoms such as silicon or germanium belong to column IV of the periodic table and have four electrons on their outer shell. In a crystal these atoms form four covalent bonds with neighboring atoms in order to complete their outer shell. Each atom is thus in the center of a tetrahedron, the corners of which are occupied by other similar atoms (Figure A.1).

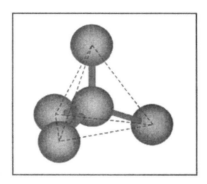

Figure A.1: Silicon atom forming covalent bonds to other silicon atoms.

The atoms in a crystal form a pattern that is repeated in the three directions of space with perfect regularity. That pattern is called the "unit cell". Silicon and germanium have the diamond lattice structure. This structure can be viewed as two interweaving face-centered lattices. In this case the unit cell is a cube (Figure A.2). The length of each cube side is called the "lattice parameter", which is equal to 5.43 and 5.64 Å in silicon and germanium, respectively.

In the unit cell presented in Figure A.2 atoms labeled "1" are completely enclosed in the unit cell. Atoms at the center of each of the six sides of the cell and labeled "1/2" belong half to the unit cell and half to an adjacent cell. Atoms located at the corners of the cube and labeled "1/8" have one-eighth of their volume included in the unit cell and contribute to seven other cells. Therefore, the unit cell contains $4 \times 1 + 6 \times 1/2 + 8 \times 1/8 = 8$ atoms. Semiconductors formed using elements from columns III and V of the periodic table, such as gallium arsenide (GaAs), have the zincblende crystal structure. The GaAs lattice cell can be viewed as two interpenetrating face-centered lattices, one containing gallium atoms, and the other containing arsenic atoms. It is also represented by Figure A.2 where atoms labeled "1" are gallium and atoms labeled "1/2" and "1/8" are arsenic (and vice-versa). The lattice parameter of GaAs is 5.65 Å.

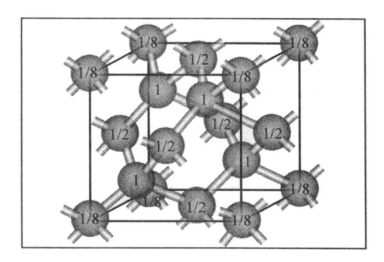

Figure A.2: Atoms in the unit cell of silicon (diamond lattice structure).

The most basic property of a crystal is that the same pattern of atoms is repeated over and over again in the three directions of space. The position of any cell in the crystal is given by a vector **l** defined by:

$$\mathbf{l} = m\mathbf{a} + n\mathbf{b} + p\mathbf{c} \qquad (A4.1)$$

where m, n and p are integer numbers, and **a**, **b** and **c** are the vectors of the lattice parameters of the unit crystal cell (Figure A.3). In most semiconductors the cell is cubic and **a**, **b** and **c** have the same length.

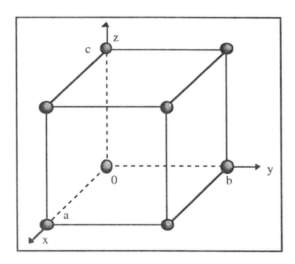

Figure A.3: Unit cell of a cubic crystal lattice.

One can define three new vectors:

$$a^* = 2\pi \frac{b \times c}{a \cdot b \times c} \quad , \ b^* = 2\pi \frac{c \times a}{a \cdot b \times c} \quad , \ c^* = 2\pi \frac{a \times b}{a \cdot b \times c} \qquad \text{(A4.2)}$$

Vectors a^*, b^* and c^* belong to what is called the "reciprocal lattice". While vectors a, b and c belong to real space and are measured in meters or centimeters, vectors a^*, b^* and c^* belong to a space where the measurement unit is meter^{-1} or centimeter^{-1}, which is called the "reciprocal space". Note that $a \cdot a^* = b \cdot b^* = c \cdot c^* = 2\pi$ and $a \cdot b^* = a \cdot c^* = b \cdot c^* = 0$; a^* is thus parallel to a and perpendicular to b and c, if there is such a thing as being parallel or perpendicular to a vector belonging to another space.

Figure A.4 represents vectors a^*, b^* and c^*. They are perpendicular to crystal planes (100), (010) and (001), respectively. Vectors perpendicular to planes (110) and (111) are represented as well. Any vector k in the reciprocal space obeys the following equation:

$$k = fa^* + gb^* + hc^* \qquad \text{(A4.3)}$$

where f, g and h are integer numbers.

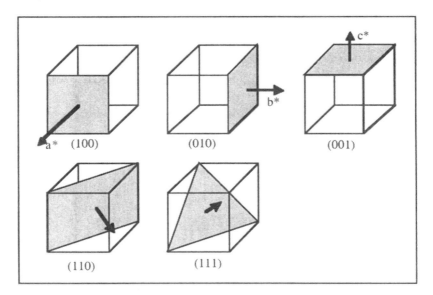

Figure A.4: Main crystal planes of a semiconductor having a cubic lattice. Vectors a*, b* and c* belong to the reciprocal space and are represented here in the real-space unit cells for a visualization purpose only.

Problems

Problem A4.1:
a: Calculate the number of atoms in a cubic centimeter of silicon and germanium.

b: Calculate the number of atoms per square centimeter at the surface of an (100)-oriented silicon sample.

 Problem A4.2:
Using Matlab place silicon atoms in the silicon unit cell in order to produce a 3D plot similar to Figure A.2. View it from different directions: random, (100), (110) and (111). The lattice parameter is 5.43 Å. Use commands [sx,sy,sz]=sphere(20) and surf1(sx,sy,sz) to draw the atoms. Use command line([X1 X2],[Y1 Y2],[Z1 Z2]) to plot the bonds between the atoms.

 Problem A4.3:
Using Matlab place silicon atoms in 3×3×3=27 silicon unit cells in order to produce a 3D plot of the lattice. View it from different directions: random, (100), (110) and (111). The lattice parameter is 5.43 Å. Use commands [sx,sy,sz]=sphere(20) and surf1(sx,sy,sz) to draw the atoms. Use command line([X1 X2],[Y1 Y2],[Z1 Z2]) to plot the bonds between the atoms.

A5. Getting Started with Matlab

◊ Matlab contains a powerful and user-friendly HELP function. For example:

```
help help
help graphics
help *    or    help +
```

will display a general help message, help on graphic functions, and help on operations such as multiplication and addition, respectively.

◊ Matlab is based on matrix operations. The following commands:

```
1 |   a = 1
2 |   b = a + a
```

will of course produce b=2 as a result, but internally both *a* and *b* are treated as 1×1 matrices, such that a = [1] and b = [2].

◊ Characters preceded by a percent sign (%) are treated as comments. Here is an example of commands:

```
1 | clear            % Clears all variables
2 | A=[1 2;3 4]      % Build a 2x2 matrix
3 | B=A/A            % Divide the A by itself
4 | C=A*A            % Multiply A itself
5 | D=A .*A          % Multiply the elements of A by themselves
6 | E=A ./A          % Divide the elements of A by themselves
7 | a=1:2:12         % Generate a vector
8 | b=a'             % Transpose it
```

The resulting matrices and vectors are:

$$A = \begin{bmatrix} 1 & 2 \\ 3 & 4 \end{bmatrix} \quad B = \begin{bmatrix} 1 & 0 \\ 0 & 1 \end{bmatrix} \quad C = \begin{bmatrix} 7 & 10 \\ 15 & 22 \end{bmatrix} \quad D = \begin{bmatrix} 1 & 4 \\ 9 & 16 \end{bmatrix} \quad E = \begin{bmatrix} 1 & 1 \\ 1 & 1 \end{bmatrix}$$

$$a = [1\ 3\ 5\ 7\ 9\ 11] \quad b = \begin{bmatrix} 1 \\ 3 \\ 5 \\ 7 \\ 9 \\ 11 \end{bmatrix}$$

Note the important difference between "*" and ".*" or "/" and "./" !

◊ Using Matlab graphic results can be produced very easily. Here are some examples:

Plot sin(x) and cos(x)

```
1   clear %Clear all variables
2   X=0:0.1:2*pi; % x varies from 0 to 2π in steps of 0.1
3   SINE=sin(X);COSINE=cos(X);
4   plot(X,SINE,'-r',X,COSINE,'--b');
5   title('Sine and Cosine functions')
```

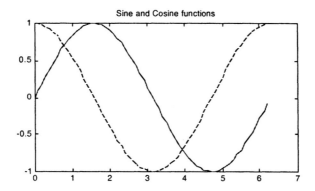

Note that x, sin(x) and cos(x) are vectors. There is no need for FOR or DO loops!

Plot a spiral

```
1   clear;clf % Clear all variables; clear figure
2   R=0:0.1:5*pi;  % R varies from 0 to 2π in steps of 0.1
3   SINE=sin(R);COSINE=cos(R);
4   plot(SINE .*R,COSINE .*R, '-b')
5   axis square
6   title('Spiral')
```

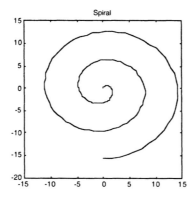

Plot a two-dimensional "Mexican hat"

```
1    clear;clf;  % Clear all variables; clear figure
2    t=50; % number of mesh points in each direction
3    A=zeros(t); % build a 50x50 matrix array
4    for i=1:t;
5         for j=1:t;
6       % Distance from center of matrix
7             r=sqrt(((i-t/2)/2)^2+((j-t/2)/2)^2);
8       A(i,j)=sin(r)/r;
9         end
10   end
11   A(t/2,t/2)=1; %center point of matrix is equal to 1
12   surfl(A) % Plot the 2D graph
13   shading interp;
14   colormap(pink);
15   title ('"Mexican hat function"')
```

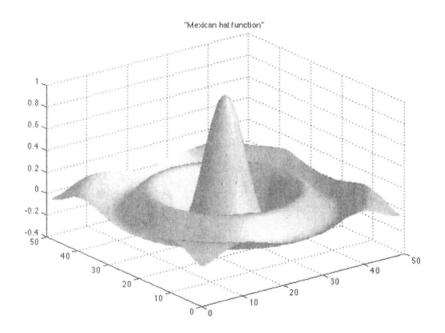

◊   Matlab can be used to conveniently solve many matrix problems. Here is a simple example. Consider the circuit below. We need to find the value of currents I_1 and I_2, as well as voltage V_1.

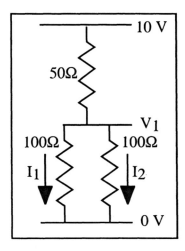

Using Kirchoff's voltage law we can write:

$$100\Omega\, I_1 + 50\Omega\, (I_1 + I_2) = 10 \text{ V}$$
$$100\Omega\, I_2 + 50\Omega\, (I_1 + I_2) = 10 \text{ V}$$
$$100\Omega\, I_2 - V_1 = 0 \text{ V}$$

or, in a matrix form:

$$\begin{bmatrix} 150 & 50 & 0 \\ 50 & 150 & 0 \\ 0 & 100 & -1 \end{bmatrix} \begin{bmatrix} I_1 \\ I_2 \\ V_1 \end{bmatrix} = \begin{bmatrix} 10 \\ 10 \\ 0 \end{bmatrix}$$

Using this simple program:

```
1    clear
2    A=[150 50 0;50 150 0;0 100 -1];
3    B=[10 10 0]';
4    IV=A\B
```

The solution is IV = $\begin{bmatrix} I_1 \\ I_2 \\ V_1 \end{bmatrix} = \begin{bmatrix} 0.5 \\ 0.5 \\ 5.0 \end{bmatrix}$ from which we infer $I_1 = I_2 = 500$ mA and $V_1 = 5$ V.

◊   Here are some Matlab functions that can be useful to solve some Problems from this Book:

Concatenation and iterative equation solving:

If $A = \begin{bmatrix} 1 \\ 2 \\ 3 \end{bmatrix}$ then writing B = [A A A] yields:

$$B = \begin{bmatrix} 1 & 1 & 1 \\ 2 & 2 & 2 \\ 3 & 3 & 3 \end{bmatrix}$$

The following example solves the equation x=cos(x) iteratively and uses concatenation to plot the values of *x* at each iteration:

```
1    clear
2    test=1;x=0;graph=[];
3    while test>1e-4
4      x2=cos(x);
5      test=abs(x2-x);
6      graph=[graph x];
7      x=x2;
8    end
9    ('the solution is')
10   x
11   plot(graph)
12   xlabel('Iteration number');ylabel('X value');
```

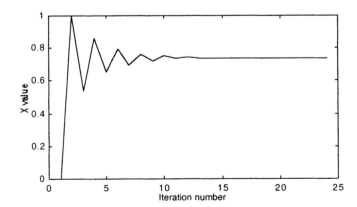

Relaxation factor:

If one tries to solve x= 2cos(x) using the iterative method described above, convergence will not be reached. Convergence can be improved by introducing a relaxation factor, α, used during each evaluation of a new x value. The value of α ranges between 0 and 1.

Instead of writing   `x2=cos(x)`
one can write   `x2= x*(alpha-1) + alpha*cos(x)`

such that *x2* is some average value between the old *x* value and the newly calculated value for *x*.

The program below uses the values 0.2, 0.4, 0.6 and 0.8 for α. Convergence is obtained for the lower α values, but not for α=0.8. Not using a relaxation factor is equivalent to writing α=1, for which there is no convergence.

```
1    clear;clf
2    graph2=[]
3    for alpha=0.2:0.2:.8
4       x=0;graph1=[];x=0;
5       for counter=1:12
6          x2=2*cos(x);
7          test=abs(x2-x);
8          graph1=[graph1 x];
9          x=x*(1-alpha)+alpha*x2;
10      end
11      graph2=[graph2 graph1'];
12   end
13   plot(graph2,'-k')
14   xlabel('Iteration number');ylabel('X value');
```

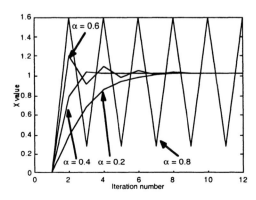

Diagonal matrices: The following program

```
1│   clear
2│   t=6;
3│   A=diag(ones(1,t),0)
4│   B=diag(ones(1,t-1),1)
5│   C=diag(ones(1,t-1),-1)
6│   A=-2*A+B+C
7│   A(1,1)=1;A(1,2)=0;A(t,t)=1;A(t,t-1)=0
```

yields:

$$
A = \begin{bmatrix}
1 & 0 & 0 & 0 & 0 & 0 \\
1 & -2 & 1 & 0 & 0 & 0 \\
0 & 1 & -2 & 1 & 0 & 0 \\
0 & 0 & 1 & -2 & 1 & 0 \\
0 & 0 & 0 & 1 & -2 & 1 \\
0 & 0 & 0 & 0 & 0 & 1
\end{bmatrix}
$$

A similar matrix is used in problems based on a numerical (finite-differences) simulation technique.

Numerical integration and differentiation:

The following program integrates and differentiates $y = x^2$:

```
1 │dx=0.01;
2 │x=-5:dx:5;
3 │y=x.^2;
4 │integral=sum(y)*dx %Definite integral (from x=-5 to x=5)
5 │integral_curve=cumsum(y)*dx; % Integral curve
  │% derivative=diff(y)./diff(x);
  │% Since the differentiation of an n-element
  │% vector produces an (n-1)-element vector we add
  │% a dummy "Not a Number"(NaN) at the end of the
  │% derivative vector, such that it has the same
  │% length as the x-vector:
6 │derivative=[derivative NaN];
7 │plot(x,y,'-b',x,integral_curve,'--r',x,derivative,'--k')
9 │text(-4,80,'BLUE:   y=x^2')
10│text(-4,70,'RED:    integral of y')
11│text(-4,60,'BLACK: dy(x)/dx')
```

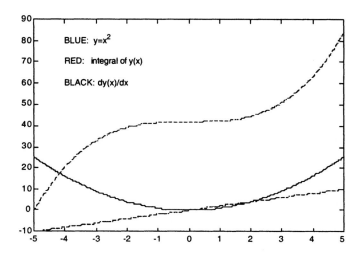

Note 1: On some computers some versions of Matlab may give you frustrating problems if you use uppercase letters in file names. So, it is good practice to use file names such as "test.m" instead of "Test.m", for example. The Problems in this Book were designed using the Student Edition of Matlab, version 5.0 for Macintosh, and version 5.3 for PC.

Note 2: Some people may find the font size in Matlab plots too small for easy reading. Plot properties such as font size and line width can be modified using the following commands:

```
set(0,'defaultaxesfontsize',14) sets the axes font size to 14
set(0,'defaulttextfontsize',14) sets the text font size to 14
set(0,'defaultlinelinewidth',2) sets the plot linewidth to 2
set(0,'defaultaxeslinewidth',2) sets the axes linewidth to 2
set(0,'defaultaxesfontname','Arial') sets the axes font
    name to Arial
set(0,'defaulttextfontname','Arial') sets the text font name
    to Arial
```

A6. Greek alphabet

LETTER	LOWERCASE	UPPERCASE
Alpha	α	A
Beta	β	B
Gamma	γ	Γ
Delta	δ	Δ
Epsilon	ε	E
Zeta	ζ	Z
Eta	η	H
Theta	θ	Θ
Iota	ι	I
Kappa	κ	K
Lambda	λ	Λ
Mu	μ	M
Nu	ν	N
Xi	ξ	Ξ
Omicron	o	O
Pi	π	Π
Rho	ρ	P
Sigma	σ	Σ
Tau	τ	T
Upsilon	υ	Y
Phi	ϕ	Φ
Chi	χ	X
Psi	ψ	Ψ
Omega	ω	Ω

A7. Basic Differential Equations

In the examples below, A and B are given constants, and C_n (n=0,1,2,3,4) are integration constants. Integration constants can be numerically determined by applying boundary conditions to the general solution of the equation.

◊ To solve:
$$\frac{dF(x)}{dx} + Ax + B = 0$$

using separation of variables we write:
$$dF(x) = - (Ax + B) \, dx \quad \Rightarrow \quad \int dF(x) = - \int (Ax + B) \, dx$$

which yields the general solution: $F(x) = -\frac{A}{2} x^2 - Bx + C_1$

◊ To solve:
$$\frac{dF(x)}{dx} + A \, F(x) + B = 0$$

we write:
$$dF(x) = - (A \, F(x) + B) \, dx$$

or:
$$\frac{dF(x)}{AF(x) + B} = - dx \quad \Rightarrow \quad \frac{A \, dF(x)}{AF(x) + B} = - A \, dx$$

Noting that $d(AF(x) + B) = A \, dF(x)$ and using a change of variables where $AF(x) + B = y$ we can write:
$$\frac{d(y)}{y} = - A \, dx \quad \Rightarrow \quad \int \frac{d(y)}{y} = - A \int dx$$

The integration results in: $ln(y) = ln(AF(x) + B) = -Ax + C_0$
Therefore, the general solution is:
$$F(x) = \frac{exp(-Ax + C_0) - B}{A}$$

or, noting $C_1 = \frac{exp(C_0)}{A}$:
$$F(x) = C_1 \, exp(-Ax) - \frac{B}{A}$$

◊ To solve:
$$\frac{d^2 F(x)}{dx^2} = A$$

we integrate a first time to find: $\frac{dF(x)}{dx} = Ax + C_1$

and then integrate a second time to obtain the general solution:

$$F(x) = \frac{A}{2} x^2 + C_1 x + C_2$$

◊ To solve:
$$\frac{d^2 F(x)}{dx^2} = A\, F(x) \quad \text{with } A > 0$$

we must find a function that is equal to its second derivative, multiplied by a positive constant. The only function satisfying this condition is the exponential function, since:

$$\frac{d^2 (C_1 \exp(Bx))}{dx^2} = \frac{d}{dx} \frac{d(C_1 \exp(Bx))}{dx}$$

$$= C_1\, B \frac{d(\exp(Bx))}{dx} = C_1\, B^2\, \exp(Bx)$$

and

$$\frac{d^2 (C_2 \exp(-Bx))}{dx^2} = \frac{d}{dx} \frac{d(C_2 \exp(-Bx))}{dx}$$

$$= - C_2\, B \frac{d(\exp(-Bx))}{dx} = C_2\, B^2\, \exp(-Bx)$$

Comparing the initial differential equation and the possible solutions, we find that $A = B^2$. Therefore, the general solution is:

$$F(x) = C_1 \exp(\sqrt{A}\ x) + C_2 \exp(-\sqrt{A}\ x)$$

Since $sinh(y) = \dfrac{\exp(y) - \exp(-y)}{2}$ and $cosh(y) = \dfrac{\exp(y) + \exp(-y)}{2}$ we can also write:

$$F(x) = C_3 \sinh(\sqrt{A}\ x) + C_4 \cosh(\sqrt{A}\ x)$$

◊ To solve: $$\frac{d^2F(x)}{dx^2} = -A\ F(x) \quad with\ A > 0$$

we must find a function that is equal to its second derivative, multiplied by a negative constant. The only functions satisfying this condition are the sine and cosine functions since:

$$\frac{d^2(C_1\ sin(Bx))}{dx^2} = \frac{d}{dx}\frac{d(C_1\ sin(Bx))}{dx}$$

$$= C_1\ B\frac{d(cos(Bx))}{dx} = -\ C_1\ B^2\ sin(Bx)$$

and

$$\frac{d^2(C_2\ cos(Bx))}{dx^2} = \frac{d}{dx}\frac{d(C_2\ cos(Bx))}{dx}$$

$$= -\ C_2\ B\frac{d(sin(Cx))}{dx} = -\ C_2\ B^2\ cos(Cx)$$

Comparing the initial differential equation and the possible solutions, we find that $A = B^2$. Therefore, the general solution is:

$$F(x) = C_1\ sin(\sqrt{A}\ x) + C_2\ cos(\sqrt{A}\ x)$$

Using $cos(y) = \dfrac{e^{jy} + e^{-jy}}{2}$ and $sin(y) = \dfrac{e^{jy} - e^{-jy}}{2j}$ we can write:

$$F(x) = \frac{C_1}{2}\left(exp(j\sqrt{A}x) + exp(-j\sqrt{A}x)\right) - \frac{jC_2}{2}\left(exp(j\sqrt{A}x) - exp(-j\sqrt{A}x)\right)$$

$$= \left(\frac{C_1}{2} - \frac{jC_2}{2}\right)exp(j\sqrt{A}x) + \left(\frac{C_1}{2} + \frac{jC_2}{2}\right)exp(-j\sqrt{A}x)$$

or:

$$F(x) = C_3\ exp(j\sqrt{A}\ x) + C_4\ exp(-j\sqrt{A}\ x)$$

INDEX

-A-

absorption coefficient 77
acceptor atom 31, 33
acceptor level 77
accumulation 170
accumulation layer 171
activation energy 383
amorphous silicon 382
anisotropy 389
Auger recombination 78
avalanche 298
avalanche multiplication 117, 298, 299

-B-

ballistic electron 348
band curvature 140
band discontinuity 317
band-to-band recombination 74
band-to-band tunneling 335
bandgap 15, 38, 39, 63, 325
bandgap engineering 316
base 252
BiCMOS 399
bipolar transistor 251
bird's beak 380
BJT 251
Bloch theorem 9
body effect 194
body factor 194, 196, 205, 229
Boltzmann relationships 42, 65
Born-von Karman boundary conditions 5, 338
breakdown voltage 117
Brillouin zone 22, 25
built-in potential 97
buried collector 257, 399
buried oxide 228

-C-

capture cross section 82
carrier freeze-out 36, 48
carrier lifetime 80, 85
CBiCMOS 399
channel 154, 160, 168
channeling 369
charge sheet 140
charge storage 123
CMP 381, 391
collector 252
common-base gain 255, 264, 265, 270
common-emitter gain 256, 270
conduction band 15, 27, 74
conductivity 56
continuity equations 64, 65, 68
Coulomb blockade 355, 357, 359
Coulomb gap 357
Coulomb oscillations 353
critical field 200
current gain 256
current mirror 311
cutoff frequency 148
CVD 381
cyclic boundary conditions 5
Czochralski growth 364

-D-

damascene process 391
Deal-Grove model 376
Debye length 172
deep depletion 181
deep level 33
degenerate semiconductor 40
density of states 25, 336, 344, 346
depletion approximation 99, 142, 176, 318

transport factor in the base 269, 281
transport model 274
trichlorosilane 364
triode 191
triode regime 191
tunnel diode 331
tunnel effect 117, 331, 333
tunnel junction 353
Two-Dimensional Electron Gas (2DEG) 323

-V-

vacancies 51
valence band 15, 74
velocity saturation 200
VLSI 391

-W-

wafer stepper 385
wave function 338, 340, 342, 411
wave number 2, 12
wave vector 3, 15, 74
weak injection 107, 108
weak inversion 179
wet etching 388
work function 139, 184, 185, 317

-Z-

Zener breakdown 117
Zener diode 118

Lightning Source UK Ltd.
Milton Keynes UK

178587UK00003B/88/P